ROBERT E. KRIEGER PUBLISHING COMPANY INC.

MALABAR
FLORIDA 32950

A MODEL-MANAGEMENT FRAMEWORK FOR MATHEMATICAL PROGRAMMING

EXXON MONOGRAPHS

BIMETALLIC CATALYSIS: DISCOVERIES, CONCEPTS, AND
APPLICATIONS
Sinfelt

A MODEL-MANAGEMENT FRAMEWORK FOR MATHEMATICAL
PROGRAMMING
Palmer

A MODEL-MANAGEMENT FRAMEWORK FOR MATHEMATICAL PROGRAMMING

KENNETH H. PALMER, *Principal Author*
N. KENNETH BOUDWIN
HELEN A. PATTON
A. JOHN ROWLAND
JEREMY D. SAMMES
DAVID M. SMITH

An Exxon Monograph

JOHN WILEY & SONS
New York   •   Chichester   •   Brisbane   •   Toronto   •   Singapore

Library of Congress Cataloging in Publication Data:

Palmer, Kenneth H.
   A model-management framework for mathematical
programming.

   "An Exxon monograph."
   Includes index.
   1. PLATOFORM (Computer system)   I. Title.

T57.74.P34   1984   658.4'033   83-23536
ISBN 0-471-80472-X

Printed in the United States of America

10 9 8 7 6 5 4 3 2 1

FOREWORD

Exxon Corporation's Communications and Computer Sciences Department has for many years conducted a program aimed at bringing advances in mathematical programming into practical use in business. This monograph describes Exxon's solution to a topic that has caused much difficulty over the years—the management of large models.

PLATOFORM has been a success story within Exxon. It has allowed us to take linear programming from the esoteric realm to the domain of practical support for more than 100 diverse applications. These applications include supply planning, petroleum product blending, and investment decisions. Such applications, in routine use throughout Exxon, have saved many millions of dollars.

PLATOFORM has become an Exxon standard. It is flexible enough to support any linear programming application. Practitioners can easily share models and concepts. Furthermore, standardization has allowed central maintenance and enhancement of a single system. Thus, although PLATOFORM is more than ten years old, it continues to employ the latest hardware and software concepts.

This monograph explains how PLATOFORM was built and illustrates how it can be used. The principal author, Ken Palmer, as well as contributing authors Ken Boudwin, Helen Patton, John Rowland, Jeremy Sammes, and Dave Smith, have all participated in the development of PLATOFORM and have made valuable contributions to the use of mathematical programming. We hope others will build on their concepts to create even more capable and flexible tools.

R. H. Harvey
Vice-President
Communications and Computer Sciences
Exxon Corporation

PREFACE

In the April 1982 issue of *Operations Research Letters,* Dr. George B. Dantzig, inventor of the simplex method for solving linear programming problems, reminisces about the origins of linear programming. His story is filled with names of giants in the field: Koopmans, Kantorovich, von Neumann, Kuhn and Tucker, Charnes and Cooper, and many more. Those were exciting days, when the pressures of the war effort, the potentialities of digital computers, and the ingenuity of dedicated men brought forth a mathematical technique to formulate and solve problems far beyond the reach of hand calculation procedures.

Dantzig devotes only a short paragraph to the industrial sequel of this epoch, wherein the war-born technique was adapted to the activities of peace. The names of contributors to the adaptation process are not so well-known, mainly because they come from an industrial environment in which the emphasis is on applications rather than on publishable results.

This book covers one aspect of the application of linear programming to real problems in real industries, an aspect that has caused the most difficulty over the years: the care and feeding of large formulations. When linear programming was first inherited from the war effort, the formulations were written out on large sheets of paper and checked by hand. Even if each equation were written only a quarter of an inch high, a 300 equation formulation would paper a whole wall! Today, formulations ten times as big are commonplace, and the problem turned out to be one of team management as well as of data management. Nevertheless, an ingeniously designed software system has contributed greatly to the rapid development and smooth utilization of such large formulations.

This book describes that software system, called PLATOFORM (PLanning TOol written in DATAFORM), which has evolved to support more than one hundred applications within Exxon. Written from the inside by people who were involved in basic system design and growth, it analyzes the key factors that led to the success of the system, including its ability to manage the vast amount of data required to build large matrices and to handle multiple cases without multiple data bases. The Introduction summarizes and places in context the subject matter of each chapter, from historical development in Chapters 1 and 2 through the example described in detail in Chapter 12.

The only prerequisite to understanding the subject matter is a knowledge of linear programming. We believe that the material we present here will be of use and interest to: practitioners of linear programming who need to know the problems that arise in its application and ways to solve them; developers of linear programming software, so that their products match the requirements of actual business problems; and those in academia who teach and study these topics.

WILLIAM P. DREWS
Manager for Research*
Communications and Computer Sciences
Exxon Corporation

Florham Park, New Jersey
April 1984

---

*Mr. Drews (now retired) held this position during the preparation of this monograph and was instrumental in its realization.

ACKNOWLEDGMENTS

In the preparation of a monograph involving a diverse and widely scattered team of authors, extraordinary diligence and dedication are expected of those to whom falls the task of pulling everything together. Thanks to the good offices of John J. O'Keefe, Inc., we were able to secure the invaluable services of Jacqueline Isherwood-Martin, who is able to smooth out everything from the alignment of an arrow in a figure, or the tangled grammar of a convoluted sentence, to the ruffled pride of an author. Barbara Janowski and Jean Hurley, each on her own side of the Atlantic Ocean, patiently performed marvels in typing and retyping miles of text and in keeping track of innumerable chapter drafts on the word-processing system. Credit for most of the graphical work goes to the staff of Gallagher Varityping Services, Inc.

W.P.D.

ABOUT THE AUTHORS

Kenneth H. Palmer (Principal Author) graduated from Imperial College, London, in 1954 with an honors degree in chemical engineering. He joined Esso Petroleum Company at the Fawley Refinery in September of that year. After four years as a process engineer and planning analyst, he became involved in an early application of linear programming for the Fawley and Milford Haven U.K. refineries. Since then, he has been closely associated with the development and growth of mathematical programming in Exxon and headed the team responsible for PLATOFORM development from 1968 through 1972. His present responsibilities include the supervision of the group maintaining PLATOFORM in Esso Europe Inc., London. (Exxon's European Regional Office).

N. Kenneth Boudwin, Jr., joined Exxon in 1971. He received his doctoral degree from the Graduate School of Business at Stanford University in 1972. His experience with Exxon has included a wide range of Operations Research applications. Since 1975, he has had a significant role in the development and support of PLATOFORM and its applications, both in the United States and Europe.

Helen A. Patton became associated with Exxon in 1976. Her education is in mathematics, with a Bachelor of Arts degree from St. Mary's College, Notre Dame, Indiana, and a Master of Science degree from Xavier University, Cincinnati, Ohio. Prior to coming to Exxon's Communications and Computer Sciences Department, she was Vice President of Management Science Systems (MSS), responsible for the design and implementation of DATAFORM. Her previous experience was with the General Electric Computer Department, where she participated in the development of the LP/600.

A. John Rowland is the Operations Research Advisor in Esso Europe. He has an extensive background in the practical application of linear programming models in the oil industry. His former positions include Group Head for Mathematical Programming Applications in the Communications and Computer Sciences Department of Exxon Corporation, and Data Coordination for European supply, refining, and distribution models. He holds a Bachelor of Science degree (Honors) from the University of Manchester, England.

Jeremy D. Sammes joined Esso Petroleum Co., Ltd., in the United Kingdom in 1971 after completing his doctoral degree in organic chemistry at Imperial College, London. He contributed to the design and development of the PLATO-FORM system, first with Esso Petroleum and later with Esso Europe, Inc., also in London. In 1979, he joined Exxon's Communications and Computer Sciences Department in New Jersey to head the PLATOFORM group. Since 1981, he has worked on the architecture of an interactive PLATOFORM facility.

David M. Smith has been a consultant in Exxon Corporation's Communications and Computer Sciences Department since 1970. He received a Bachelor of Chemical Engineering degree from Cornell University in 1952. Prior to returning to Exxon, he held positions as Vice-President of Orchard-Hays and Company; Assistant Director, Information Processing Technology Division of CEIR; and Engineer in the Esso Standard Oil and Esso Research and Engineering Companies. He was president of SHARE, Inc., in 1971–1972 and was the founding chairman of the NSI, SPARC Data Base Study Group.

CONTENTS

Kenneth H. Palmer

INTRODUCTION

This Exxon Monograph describes a support system for mathematical programming applications, not from the user's point of view, but from the inside. All six of its authors have been closely involved with the basic design and subsequent growth of a system that started with modest ambitions and developed into an indispensable tool used throughout one of the largest international corporations. Were it not for this success, our story would not be worth telling.

Since the introduction of linear programming (LP) in the mid-1950s, Exxon developed many systems (most of which were short-lived) to support the use of LP to improve its business decision-making. Following the emergence of PLATOFORM (the subject of our book) in 1972, no new support systems have been developed. Today PLATOFORM supports more than a hundred diverse applications, and although ten years old, it is at least as effective and vigorous as it was in its early life.

PLATOFORM, then, is a system fundamentally different from its early predecessors: a system that is truly general enough to support any mathematical programming application. In writing this book, we have tried to analyze the key factors that have contributed to this difference, in the hope that they will provide some useful pointers both to future developers of similar systems and to suppliers of related software.

PLATOFORM was developed using the Enhanced Mathematical Programming System (EMPS) and its associated language DATAFORM. These were originally products of Management Science Systems, and they are currently marketed as MPS III by Ketron, Inc. The detailed structure of PLATOFORM, as described in Chapters 4 through 11, was inevitably influenced by this choice of supporting software, but the basic concepts have always been more fundamental. This is not to say that the choice of software is not important. Certainly, such a system as PLATOFORM could not easily have been implemented using earlier mathematical programming packages, but today's technical advances have supplied a number of alternative competing packages.

In describing the structure of PLATOFORM, we have tried to draw attention to those features of the supporting Mathematical Programming System (MPS) that have proved essential to the production of a successful system, that is, one that could be entirely "data driven" and could also combine all activities associated with linear programming into a single integrated system. From this discussion, it becomes apparent that the level of indirect references and the ability to communicate with the host MPS are key ingredients of any implementation language. Chapter 3 provides sufficient details of the DATAFORM language to illustrate how these key features are incorporated into EMPS.

In Chapters 1 and 2, we have set the scene by tracing the early history of LP in Exxon that led to the development of PLATOFORM during 1968–1972. These chapters also analyze the reasons for PLATOFORM's subsequent success by describing its evolutionary growth into applications far removed from those of its initial design.

The traditional tasks associated with LP concern the generation of a matrix, the finding of an optimal solution, and the subsequent analysis of this solution by means of report writers. While these important activities were addressed by many of the earliest systems, PLATOFORM, at the time of its development, incorporated a unique feature that further distinguished it from its predecessors —that of managing the vast amount of data required to build large matrices. In this respect, PLATOFORM was well ahead of its time by including a logically structured data base that incorporated many of the relational features seen in modern data-management systems. The story of the design of this data structure is told in Chapter 4, which also shows how PLATOFORM solved one of the most intractable problems associated with the routine use of LP models: the handling of multiple cases without a multiplicity of data bases. This story of data sharing is taken up again in Chapter 8, which explores the relationship of data organization to the structural organization of support personnel within a multiaffiliate application.

In Chapter 5, we look at the system design of PLATOFORM that enabled it to handle all of the activities associated with LP within a single system and yet to retain the flexibility to execute these activities in any sequence. The consequent modular structure of the overall system leads to certain specialized maintenance considerations. Furthermore, its open-ended nature needs to be controlled to ensure a proper migration toward a system that would support more and more diverse applications. Within the complex interrelationships that exist in PLATOFORM, a most important concept emerges that emphasizes how system-input requirements can, and should, be separated from user-input preferences. A translation is accomplished by means of an independent translation module, and Chapter 6 illustrates how this module affords the user a more "friendly" view of PLATOFORM.

LP practitioners use a simple syntactical "language" to control all aspects of their model-related work. This same language is used in building and modifying the data base, controlling the use of the matrix generator, selecting appropriate solution strategies, and calling any required solution reports. Side cases are easily set up and solved. Experience has shown that new users can learn this language—the only required interface with PLATOFORM—with about half a day of instruction. Becoming familiar with the organization of the data tables takes longer, since these can be designed independently for each different application.

One of the major modules of PLATOFORM is the Data-Base Manager (DBM). This module is used by every application and is intrinsically geared to the basic data structure of the system. In Chapter 7, the reader discovers module features that allow the user to construct, modify, and display the data base. This chapter contains illustrations of some particularly powerful operations; for example, changing the code for a given entity throughout an entire data base by means of a single command. Examples are given of the programming steps required to do this. Such operations are made possible by the logical relationships of the data.

Whereas the DBM is common to all applications, is centrally maintained, and is not allowed to be changed by local system support groups, matrix generators are more specific and depend on the type of application being modeled. Occasionally, a new application will require the development of an entirely new PLATOFORM generator module, but such a module can usually be made sufficiently general to serve a wide variety of applications. One such general module is the generator developed for the initial PLATOFORM application. This module now supports nearly all applications relating to the refining of crude oils and the subsequent distribution of oil products to the marketplace. In Chapter 9, we have chosen this initial module to illustrate the essential concepts of building matrix generators. Throughout this chapter, our emphasis is on the means by which these generators can be data driven and, hence, can support widely differing applications with minimal maintenance.

Integration of the optimization algorithms into PLATOFORM proved to be one of the most challenging tasks. Here, the links between the implementation language and the host MPS were of paramount importance. It was not until comparatively late in the life of PLATOFORM that a fully general-purpose "optimization" module evolved that was capable of coping with all required optimizing strategies—linear, nonlinear, network, and parametric algorithms—and of automatically handling the vast range of off-normal conditions that could occur, and all this without the user having to know anything about the basic MPS syntax. Here again, the reader discovers in Chapter 10 how the underlying data structure of PLATOFORM was used to advantage.

The final part of our story is told in Chapter 11. Once again, we describe a relatively late development, but one that finally made PLATOFORM a truly "off-the-shelf" system. The generalization of the report-writing module gave PLATOFORM users the ability to set up entirely new applications without the need for any system support in developing new software. Assuming the new application could make use of an existing matrix generator module, the only remaining tasks were the organizing and inputting of the data base, part of which included defining report requirements. The fundamental data structure of

PLATOFORM proved to be the key to the development of this "do-it-yourself," data-driven report writer.

After reading Chapters 4 through 11, the reader will have been exposed to the internal construction of a quite complex system. The purpose of our final chapter (Chapter 12) is to look at this system from the outside—the user's point of view. Here, we have taken a necessarily simplified oil-refining problem to show how PLATOFORM would be used to set up, solve, and report on such a model. Examples are given of carrying out sensitivity analyses and side cases using parametric techniques. The reader will notice that these tasks are achieved using just the simple PLATOFORM language, without recourse to DATAFORM, Program Control Language (PCL), or even Job Control Language (JCL).

Most of the examples used to illustrate the text are based on an oil-refining application. This was done to provide continuity among the chapters and to place the final chapter in perspective. Such a policy, of course, also runs the risk of leaving the reader with the impression that PLATOFORM, or any similar system, is good for solving only a single class of problems. In reality, the essence of such systems is to be able to adapt to any application. Within Exxon, PLATO-FORM has been used to model such diverse activities as oil rig scheduling, warehouse location, and portfolio selection, as well as the more traditional refining and supply problems. Such a system is in no way confined to oil industry use. All data used in the examples throughout the book are for illustrative purposes only. Although deliberately made to appear realistic, they are in no way based on actual data used by Exxon in its planning activities.

In some of the more technical chapters, we have provided extensive explanatory notes to the figures for those readers sufficiently curious to discover how some of the system features have been implemented. In these instances, the text concentrates more on what has been done, while the figures add substance to the detailed methods used.

The initial development of PLATOFORM (in parallel with that of EMPS) spanned a period of about four years and represented some ten person-years of effort. There followed a period of active, evolutionary development. Today the system is supported, mostly centrally, by three to five systems analysts. This represents a level no more than one-tenth of that which would otherwise be required to support the hundred or more applications that currently use mathematical programming.

While there were some initial concerns that a system as ambitious as PLATO-FORM, based on a powerful, high-level language such as DATAFORM, might run into execution-time problems, such problems have never been encountered. The reasons for this have been twofold. First, DATAFORM's close linkage with lower level languages such as FORTRAN has always provided a way out. The ADHOC report writer described in Chapter 11 contains many FORTRAN functions. Complex reports typically execute in one to five CPU seconds on an IBM 3033. Here, DATAFORM plays an essential role in linking the FORTRAN functions and the rest of the system architecture. The second reason concerns DATAFORM's underlying file-handler TILLER. This has received continual attention from the vendors and, today, is established as an extremely efficient data-retrieval tool. TILLER has enabled DATAFORM to be used directly as an interactive language to support recent on-line extensions to PLATOFORM. Although not discussed in this book, these extensions are allowing PLATOFORM to become a competitive, on-line data-management system for applications quite removed from mathematical programming.

David M. Smith

1. LINEAR PROGRAMMING IN EXXON

1.1 ADVANTAGES OF A MODEL-MANAGEMENT SYSTEM

The process industries, and Exxon in particular, are enthusiastic users of mathematical programming techniques to support a variety of planning decisions. These companies operate in a highly dynamic environment where wrong decisions can add dramatically to costs. Right, or optimal, decisions require the support of very complex calculations. The differences between a "good" and a "bad" operations plan for an Exxon region or affiliate often can be measured in hundreds of millions of dollars per year. However, these savings cannot be realized unless an operable modeling system is used, the model adequately represents reality, and the people reviewing the plan find the results convincing.

The development and widespread acceptance of PLATOFORM, Exxon's present model-management system, was a milestone in the corporation's use of linear programming (LP). The rapid increase in effective formulation capability and model management stimulated a corresponding growth in use of mathematical programming. The corporation now has well over a hundred models in regular use. Most of these applications regularly yield high returns, but some are used for only a single study. With PLATOFORM, new problems can be solved in a very short time—as little as two to four weeks—so that a solution is obtained while a problem is still of interest.

Before PLATOFORM, mathematical programming in Exxon was limited almost exclusively to operational planning studies, allocating crude to refineries, approximating refinery running plans, and distributing finished products. A few case studies, justifying investment in plant equipment supporting these activities, were also carried out. With PLATOFORM, the range of problems has broadened to span such complex procedures as energy conservation, chemicals manufacturing, oil exploration and production, ship and plant scheduling, warehouse or bulk plant location, and many other business applications.

Presently, broader and more effective LP use in Exxon is limited by needs for simpler user training, improved ease of use, and better understanding of the models. A program is now underway to develop a prototype system with interactive capabilities addressing these needs, and the next major breakthrough is expected to come from incorporating much of the knowledge about the system within the system. Thus, users will be provided with expert advice while at their terminals, significantly improving the human factors of making runs.

1.2   EARLY DEVELOPMENT

The simplex solution procedure (or algorithm) was invented by G. B. Dantzig in 1947, and the concepts and methods of LP became generally available to industry through the medium of Koopmans's 1951 monograph.[1] Soon after this, several Exxon employees began experimenting with the simplex procedure and speculating about the costs and benefits of potential company applications. Intracompany memoranda were exchanged early in 1953, and some papers were presented at industry and professional meetings later that year.

First, a few successful trials were conducted in the mid-1950s with internally written, elementary computer programs. Then, Exxon organized and supported its LP methodology with training courses, application projects, ancillary program development, and research projects that improved effectiveness and broadened applicability. Improved tools were adopted as they became available in the public or commercial domain, often replacing programs written internally.

Exxon encouraged the development of needed tools that were not yet being marketed, employing various forms of cooperative effort from demonstration prototypes to joint ventures and direct contract purchases. This positive, long-term supporting stance for LP was undertaken because the company perceived that its economic value would greatly exceed its cost. This perception became the justification for a substantial investment in research and in the development or acquisition of advanced tools—a view eventually supported by practical experience and now widely accepted.

Admittedly, there were some failures, generally in ambitious efforts attempted too early. A failure could usually be traced to attempts that exceeded the capabilities of computers, programs, or organizational support then available, rather than to inadequate improvements in decision-making. As expected, the percentage of success in applications requiring advances in research and development has been lower and the timing later than with more conservative applications. Nevertheless, many otherwise impractical applications earned high benefit ratios when a breakthrough in the supporting technology became available at the crucial moment. Thus the application, research, development, and acquisition of advanced LP methodology have proved mutually reinforcing.

The assimilation pattern, in retrospect, seems to have paralleled other successful developments in communications and computer science. The initial idea and the first methodologies came from outside the company. They were then tried by a small cadre of Exxon enthusiasts who showed that the application of LP demonstrated large potential benefits. This encouraged additional internal

---

[1]Koopmans, Tjalling C., et al. *Activity Analysis of Production and Allocation,* Cowles Commission Monograph No. 13. New York: John Wiley (1951), p. 404.

contributors to develop and use computer tools successfully to solve actual problems. Thus began an institutionalization process that continues today.

This process has two complementary aspects. In one, the number, size, and complexity of applications are expanded until they reach a major obstacle. In the other, research and development or acquisition are directed toward recognizing and relieving such limitations. The process constantly repeats itself; breaking through one limit always uncovers another, possibly unanticipated, barrier. This repetitiveness is particularly evident as the underlying computer technology improves by orders of magnitude per decade, dramatically changing economic feasibility, design trade-offs, and make-or-buy decisions.

1.3 LIMITATIONS AND BREAKTHROUGHS IN LP

Keeping the tools of LP as sharp as computer technology permits has maintained a usage growth rate nearly as high as that of computing itself. The pace of assimilation has not been completely smooth; limitations were encountered that slowed or almost stopped growth before they could be relieved. The particular limitations perceived as most constraining varied over time. The table in Figure 1.1 chronicles these limitations along with the accompanying products that enabled a breakthrough to the next limiting factor.

1.3.1 Computational Feasibility

Dantzig's simplex procedure avoids complete enumeration of a very large number of solution candidates by stepping from one basic solution to a "better neighboring one," by moving a selected "good" nonbasic variable from its

TIMEFRAME	PRIMARY LIMITATION	BREAKTHROUGH
1953-1955	Computational Feasibility	Exxon Baton Rouge's Simplex Programs for IBM 650/705
1956-1961	Operational Problem Size	CEIR's LP/90 Program for IBM 7090
1960-1970	Formulation/Analysis	MSS's DATAFORM Language for MPS-III
1967-1973	Model-Case Handling	Esso Europe's PLATOFORM
1960-1978	Non-Linear Modeling	Exxon's SLEUTH Extension to PLATOFORM for Recursive LP
1975-?	User Training, Procedures, Understanding/Insight	?-Exxon's Fully Interactive PLATOFORM Support System-?

Figure 1.1   LP's initial limitations and the breakthroughs that led to their resolution.

bound until it is exchanged for a basic variable that has just reached its bound. Even very simple selection and exchange rules yield amazingly good results. Although the number of "basic" solutions is astronomical, this stepwise procedure looks at very few of them.

Manual solutions, however, were still impractical for all but tiny problems since each change of basis could take as many calculations as there are numbers in the matrix representing the problem. Even the smallest LP problems, such as the 30 $\times$ 40 size needed to accommodate gasoline-blending models, could require weeks to optimize manually. In contrast, a skilled analyst, doing trial and error on a desk calculator, could derive fairly good (but not optimal) blend compositions in a day. The operational application of LP in Exxon had to await the acquisition of adequate computers.

A breakthrough occurred in 1955 and 1956, when Exxon obtained a few IBM computers (models 650 and 705), and an engineer at Baton Rouge Refinery wrote LP computer programs for them. These programs could handle problem sizes up to 40 $\times$ 60 and 60 $\times$ 100. They were the milestone that established the computational feasibility of LP in Exxon.

1.3.2 Problem Size

For the next several years, practical use of LP was directed toward the solution of small blending problems for a few products at plants that had IBM 650 or 705 computers. Preparing accurate input data, analyzing the solution output, and recasting it in the formats used for reporting blending results were tedious tasks. This tedium was relieved by constructing special-purpose, blending matrix generators and report writers to aid problem formulation and analysis. The most severe limitation on the use of LP in Exxon at that time was the lack of a robust, large, and fast LP computer program capable of optimizing the running plan for a whole plant—a complete and consistent set of the many feedstocks, operating conditions, and offtakes for each process unit in the plant.

In conjunction with the larger and faster second-generation computers introduced by 1961, the product form of the simplex algorithm and related procedural improvements provided a breakthrough. Solutions of complete refinery planning problems, rather than just the blending of a few products, became operationally practical. (Problem size has not recurred as a primary limitation since 1961, because progress in both computers and algorithms has generally kept pace with the assimilation of LP.)

From the mid-1950s into the late 1970s, increasingly sophisticated methods were developed to take full advantage of:

Sparsity. LP models depicting real-world problems have few nonzero elements per matrix column, since most activities are governed directly by less than a dozen rules, regardless of problem size.

Stability. Only a small fraction of the nonzero elements in a given model change from one case to the next.

Starting with "good" basis affords a further significant reduction in the computer resources needed to solve an LP problem. If the starting basis is nearly optimal, only a few changes of basis are likely to be required. However, guessing at a starting basis is almost certain to be fruitless without computer aids.

By 1958, Exxon had independently developed a larger, faster LP program for the 705 computer using the product form of inverse, but it was still inadequate to solve refinery planning problems. Shortly afterward, another unit of Exxon acquired the RSLP1 program from SHARE Inc. for their new IBM 704.

PLATOFORM's principal solution algorithms and start-up procedures are the direct descendants of these Exxon precursors. More important, in these two computer programs the matrix rows and columns are identified by names rather than index numbers. *This provision for user-defined symbolic identifiers is perhaps the most fundamental concept of PLATOFORM.* The system certainly would be different, and might not have been possible at all, without the concept of mnemonic identifiers and the means to manipulate them that were developed over the next decade.

In 1959, an entire refinery running plan was successfully optimized using the RSLP1 program. This led to a forecast of increased usage of LP techniques in the company. Consequently, starting in 1960 Exxon negotiated the purchase of a series of faster, larger versions of the RSLP1 system. This work was done under the guidance of the original principal author, William Orchard-Hays, who was then with CEIR. This series of programs, called SCROL on the IBM 704, or LP/90 on the "second generation" IBM 7090 computer, was written by CEIR on behalf of sponsor groups and CEIR's own service bureaus. This was a very early public sale (perhaps the first) of "software," a generic computer program of broad applicability.

The LP/90 matrix capacity was soon extended to 1,023 rows. This program ran about twenty times faster on a 7090 than its predecessor on the 704. It also had many superior operational and ease-of-use characteristics. Consequently, routine solution of refinery problems within the company finally commenced in 1961, when a refinery model of several hundred rows could be solved overnight, using only a few hours of computer time.

Even with the installation of an IBM 7090 with LP/90 at the Exxon Engineering headquarters in New Jersey, there remained other obstacles to the widespread use of LP in Exxon. None of the refineries possessed computers as large as IBM 7090 or 7094 computers, and the telecommunication of data at that time was rudimentary, to say the least. Local computing capability was needed.

This limitation was relieved by Exxon's purchase of the LP/1 program from

Haverly Systems, Inc., in 1962. LP/1 used essentially the same algorithms as LP/90, but was designed to run on the much smaller IBM 1400 series of computers. Some other Exxon plants used programs purchased from Bonner & Moore, or obtained from user groups, to solve small problems more conveniently or cheaply than was possible with LP/90. Exxon wrote only one more major internal LP program for the IBM 7070 at Baton Rouge, since publicly available programs for that computer were inadequate.

1.3.3 Continued Improvement

In 1964, IBM announced a set of revolutionary new products, for delivery two years later: a series of computers, System 360, which offered major improvements in price, performance, and speed. In addition to the usual operating systems and languages, IBM offered at no charge some applications software, including a full LP program called MPS/360, and a matrix generator called MARVEL. These programs were intended to replace and upgrade the LP/90 capabilities then available on the 7090 computers through SHARE.

The new MPS/360 LP program permitted a maximum matrix size of 8,000 rows, and it made use of the random-access capability of disk storage. The internal logic of LP/90 was improved as well. IBM also provided a compiled Program Control Language (PCL), executed under a standard operating system (OS/360), in place of the special operating system (OS) and sense-switch-invoked agenda control cards of LP/90.

While MPS/360 was accepted by most large sites in Exxon, MARVEL was not. Although it ran on a faster computer, MARVEL took longer to execute than Exxon's previous systems. Recognizing the crucial role of the model-management system to practical-sized industrial LP applications, and responding to the recommendations of a special LP Task Force, Exxon worked jointly with Management Science Systems (MSS) to develop specifications for an alternative. These specifications embodied MSS proprietary algorithms, a very sparse reinversion, and a generalized upper-bound technique, as well as the first working definition of DATAFORM, a comprehensive model-management language exploiting the concept of mnemonic identifiers. Subsequently, Exxon sponsored the development of this system by MSS. It is now marketed as MPS III by Ketron, the corporate successor to MSS. Within Exxon, this combination of algorithms and analysis capabilities is called the Enhanced Mathematical Programming System (EMPS).

Benefiting from the tenfold increase in computer speed provided by the new computer generation, EMPS now routinely solves most linear programs of up to 4,000 rows. Even sizable nonlinear or integer mathematical programming problems are solved with modest analyst and computer effort, through PLATOFORM extensions to EMPS.

1.3.4 Formulation/Analysis

The earliest general attempt to aid the LP formulator/analyst was the provision in the RSLP1 program of 1956 of column names instead of numerical indices to denote the position of an element. *This seemingly trivial extension, made mostly in the interest of "user friendliness," eventually was to have profound implications on the assimilation of LP.* The next step—providing row names as well as column names in LP codes—appeared equally trivial. However, the PLATOFORM System probably could not have been developed without the underlying structure of row- and column-*naming* conventions that evolved over the succeeding decade.

Even aided by the most sophisticated naming conventions, the analyst still had to devise the matrix representation, derive the value of each element, and interpret each answer manually. Ancillary supporting programs were needed to generate the matrix elements for a given formulation and to display the results in a meaningful report.

The first such support programs were very specific. Often written in an assembler language, or sometimes in FORTRAN, they were designed for a particular plant location and planning problem. This kind of support program was easy to use but hard to maintain, and each model required a separate development effort.

The next attempts at LP support provided ancillary programs that were very flexible, but quite tedious to use. They required a separate formula to calculate each matrix element and each numerical value in the report. Row and column names for the matrix elements were also supplied by the user: such names were used as solution references in the report formats and formulas. This approach gave great flexibility in formulation and analysis, but required careful and detailed matrix and report specification in special languages with very rudimentary syntax.

As the size and complexity of computable models increased, more general and effective support programs were needed to enable formulation efforts to keep pace with optimization. By the early 1960s, the LP/90 system could handle models of more than 300 rows. Using the flexible but simple support programs then available, an analyst might spend ten to twenty hours preparing and checking updates to the data and support programs for a one-hour computer optimization run. Under these circumstances, each plant LP model took months to develop and required more than one full-time person to support.

Thus formulation and analysis became the new limiting factors in the growth of LP. This situation led to research efforts seeking more effective ways to develop and support LP models. Three main approaches or points of departure were tried:

• Generalizing support programs by more extensive use of data describing model structure and employing higher level languages to reduce development and maintenance efforts.

• Introducing looping and/or subroutine capabilities in the support programs.

• Developing more suitable support architecture by extending the naming conventions of LP element input to generalized arrays.

In the mid-1960s, Exxon purchased and used MaGen, which had been developed by Haverly Systems, Inc. This software showed that general-purpose software could integrate the generation of LP models and the preparation of corresponding reports in a data-driven mode.

By 1970, enough progress had been demonstrated with the third approach, developing more suitable support architecture, for it to become dominant in Exxon's work. Indeed, the first efficient support programs that were also generalized and concise were later written in the DATAFORM portion of EMPS, which is based on a subset of generalized array architecture. The key issue was to generalize the formulas for computing matrix or report elements in such a way that a single statement would produce a set of the elements using that particular formula. The difficult part was to denote precisely which ones were to be produced and to define exactly how to substitute original input data items into the formula.

Exxon did make some successful prototype trials of generalized array concepts in conjunction with LP/90, most notably with a program called MAGIC. However, usage was restricted by the limited size of directly addressable memory and the lack of adequate disk storage to support large tasks efficiently. Nonetheless, MAGIC demonstrated the generality and power of array-like objects and operations. It even foreshadowed, in some significant aspects, the Relational Data Model proposed by Codd in 1971,[2] which is just now becoming commercially available and which may well become very popular as the 1980s progress.

The matrix generators written in MAGIC took far fewer statements to program than if they had been written in FORTRAN. Most meta-data (data about data) required by FORTRAN, such as the dimensions of arrays, was not needed by MAGIC. Moreover, basic control information, such as the admissibility of crude oils to refineries or the relevance of a specification to a certain grade of product, usually was kept in arrays rather than being embedded in procedural code. This data-driven architecture, facilitated by MAGIC, demonstrated appreciable savings in generator maintenance. Unfortunately, MAGIC also consumed considerable computer resources; therefore, the effective problem size was restricted by

[2]Codd, E. G. "A Relational Model of Data for Large Shared Data Banks." *Communications of the Association for Computing Machinery.* **13**(6): 377–387; June 1970.

the limited capability of the 7090 computer. And, although many of the concepts of MAGIC were later incorporated into the architecture of MARVEL for the System 360's new computer hardware, the MARVEL product proved inadequate, and was eventually withdrawn by IBM to be replaced with a more conventional matrix generator and report writer.

In the meantime, Exxon was left in a quandary. Neither MAGIC nor most of the other ancillary programs supporting LP/90 in the corporation could run on System 360, because they were written in 7090 Assembler and thus were not transferable. MARVEL was too slow to be operationally effective, yet IBM had no reason to search out a remedy for this since less than 1% of their customer accounts had any significant use of LP. What should be done to provide support for the many LP applications then in use throughout the corporation?

1.3.5 Origins of DATAFORM

In view of this situation, Exxon management formed the Mathematical Programming System Task Force in 1968 to propose a long-range plan for the support of LP and its extensions in the corporation. The Task Force found that LP was consuming about 10% of the company's computing resources. There was good growth potential, but actual usage was constrained by the lack of formulation and analysis tools. The Task Force proposed strengthening internal support and procuring major enhancements to MPS/360, particularly in matrix generation and report writing. Exxon selected the proposal embodying DATAFORM specifications from MSS in 1969. The contract called for strong Exxon participation in refining the specifications as well as some joint development.

The first phase of an improved DATAFORM system (called EMPS within Exxon) was delivered for testing in 1970. It achieved production status in 1971, after which it was marketed publicly as MPS III by MSS and its corporate successor, Ketron. EMPS continues to be enhanced regularly by MSS/Ketron and Exxon. The ease of this enhancement process, considering the cumulative magnitude of the improvements, is a tribute to the foresight of the original designers.

DATAFORM could now support quite generalized LP model formulation and analysis, but it required programming specialists and skilled system designers to lay out user interfaces, particularly for complex case study environments. Thus while one bottleneck was relieved by DATAFORM, another was uncovered: that of model management and case handling.

1.3.6 Model Case Handling

By the time DATAFORM was being developed, Esso Europe's LP support system had become almost unmanageable. Too many changes, from too many organizations, came in during a planning cycle. Each change required many meticulous updates to the model in order to maintain data consistency. Consequently, Esso Europe advanced persuasive arguments during and following the development of DATAFORM to include extensive provisions for model and case management. DATAFORM thus was made adequate to implement a comprehensive system of data management, model formulation, solution control, and analysis for new plans and case studies.

At about this time, the organizational planning environment had become rather complex, particularly in Esso Europe, where about seventeen plants were owned by a dozen affiliates in as many countries. Over time, most of these plants had developed sets of their own local LP models along with both short- and long-range representations at Esso Europe headquarters. While legitimate differences among the models were required to support variations in scope or intent, inadvertent or unnecessary discrepancies in formulation plagued the analysis of results.

Consequently, the Esso Europe Logistics Department established common data definitions and a central repository of reference data. Using standardized methods wherever possible, analysts in each affiliate then formulated their models, based, for the most part, on these common definitions and reference values. The rest was determined locally. PLATOFORM was developed to support this combined local–central planning, with as much added system flexibility and generality as could be incorporated at a reasonable cost.

PLATOFORM was immediately successful. In addition to being fully adopted in the Esso Europe refining circuit, it spread to support essentially all Exxon LP models. It can be used either locally or centrally, and can also be used to communicate between local and central planning groups. Two key elements in PLATOFORM's success are this hierarchical flexibility and the fact that abstract process formulations are almost totally driven by external data tables. This idea of separating the conceptual process model from the specific case data is important. The exploitation of this idea has improved over time.

Today, very few local variations or even models of totally new processes need to be programmed in DATAFORM. Generally, such a new situation can be formulated simply by modifying the data tables. Therefore, most new models can be formulated in months rather than years, and sizable case studies often take under a week. Model maintenance and analysis with PLATOFORM is reduced greatly and consists almost entirely of changes to external data rather than modifications to processing modules.

1.3.7 Nonlinear Modeling

PLATOFORM has proved useful in the field of nonlinear models as well. Since the 1960s, theorists' attempts to extend LP solution methods to intrinsically nonlinear continuous, as well as integer-valued, problems had only limited success, for example, with piecewise-linear convex problems or separable, mildly nonconvex problems. PLATOFORM model-manipulation tools provided analysts with an effective new medium to control successive approximations or recursion procedures, according to their insights into the problem structure.

Procedures for quick communication between matrix modification and optimization modules were developed. These extensions to PLATOFORM were especially suited to mixed-integer methods and successive linear approximations. Successful applications included fixed-charge problems such as optimum warehouse location, some scheduling problems, and the "pooling" problems that pervade refinery processing models. (In a pooling problem, plant streams are mixed or pooled in production proportions before blending or further processing. The quality of the mixture cannot be determined in advance of the solution, as it depends on the level of activities to be optimized, and yet the numerical value of the quality must be used in representing other constraints in the model.)

1.4 SUMMARY

This chapter has sketched the assimilation of LP in Exxon from its inception in the 1950s to the expectations for the 1980s, together with PLATOFORM as a key development in this evolution. The PLATOFORM system itself dramatically reduced the analyst effort required to formulate and maintain models. It represents a major consolidation and refinement of earlier concepts, and it is still being extended and made easier to use. But it already is the dominant vehicle for linear and mathematical programming in Exxon.

David M. Smith

2. THE PLATOFORM CONCEPT

2.1 WHAT PLATOFORM IS

The PLATOFORM system can be broadly characterized as an information support system for the planning activity that emphasizes formulating, solving, analyzing, and maintaining mathematical programming models. The planning activity usually includes a study of several cases, or alternatives, to span a reasonable degree of uncertainty about the future and the accuracy of the mathematical model.

As shown in Chapter 1, the solution of mathematical programming models depends on sophisticated software systems and powerful computers. Concise, yet flexible, ancillary software is required to automate the mechanics of model synthesis and analysis. In addition, considerable support in managing the model is essential for a system to be effective in a complex planning environment. This need is particularly evident in coordinating the highly decentralized operations that are typical of Exxon Corporation.

The central idea behind PLATOFORM was to construct an integrated framework of capabilities for managing mathematical programming models, one that could be easily used at a number of locations, functions, and organizational levels. The framework had to strike a reasonable balance between flexibility for local tailoring and control of a uniform core of centrally supported procedures and agreed-on reference data. This contrast between local flexibility and central control is reflected in the nature of the communication among the modular components in the PLATOFORM system. However, the flexibility of PLATOFORM also makes it rather hard to describe in a narrative fashion, because the system can be used in many modes or be viewed from several angles.

In overview, PLATOFORM appears as a rich collection of named objects kept in a large computer. The objects range from data elements and metadata (data about data) to formulation, solution, or analysis procedures. Such procedures include the languages to describe the results desired or to prescribe actions to be taken. Together, these procedures provide users with a comprehensive support environment for employing mathematical programming techniques.

2.2 WHAT IT DOES

The analysts/users of PLATOFORM can employ its command structure without writing any procedural programs. They can define a data-base schema, manage base case and side study data, formulate a mathematical programming model, generate partial or complete model matrices, solve an instance of the model starting from the basis saved in a prior run, save the solution, and prepare a set of analysis reports on the original data and a collection of the resulting solutions. Procedural programming is needed only to implement new formulation methods or solution algorithms.

By analogy, PLATOFORM is like a small, well-equipped workshop that contains hand tools, power tools, premade jigs, and subassemblies, or unfinished goods. The worker can create a new tool, a jig, or just assemble or finish some work in progress, depending on the need at the time. In this sense, PLATOFORM creates an environment, rather than delivering an application package. Carrying the analogy further, each group using PLATOFORM shapes this environment to its needs, just as workers acquire and arrange their tools and jigs according to their current projects.

This broad set of capabilities has improved analyst/user productivity tenfold. In about a month, an entirely new application can be developed that otherwise would have taken over a year, and a side study on an existing model can be produced in days rather than weeks. Moreover, consistency can be maintained among separate models supported by different groups, appropriately using the system's model and data-management facilities.

The total cost of using PLATOFORM (for model development, support, and usage, including both systems and user personnel, together with all computer charges) is significantly less than that for conventional matrix generator/report writers. Still, the primary driving force behind PLATOFORM's rapid assimilation and growth is more likely the tenfold reduction in elapsed time required to develop a model or to obtain a valid case study. It is this latter characteristic that lets ad hoc problems be solved while the answers are still relevant. Taken in combination, these attributes have led to a rapid acceptance of PLATOFORM throughout the corporation, including both the development of many new applications and the conversion of most previously existing models.

2.3 HOW PLATOFORM WORKS

PLATOFORM is a modeling system, rather than a transaction-processing system. Its emphasis is on analysis and synthesis capabilities for modeling together with a planning data base. This emphasis may be contrasted with the transaction-processing, simple inquiry, or consolidation capabilities of a conventional business system.

The data organization supported is a subset of the full generalized-array model, which is based on set theory and was developed by the SHARE Theory of Information Handling Committee in 1961–1962. The PLATOFORM database manager is equivalent in capability to most realizations of the relational data model in batch systems, although the command syntax is quite different. In addition, it has substantial case handling and symbol manipulation capabilities that are not yet available in other systems.

The overall PLATOFORM system is highly modular and quite general. It permits local additions, or sometimes substitutions, of capabilities relevant to a particular application, without losing the advantage of central support. This extensibility is readily manageable, so the more useful modifications are often promoted to the centrally supported core system, preventing an uncontrolled proliferation of incompatible variants. This has led to an orderly evolution to substantially increased generality, power, and effectiveness that was accomplished with very low support costs.

Three main concepts underlying the PLATOFORM architecture are responsible for its flexible and powerful capabilities. First, the objects of discourse handled by the data manager are high level, matching the analyst's conception of the problem. In addition to tables of ordinary data, they include extensive meta-data, such as catalogs, dictionaries, and masks to define relationships and ascribe meaning to the data. Next, a stored program may be treated either as a procedure to be executed or as a table of control elements that can be updated. This dual nature is exploited to provide appropriate standard and off-normal actions, during the course of a PLATOFORM run, with a minimum of "nonprocedural" specifications. Lastly, the system components are integrated, but loosely coupled, via well-defined linkages. Abstract procedures are bound to parameters or data objects by name, rather than position, at execution time.

In 1970, when the PLATOFORM development was started, only transaction-oriented tools like IMS/DL-1 were available. Providing the capabilities described above at that time meant that PLATOFORM required development of its own data-base management and manipulation facilities. The PLATOFORM Data-Base Manager (DBM) was implemented almost entirely in the DATAFORM language, which originally had been intended for linear programming (LP) matrix generation and report writing alone.

2.4 HOW IT CAME ABOUT

In 1966, as a part of a corporate reorganization, Esso Europe, Inc., was created as a major geographical unit reporting to the Exxon Board of Directors. Esso Europe held line responsibility for a number of European affiliates. These were the operating companies that acquired crude oil, refined it, and marketed the end products in about a dozen Western European countries. Most of these affiliates, but not all, were wholly owned. In aggregate, they operated seventeen refineries, and there was scope to optimize crude distribution, product movements, and so on. The individual affiliates retained operating responsibility and scheduled their local operations.

Before the reorganization, the larger plants and affiliates had developed their own LP models for planning. There also were a few highly aggregated interaffiliate LP models that were transferred to the Esso Europe Logistics Planning Department when it was formed. These models were developed at various times by different project teams and were supported almost independently by each organization on its own computers. They did not form a coherent system. In addition to obvious differences in scope and detail, there were substantive variations in formulation style and even some inconsistencies in basic data elements, such as process yields.

Maintaining and validating the aggregated interaffiliate (regional) planning models received increasing attention as discrepancies and inconsistencies among local models became apparent. A given instance, or revision, of a regional model could be solved fairly readily. However, a very large number of revisions were needed in each planning cycle to reach a formulation yielding results that could be agreed upon by both affiliate and regional planners. In essence, the regional model was rather easily solved when properly posed, but it proved difficult to establish a representation that was sufficiently valid to satisfy the users.

Moreover, the mechanics of revision, verification, and analysis of the regional models were hampered because no suitable computer tools were available to aid their formulation and analysis. The individual affiliate and plant models were in only slightly better shape. Their various matrix generator/report writers were both too specific and too verbose for easy maintenance.

2.4.1 PLATOFORM Project Initiation

In 1970, Esso Europe was feeling a pressing need to improve the operability of its LP planning systems. At that time, with DATAFORM implementation just well underway at Management Science Systems, Inc. (MSS), only the most rudimentary capabilities of the new system could be exercised; it was impossible to estimate accurately what its performance would be at final delivery. Nevertheless, Esso Europe formed a design group to draw up the broad specifications and justification for the DATAFORM-based model-management system to be called PLATOFORM. The earliest estimate for an operational target date was 1972–1973.

A two-pronged strategy was selected. To carry Esso Europe through the 1971–1972 planning cycles (at least), a previous FORTRAN model generation system was converted from the IBM 7090 to the System 360. This system would afford a fallback capability in the event that MSS's DATAFORM should prove unsuitable to support the PLATOFORM development. On the other hand, initial work on PLATOFORM was approved as a test of DATAFORM's capabilities, but with the proviso that it proceed deliberately.

The strategy worked. The availability of the earlier system helped the next few planning cycles, but the strain on the system support work force was not relieved. Moreover, the long-term technical problems remained. These were:

- Inconsistencies in reference data and discrepancies in modeling style.
- Tedium of formulation, data management, and reporting chores in model support.
- Large effort to maintain the generator/analyzer modules.
- Long elapsed time to produce a new plan or case study.

By late 1971, full PLATOFORM development was approved, based on more extensive but still preliminary DATAFORM test results. The system became fully operational in 1973 and proved so successful that it soon came to be used for almost all LP applications in Exxon.

2.4.2 Functional Capability

The PLATOFORM design team, assembled by Esso Europe from affiliates and from headquarter staffs, had an almost classical breadth of experience. Users provided knowledge of the needs of the planning process. Engineers contributed the best current process technology. Managers supplied experience with the interrelated planning cycles and communications. And, significantly, experts in LP, data-base technology, and computing systems could assess what was available, what was likely to become available, and what could become available only if Exxon funded development. Fortunately, most team members embodied more than one of these characteristics, so the team was kept small enough to be effective.

The overall goal was to provide a general, data-driven support system that would permit the company to realize the operational benefits demonstrated for LP models. The immediate objectives were to make the regional Esso Europe supply/refining models more manageable, and to achieve compatibility with their satellite affiliate models under a common system framework. The broad functional requirements adopted for PLATOFORM were quite ambitious, considering the state of the art for model management in 1970, the conservative expectations for computer hardware improvements, and the unsatisfactory performance of MARVEL. The team identified four basic design targets:

1. *A single, integrated system was required.* It would combine major modules for data definition, model formulation and revision, complete or partial matrix generation, LP solution from a prior starting basis, and recurring or ad hoc reporting and analysis of solutions. A few "bridges" were expected to provide bulk data acquisition from other sources, such as market demand forecasts from the corporate Uniform Reporting System, crude oil properties from the crude assay library, or process yields from engineering simulations.

However, the PLATOFORM data base was to be the central repository for all the data, together with suitable entry, update, and display facilities. This implied a need for a comprehensive storage structure and a communication mechanism between modules far superior to any available at the time. It appeared that the Enhanced Mathematical Programming System (EMPS)/ DATAFORM combination, together with its convenient links to FORTRAN, would be adequate for the required level of storage and communications facilities.

The requirement for integration was given high priority, since experience with previous systems taught that completely independent modules soon became desynchronized unless extensive and continuing efforts were made to keep them in step. However, when modules were too tightly coupled, a change in the output from an early module would require compensating changes in all subsequent ones. This meant that the communication and storage structure trade-offs would have to be very carefully drawn.

2. *A higher level input language was proposed.* The existing MPS/360 input was quite complicated. Its Program Control Language (PCL) prescribed in meticulous detail the exact solution process to be used in a given run, as well as actions required for the various conditions where an optimal solution is not attained. Input to EMPS was expected to be even more complex. In addition to specifying the desired solution steps in detail, the PCL program would invoke compilation of any new or revised DATAFORM modules. This had to be done for each run.

Even expert system programmers had difficulty in assembling such a complex input, which included the preparation or revision of hundreds of PCL statements without error. Such detail would clearly impose an unacceptable burden on a planning analyst with little or no programming experience. Designing a simple, yet flexible, input stream and translating it into the precise sequence of commands demanded by EMPS proved difficult. But the ultimate success in this area was a critical factor in PLATOFORM's acceptance.

3. *A comprehensive facility for model management was greatly needed.* This requirement was proposed, not only to ensure consistent formulation practices, but also to minimize the need for linear programming or DATAFORM expertise in the user community. The data tables would be clustered into geographical and/or functional areas corresponding to the organizational responsibilities of the planning groups and system support staff. These groups were well coordinated and had cooperative relationships. Nevertheless, a powerful case study mechanism was needed to avoid having one user inadvertently change another's data. At the same time, however, this mechanism needed to permit "universal" updates to propagate to all relevant cases.

4. *A modular approach had to be taken.* Perhaps the most challenging requirement was to modularize and package the system so that it would be extensible centrally and also easily tailored locally. The objective was to minimize long-term development and maintenance efforts. This dictated a careful assessment of modularity requirements to provide adequate flexibility for common use while avoiding excessive complexity. A minimum set of abstract procedures to generate model/matrix segments had to be written in such a way that the segments would assemble correctly into a complete formulation for nearly any situation. Thus, the procedure representing a given process could not contain data, or even assumptions, about connections to preceding or succeeding processes. All such information would be extracted from data and control tables at execution time.

This set of functional requirements was largely achieved even in the initial version of PLATOFORM. The system was unified to the extent that all but a very few bridge modules operated within a single job step. Simple user input format required only a knowledge of PLATOFORM nomenclature and control conventions for standard runs. The model-management capabilities were quite effective. Program development and maintenance were reduced substantially. Most of the required formulation could be generated from standard process modules, rather than requiring model- or location-specific procedures.

Nevertheless, it took several years to deliver a sufficiently flexible and general report writer to supplant individual DATAFORM analyzers for each model. Subsequent system enhancements, coupled with judicious local program modifications and extensions, have brought the current system even closer to its ambitious goals.

2.4.3 Initial Design Concepts

The key development decisions pertained to the logical and physical design of the data base. DATAFORM supported a number of LP-oriented object classes, including two-dimensional tables of either numeric or alphabetic information. While the tables had named rows and columns in DATAFORM, there was no implication that data about the nature of how one table would be used should be present in the content of another. The PLATOFORM concept required a more "knowledgeable" data manager, one that could recognize such meta-data and use it to impose order on the problem representation.

DATAFORM was enhanced to support case studies. Stable data could be stored once and yet be referenced by all cases in a study. More volatile data, on the other hand, would be stored as different versions of a given table and referenced via the name of the case. The PLATOFORM DBM was designed to manipulate this higher level information stored in the form of tables, such as catalogs, dictionaries, lists of controls, or comments.

The PLATOFORM DBM was one of the first extensive realizations of relational data-base concepts. In the nomenclature of PLATOFORM, the "Contents Catalog" designates the source and latest update of all data available to a case. The "Generic Table Catalog" defines directly accessible relations, with up to seven concatenated primary keys, specifying for each key the "Dictionary" supporting its domain. Comments about any meta-data or problem data table are encouraged by supporting an annotation capability tied to the table name.

> Dictionaries support a domain by enumerating a coordination table of short character codes, with the names and descriptions of all expected instances of entities in the domain, together with optional subset classifications. For example, the "Crudes Dictionary" might include the entry:
>
> AL   ARABLT   ARAB LIGHT MIXTURE   PG
>
> where AL is the code, ARABLT is the name, and PG denotes that this crude comes from the Persian Gulf. The description, ARAB LIGHT MIXTURE, is not considered a potential key field but can appear on designated printouts. A dictionary is said to support a domain rather than to define it, as data entries not in the dictionary are legal but cannot be acted upon by automatic translations or subsets.

The generator and reporting or analysis modules of PLATOFORM are "aware" of the DBM meta-data and adhere to its conventions. Also, these modules depend heavily on the DATAFORM capabilities for symbol manipulation and execution-time binding of procedures and data. That is, they compute sets of matrix or report elements in a logically invariant pattern, the scope of which is determined at execution time by control tables. The generator structures reflect engineering correlations and formulation practices, and the report structures match the usual presentation formats for plans. The numerical values come from data tables denoted in the Contents Catalog, from a designated set of solution values, or from current matrix elements.

2.4.4 Design Evolution

The two most difficult standardizations, even by such a process of abstraction, were those concerned with reporting modules and with translating user input to Job Control Language (JCL) and PCL programs.

When PLATOFORM was first introduced, specific reports for individual applications were written in the DATAFORM language. This led to a proliferation of individual DATAFORM analysis/reporting modules, which required individual maintenance and local expertise. Turnaround time to develop a new analysis report also was far too long, compared to the ease with which the formulation could be changed.

Both the turnaround and the maintenance problems were remedied a few years later using the ADHOC module. ADHOC permits easy nonprocedural specification of simple reports, often directly by users rather than specialists. It also facilitates a much faster preparation of complex reports, or even of data bridges, by experts. Most of the original individual DATAFORM reports have been replaced with ADHOC versions. Maintenance and support effort has been greatly reduced.

Just as development of ADHOC had to await the accumulation of PLATO-FORM/DATAFORM experience to be able to generalize effectively, so did the development of a flexible and robust abstract run procedure. The difficulty stemmed largely from the fact that part of the function of JCL was being usurped by lower level subsystem controls, such as PCL. Moreover, PCL was not designed to mediate the flow of control between itself and DATAFORM in a partial or recursive model-generation mode.

The adopted solution may seem unduly convoluted, but it was a pragmatic approach in light of the ground rules. These directed that PCL not be changed at all, that JCL be modified as little as possible, and that changes be strictly in accord with PCL design conventions. In effect, an inverted natural hierarchy made DATAFORM the "real" supervisor, feeding parameters to abstract JCL or PCL procedures. These parameters were held in DATAFORM tables supplied with the PLATOFORM system. The details are obviously complicated. The overall process, though, works quite reliably, with almost no tailoring or intervention needed by local operating system or mathematical programming system (MPS) support staff.

2.4.5 Future Directions of PLATOFORM Evolution

Mathematical programming applications have long been considered one of the "number-crunching" usages of computers. However, the time required to solve one of these problems has gone from days to hours to minutes, even in the relatively short time span of the last thirty years. Future speedups will carry the time requirement into the same range as the attention span of human intellectual activity, opening up the whole area of interactive computing.

Some of the same functions that PLATOFORM now facilitates will be among the first candidates for interactive computing. These include viewing and updating the data base and validating changes so made. Soon after, we will see some of the features that are making personal computing so popular, such as helping the user with online prompting and explanations. Assistance to the user in "browsing" the results is already available in rudimentary form; these facilities will be greatly enhanced.

"Browsing" of results is an essential part of the human/machine cooperative process in which an analyst seeks insight into a problem, using mathematical models and computers as tools in this search. For example, one may wish to know which variables remain active over a broad range of input assumptions. A simple search procedure would be to construct a side-by-side tabular comparison of a large number of case studies. A more sophisticated procedure would be to construct the logical intersection of the optimal basis sets for the same group of cases. This is an example of what might be called "analytical usage of mathematical programming," where the result sought is insight or generalization, rather than a particular numerical solution corresponding to a given set of input data.

Such analytical use of mathematical programming has been somewhat neglected in the push to set up routine operational applications. We will see a resurgence of interest when the burdens of data management and batch computing have been largely removed from the analyst's shoulders. The optimization algorithm will be thought of as simply another mathematical function, converting alternative data assumptions into the consequent operational changes. Insights into the inner workings of a complex group of production processes will then be sought through simulation within an optimization framework. Such activities will place new and more sophisticated demands upon the future successor of the PLATOFORM system. Among these demands will be enhanced graphic display capabilities, tracing of causal relationships among the active variables, and selection of preferred representations among a set of alternative optima. Each of these new demands implies a more intimate interaction between solution algorithms and model management facilities than now exists.

2.5 CONCLUSIONS

Designed over ten years ago, the PLATOFORM system has become widely and intensively used as an integral part of the company's planning activities. The system has expanded greatly over the years, but retains its original architecture. To a large part, it remains ahead of commercially available, state-of-the-art, information support system technology by providing a pragmatic combination of symbol manipulation and set-theoretic data model capabilities. PLATOFORM continues under active development intended to preserve this advantage.

Chapter 3 provides background on Exxon's Enhanced Mathematical Programming System, and Chapters 4 and 5 explain the design trade-offs made in data organization and overall system structure. The remainder of the book discusses the major modules in more detail, while Chapter 12 presents an example of the actual use of PLATOFORM.

HELEN A. PATTON

3. THE DEVELOPMENT OF EXXON'S MATHEMATICAL PROGRAMMING SYSTEM

3.1 INTRODUCTION

The Enhanced Mathematical Programming System (EMPS) was developed by Management Science Systems, Inc. (MSS) under contract to Exxon for use on large-scale IBM 370 or compatible equipment. The functional specifications were agreed upon in late 1969, and development proceeded through 1972. Under the terms of the contract, as mentioned earlier, MSS retained title to the system and marketed it under the name MPS III. It is currently owned and marketed by Ketron, Inc., which has offices in Rosslyn, Virginia.

EMPS and its data-management subsystem, DATAFORM, are used within Exxon as the underpinning for PLATOFORM and as a vehicle for the development and dissemination of mathematical programming tools. In this chapter, we outline those features of EMPS in general and DATAFORM in particular that are important to the development of PLATOFORM.

3.1.1 The Motivation for Building EMPS

In the late 1960s, the mathematical programming system (MPS) available on the IBM 360 equipment was IBM's MPS/360. Exxon's reasons for undertaking a separate effort were to provide increased capability in the following areas:

- Features and facilities to promote easy use.
- State-of-the-art algorithm capabilities.
- Creation of an open-ended system that could be modified and enhanced as the technology advanced.

EMPS was proposed as an extensive redesign, enhancement, and modification of MPS/360. Due to the anticipated long life of the system (forecast as five to seven years in 1969!), the gains in flexibility and capability that would permit the development of new applications were considered sufficient justification.

The significant differences between EMPS and MPS/360 were:

- The addition of a completely new data-management subsystem called DATAFORM, which provided integrated file and array management, matrix generation, matrix revision, and report generation.
- The addition of optimizing procedures that included a generalized upper bounding algorithm (VARIFORM) and a generalized matrix-inversion procedure (GENINV).

3.1.2 EMPS Genealogy

Three systems significantly influenced the design of EMPS:

1. *MPS/360.* This system formed the basis for the EMPS development and can be considered a direct parent. Indeed, some of the original MPS/360 subroutines can still be found in EMPS. EMPS accepted and used the basic procedure control and communication mechanisms of MPS/360. Many decisions regarding file formats were based on the desire to remain compatible with MPS/360 and to ease the problem of integrating the EMPS extensions into the existing system.

2. *MARVEL.* MARVEL was a language processor for the data preparation, matrix generation, and report writing functions associated with MPS/360. The design objectives of MARVEL and DATAFORM were remarkably similar, but MARVEL was never accepted by the MPS user community because of excessive running times. Certain restrictions were deliberately imposed in the design specifications for DATAFORM, based on an analysis of those features of MARVEL that contributed to its poor performance.

3. *LP/600 (LP/6000).* This system was developed for the General Electric 600 series, now the Honeywell 6000 series. Many of the ideas in DATAFORM had their beginnings in the LP/600 Matrix Generator, MATGEN. In particular, the underlying data management routine, TILLER, was in place under the name ZORCH. The LP/600 had separate languages for matrix generation (MATGEN) and report writing (DEFINE and CALC) that reflected a partial integration of the data. The DATAFORM data-base design was the next logical data integration step.

3.2 IMPORTANT DATAFORM CONCEPTS

DATAFORM is a data-management system for generating, maintaining, and reporting mathematical programming models. DATAFORM provides unique facilities that are essential for the development of data-driven optimization systems such as PLATOFORM. In the rest of this chapter, we explore the concepts behind those facilities and illustrate some of the language constructs.

The significant system-building concepts that are embodied in DATAFORM fall into two categories:

1. *The data-base design.* The important concepts in the DATAFORM data-base design are:

- The DATAFORM data base has a hierarchical structure that permits the logical breakdown of the data into sets, each associated with an individual problem. The physical organization and accessing of the data base do not concern the user, but the user does control the logical organization of the data into sets.
- DATAFORM employs random access rather than sequential access techniques. This means that DATAFORM has equal access to all the data in a problem without having to pass through any other problem. It can also access data concurrently from more than one problem.
- The DATAFORM data base can be expanded and contracted dynamically, based on the demands of the application. The user does not have to be concerned with core-storage allocation or traditional I/O statements. DATAFORM operates as though all of the data requested from a data base is available in main storage.
- The DATAFORM data base is more highly integrated than previous systems. DATAFORM expands the traditional definition of an MPS problem to include the model and associated data tables and solution values. One data base is used to store all of the application system's data, and one programming language is used to process the data for a variety of operations.

2. *The language design.* The important concepts in the design of the DATA-FORM language are:

- The DATAFORM syntax permits great flexibility in the use of expressions. The simplest expression consists of a "primary," which may be a single constant, variable, data-base, or function reference. A complex expression consists of two or more primaries combined with operators that define the operation to be performed. The DATAFORM language usually permits the use of a complex expression to produce a value required by the syntax. In addition, the names of any of the language elements can be derived from an expression and referenced indirectly. In this chapter, we illustrate that these syntax constructs permit the definition of meta-data (i.e., meta-tables that identify the application data tables and provide instructions regarding their use by the DATAFORM program). The operation of a DATAFORM program can be controlled by altering the meta-tables rather than changing the program.

- The DATAFORM language is designed to deal with a dynamic data base. The syntax contains constructs to process elements whose dimensions are continually changing. One of the major report-writing statements produces multipage reports automatically whenever the report dimensions overflow a single page. As a result, DATAFORM programs "float" on the associated data base, automatically responding to changes in the size of the data tables.

- The DATAFORM language can communicate with all the levels of control in the EMPS system, including the Program Control Language (PCL) and the Communication Region (CR). DATAFORM programs can access data at the PCL level, alter a linear programming (LP) matrix dynamically as it is being solved under PCL control, and invoke some PCL commands directly.

- The DATAFORM language contains extensive character-manipulation and name-matching facilities. These features enable the implementation of the meta-table concepts introduced above.

- The DATAFORM language can access external subroutines written in other computer languages such as Assembler, FORTRAN, COBOL, and PL/I. This opens up DATAFORM-based systems to all data files and systems accessible through the external subroutines, and preserves compatibility with previously developed technology. Furthermore, these functions are incorporated into the DATAFORM program at execution time, so they can be modified or maintained independently of the DATAFORM programs that call them.

3.3 THE DATAFORM SYSTEM

The DATAFORM system consists of:

- A compiler (CPDF) that analyzes DATAFORM programs for correct syntax and produces a compiled program.
- An executor (EXDF) that interpretively executes the compiled program.
- A set of utility programs for copying, editing, and manipulating DATA-FORM data bases.

3.3.1 DATAFORM in the EMPS System

EMPS, as a system, consists of a set of computer programs or procedures that perform specific functions. EMPS also provides a basic framework that links to the operating system, establishes a communication environment, and controls the sequence of execution of the various procedures. This facility, called the Program Control Language (PCL), can be considered a high-level language like FORTRAN. It is able to call procedures in a manner similar to calling subroutines.

The PCL program is executed under the control of a program called EXECUTOR. EXECUTOR is resident throughout the entire EMPS run, while the individual procedures are resident only when they are being executed. Therefore, there is no direct communication between procedures. Interprocedure communication is handled in two ways:

The Communication Region (CR). The CR is a common block of memory whose location is known to all procedures. It is resident for the entire run.

PCL Variables. These are variables established in the PCL program, called DC cells. The DC cells are resident for the entire run.

Figure 3.1 illustrates the memory allocation during an EMPS execution.

In order to operate as a part of EMPS, the DATAFORM procedures were designed to conform to the environment established by EXECUTOR. DATAFORM accepts the established parameter mechanisms, I/O interfaces, and core-management procedures. One of the powerful features of the DATAFORM language is its ability to access both the CR and the PCL variables. The actions of the PCL as well as other EMPS procedures can be controlled by a DATAFORM program. Conversely, the PCL or other procedures can influence the actions of a DATAFORM program.

3.3.2 Management of the Data Base

No discussion of DATAFORM is complete without an understanding of TILLER, the underlying data-management routine that combines the concepts of hierar-

EXECUTOR
Area for procedure load module
Area available to the procedure
Communication Region
DC cell Area

Figure 3.1   Memory allocation during EMPS execution.

chical data structure, random access, and virtual paging. The power and flexibility of TILLER had a profound effect on the design of the DATAFORM language, as well as the implementation of the DATAFORM compiler and executor. TILLER provides clean design and implementation procedures to handle:

- Efficient access to items in the data base.
- The ability to have equal access to all the data-base items.
- The ability to control dynamic expansion and contraction of data-base elements at any level.

TILLER Basics. A TILLER tree is composed of a hierarchy of elements residing on a direct access file called an Attribute Comparison Tree File (ACTFILE). The highest level of the hierarchy has only one element, called the "root." Except for the root, every element has one and only one element related to it on a higher level, called a "superior" element. Each element can have one or more elements attached at a lower level, called "subordinate" elements. Elements directly attached to the same superior are called "peer" elements. Figure 3.2 shows a logical schematic of a hierarchical tree. In this figure, element 1 is the root. Element 11 has element 4 as a superior element. Element 2 has elements 5, 6, 7, 12, and 13 as subordinate elements. Elements 5, 6, and 7 are peers.

A TILLER element consists of three distinct parts:

- A numeric identifier (id), which defines the element's hierarchical level. For a TILLER tree, the only restriction is that the value of the id must increase as one progresses down the tree.

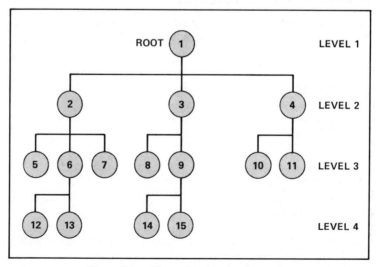

Figure 3.2   Hierarchical tree schematic.

- The element's name, an alphanumeric string that uniquely identifies this element within its set of peer elements.
- The data array, which contains data that is associated with the element but that does not influence its position within the hierarchy. This part of the element is optional and can be expanded or contracted.

An element is expressed as the combination of id and name. The interpretation of an element completely depends on its position in the logical structure; that is, in the context established by its superior elements. Consider a TILLER tree of cities in the United States grouped by state. We define the following logical structure:

Level 1   Root   id = 0, name = '0'
Level 2   State   id = 3, name = state name
Level 3   City   id = 6, name = city name

Figure 3.3 illustrates part of the city/state tree. The meaning of any element on the city level depends on its superior element on the state level. Therefore, the duplication of the name SPRINGFIELD on the city level is legal and unambiguous. This feature is important for defining the DATAFORM data-base structure.

TILLER Usage. DATAFORM uses TILLER to perform operations on ACTFILEs. The random-access I/O scheme used by TILLER allows DATAFORM to address the file as if it were wholly in main memory at all times. An initialization procedure establishes the existence of an ACTFILE and allocates memory for file control. TILLER may then be used to:

- Locate any element in the file.
- Add an element to the file.
- Delete an element and all its subordinates.
- Retrieve elements in level and sort order.
- Store data into an element's data array.
- Retrieve data from an element's data array.
- Search an element's data array.
- Modify data in a data array.
- Delete a data array.

TILLER Buffers and Paging. TILLER operates on information that is contained both in main memory storage and on an ACTFILE. An ACTFILE is a set of sequentially numbered, randomly accessible records called "pages." As a tree is created or modified, TILLER keeps track of the tree structure by posting

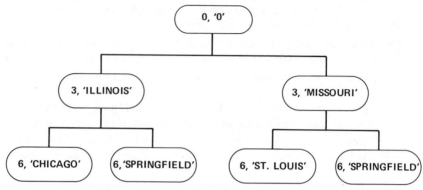

Figure 3.3   A portion of the city/state tree.

pointers that contain the page number and the intrapage location of the other elements connected to the current element. The intrapage location is referred to as a line, and each element has a unique page and line location within the tree.

As a part of its initialization, TILLER divides an area of core storage into fixed-length blocks that are exactly the right size to hold a file page. In order to access any information, pages from the file must be read into the page blocks in memory. TILLER maintains a directory for each open ACTFILE that records the current allocation of file pages to page blocks. TILLER finds its way through the branches of the tree by using the element pointers and the page directory. A page from the file may be located in any one of the page blocks. If a page is relocated to another page block for any reason, only the page directory needs to be updated to indicate the correct page block.

TILLER is forced to juggle page blocks when the number of required pages is greater than the number of available page blocks. In order to release a page block, TILLER must choose a page to write back to the file. TILLER chooses the page that has gone for the longest period of time without being accessed. If the page has been changed since it was last read, it is written to the file, the page block is freed, the new page is read in, and the page directory is updated. If the page has not been changed, then it does not have to be written back to the file. At the end of processing, the file must be closed. The file-closing procedure involves checking the page directory for file pages that are in page blocks and writing those that have been altered back to the file. Until the closing is completed, the valid contents of a file are split between the pages on the file and the page blocks in core. An abnormal termination can result in an invalid file.

Because of the manner in which TILLER operates, the user always deals with the information as though it were totally in memory at all times. Any paging that is required is performed automatically by TILLER in order to carry out the function requested by the user.

3.3.3 The EMPS Data Base

The selection of the structure and accessibility of the data is one of the most critical decisions that can be made in the construction of a computer system. These decisions set the philosophical tone of the entire system and influence the thought processes of the user population. The removal of arbitrary barriers to effective data communication is rarely accomplished satisfactorily without a major upheaval of the system.

The data structure selected for DATAFORM was very different from the structure available in MPS/360 in 1969. In fact, the level of data integration in the DATAFORM data base exceeded any other MPS system of the time. This feature is directly responsible for the flexibility of DATAFORM, and it has played a significant part in the ability of the system to evolve over its long life. To date, the original data structure has supported the desired evolutionary changes.

The basic data-processing problems that confront an MPS include managing the required data, transforming the data into the required structures, and interpreting the results. An MPS deals with three basic categories of information:

The Basic Data.   This data consists of the basic information necessary for the construction of a viable mathematical model (i.e., costs, transportation links, product qualities, capacities, etc.).

The Matrix.   This information consists of the structure of the mathematical model and the actual coefficients.

The Result.   This information is one solution of the model.

In order to understand the difference in philosophy represented by DATAFORM over the previous system, we will review how MPS/360 and DATAFORM deal with these three basic categories of information.

MPS/360 File Structure. A data flow diagram of MPS/360 as it existed in 1969 is illustrated in Figure 3.4. MPS/360 handled the three basic data categories as follows:

Basic Data.   The system facilities of MPS/360 did not deal with the basic data at all. The creation of a mathematical model from more basic information was left as an exercise for the user. Manual computation procedures gradually evolved into computer programs called matrix generators. The result, in any case, was the symbolic input file that was accepted by MPS/360.

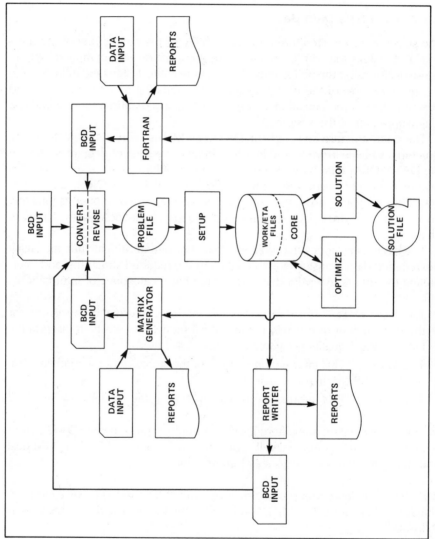

Figure 3.4   Structure of the MPS/360 system.

The Matrix. MPS/360 dealt with three representations of the mathematical model:

Symbolic input. This representation presented the matrix in a form that could be understood by the user. The information consisted of the names of the rows, columns, and right-hand sides; the values of the coefficients; and the values of the bounds and ranges. There were stringent requirements regarding the order of the input. For example, all the coefficients for a column had to be specified together. This information was processed by a procedure called CONVERT, which transformed it into:

The PROBFILE. This representation was one step closer to the requirements of the computer, but was not readily understood by the user. The PROBFILE format provided a better way of saving the data than the symbolic input file, but it could be changed only through the facilities of MPS/360. The PROBFILE was processed by a procedure called SETUP that transformed it into:

The work file. This representation was intelligible only to the computer. It was constructed in the form that best satisfied the requirements of the algorithms.

The Results. When the algorithm completed its work, the results were processed by the procedure SOLUTION, which could produce a printed report or a file, or both. The Solution File could be processed by user programs to produce management reports or symbolic input that could be used to alter the model representation.

The MPS/360 data-structure decisions represented in Figure 3.4 had the following effects on the system:

- The data is stored on distinct files with distinct formats. The user is completely responsible for keeping track of which set of input data and which set of solution values belong to which version of the model.
- Since the files do not have a uniform storage mechanism, a mixture of programming languages must be used to perform all the processing required by the application system.
- The PROBFILE is particularly isolated. No application program can access or change it. All communication with the PROBFILE matrices must be done through the CONVERT/REVISE facilities of MPS/360. Developing a system that automatically synchronizes the input and output data with the model becomes very difficult.

The Structure of the DATAFORM Data Base. Now, let us consider the flow of information provided by DATAFORM as illustrated in Figure 3.5. The DATAFORM data base consolidated all the information except the work file into one

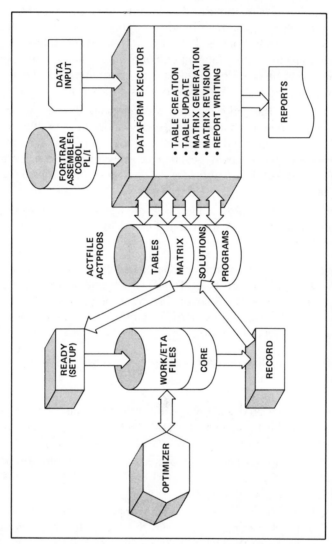

Figure 3.5 Structure of the DATAFORM system.

unit called an ACTPROB. DATAFORM handles the basic data categories as follows:

Basic Data. DATAFORM handles the basic information required for the model in the form of two-dimensional arrays called tables.

The Matrix. DATAFORM deals with two representations of the mathematical model:

The ACTFILE matrix. This representation contains all of the information that was held in the MPS/360 PROBFILE. It is processed by the procedure READY, which transforms it into:

The work file. This representation is the one required by the algorithms. It is very close to the work file representation in MPS/360.

The Results. When the algorithm completes its work, the results are processed by the procedure RECORD, which places the results back on the ACTFILE so that they can be accessed by DATAFORM.

The DATAFORM environment provides for one data base that contains all of the basic categories of information required by the MPS system. As a result, one programming language can process the data base for any desired function.

The Elements of the Data Base. Let us take a closer look at the details of the ACTPROB entities. We discuss the characteristics of each ACTPROB entity first, then show how ACTPROBs are organized into ACTFILEs, and finally illustrate the logical hierarchical TILLER tree structure. The TILLER tree chosen for DATA-FORM is one instance of a hierarchical tree that could be constructed using TILLER. This particular tree was designed to satisfy both the data-access requirements of the DATAFORM language and the data-structure requirements of the algorithms.

Data Tables. Tables are two-dimensional arrays with named rows and columns. DATAFORM recognizes two different types of tables: T-tables, which contain numeric information, and Z-tables, which contain alphabetic information. An example of each type of table follows:

Numeric table				Alphabetic table		
T:LBDEM =	DEMAND	RANGE	PRICE	Z:TIMBER =	A1	A2
MC =	90	10	3.00	NT	= NORTHERN	TIMBER
YJ =	100	10	3.50	ST	= SOUTHERN	TIMBER
PM =	110	10	3.20	OT	= OUTSIDE	TIMBER

The names of the table rows and columns are in the "rim" of the table. The column that contains the row names is called the "stub" and is referenced by the column index zero. The row that contains the column names is called the "head" and is referenced by the row index zero. The elements in the body of the table are referenced by naming the head/stub elements or through the use of a numeric index ranging from one to the maximum row/-column dimension. For example, the numeric table named LBDEM has three rows and three columns in the body of the table. The stub consists of the row names MC, YJ, and PM. The head consists of the column names DEMAND, RANGE, and PRICE. The notation T:LBDEM(1,0) refers to the first element in the stub (the zero column). In our example, this element has the value MC. The notation T:LBDEM(MC,3) refers to the element at the intersection of the row named MC and the third column. In our example, this element has the value 3.00. The Z-table is accessed in a similar fashion. All table elements, including those with a value of zero or blank, are retained. A single ACTPROB contains many tables.

The LP Matrix.   The matrix consists of a rectangular array with named rows and columns and their coefficients. The rows are the constraints and the columns are the activities of the model. In addition, the LP matrix includes special constructs such as named right-hand sides, bound sets, and range sets. Unlike tables, only nonzero intersections are retained. In DATAFORM, the matrix is not order dependent; the coefficients can be established in any order. A single ACTPROB contains only one matrix.

The Results.   The model results are placed on the ACTFILE in two sections. The column section includes the primal activity (X values), the composite input cost, the column lower limit, the column upper limit, and the reduced cost (d_j). The row section includes the slack value, the row lower limit, the row upper limit, and the dual activity (pi values). Each solution case is stored under a case name. A single ACTPROB contains many solution cases.

An ACTFILE is composed of one or more ACTPROBs. Figure 3.6 illustrates the construction of an ACTFILE as multiple ACTPROBs. The data-structure decisions represented in Figures 3.5 and 3.6 had the following effects on EMPS:

- One data base can contain all of the application system's data, so a single programming language can process the data base for all of the operations required by the application.
- DATAFORM deals directly with all the ACTPROBs on an ACTFILE and all the entities in each individual ACTPROB. Systems written in DATAFORM have the same access and can use the logical structure of the ACTFILE to maintain synchronization of all data in the system.

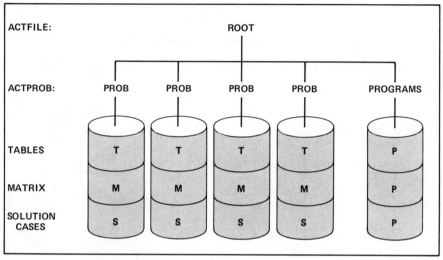

Figure 3.6 DATAFORM ACTFILE.

- A single DATAFORM program can access more than one ACTFILE. The user has great flexibility in the organization of the data into ACTFILEs and ACTPROBs.

The Logical Tree. The logical representation of an ACTPROB as a TILLER hierarchical tree is illustrated in Figure 3.7. Note that this is a logical description of the organization; the actual physical tree as implemented in DATA-FORM differs in order to minimize disk space and access time. From the user's point of view, the data base behaves as though it were constructed according to the logical tree. Figure 3.7 enables us to derive those places in the DATAFORM data base in which duplicate names are legal and unambiguous:

- In the table section, a numeric table can have the same name as an alphabetic table because the superior elements of these two kinds of tables are different. Further, a table in one ACTPROB can have the same name as a table in another ACTPROB.
- In the matrix section, a row, column, or right-hand side can have the same name since they have different superior elements. A bound set and a range set can have the same name, since the bounds are defined under the columns and the ranges are defined under the rows. As with tables, matrix elements on one ACTPROB can have the same name as matrix elements on another ACTPROB.

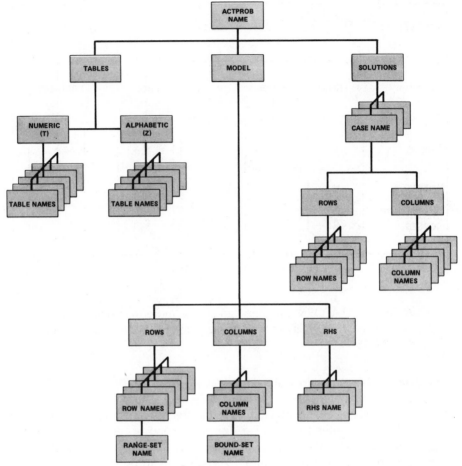

Figure 3.7   DATAFORM logical hierarchical tree.

- In the solution section, each case name in one ACTPROB must be different, but different cases can contain the same rows and columns. Case names can be duplicated across ACTPROBs.

The Working Data Base. We have seen that any element in a hierarchical tree is meaningful only in relation to its superior elements. We have also seen that a DATAFORM data base is maintained on a named ACTFILE, which is divided into named ACTPROBs. If the DATAFORM language dealt directly with all ACTFILEs, then references would have to include the two qualifiers ACTFILE and ACTPROB. This would make the DATAFORM syntax tedious to write and debug. Data-base integrity must also be considered, since an abnormal termination could result in an invalid file.

Instead, DATAFORM establishes a working area, called the working data base. The working data base is an ACTFILE whose name is SYSACT, distinguished by the lack of an ACTPROB level and by the fact that it exists only during the execution of the DATAFORM program. Most DATAFORM statements are defined to operate on data-base information on SYSACT.

For data-base elements to be operated on, they must be transferred from the proper ACTFILE/ACTPROB into the working data base by the DATAFORM statement RECALL. Once on SYSACT, elements can be created, modified, or deleted, as desired. When the operations are complete, the elements are transferred back to the proper ACTFILE/ACTPROB by the DATAFORM statement ENFILE. Figure 3.8 illustrates the relationship between DATAFORM, SYSACT, and ACTFILEs.

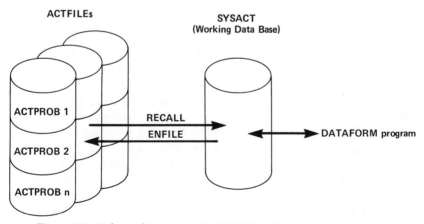

Figure 3.8 Relationship among DATAFORM, SYSACT, and ACTFILEs.

3.4   DATAFORM LANGUAGE FEATURES

DATAFORM is a programming language that was specifically created to be used as a part of an LP system. As such, it has special features for dealing with that type of application. This section illustrates some of those features.

3.4.1   Reference Levels

The DATAFORM language recognizes and can access a formidable number of data items, namely:

- Arithmetic, alphabetic, and logic variables local to the DATAFORM program.
- All the elements in the data base:

    Arithmetic-table elements.
    Alphabetic-table elements.
    Matrix coefficients, including bounds and ranges.
    Solution values.

- Arithmetic and alphabetic variables defined in the PCL.
- Arithmetic and alphabetic variables defined in the CR.
- Bit maps.
- FORTRAN, PL/I, COBOL, and Assembler functions and subroutines.

The final total came to forty separate reference types. In addition, the language had to be extensible for any new reference types. The DATAFORM solution to this problem was to designate the individual elements through the use of a prefix, a predefined character that is separated from the name of the element by a colon (:). For example:

V:name

is the syntax required to reference a local arithmetic variable. Because of the number of references, two characters were required for some prefixes.

The use of prefix notation means that the element type and its source are immediately known. Since DATAFORM can reference such a wide diversity of elements, this eliminates the need for a complex DECLARE statement. It also eliminates confusion over duplicate names since the prefix defines the reference category. It is perfectly all right to have a compute variable (V:ABC), a name variable (N:ABC), and a logic variable (S:ABC) with the same name. As the element references are extended, new prefixes must be defined and integrated into the language.

3.4.2   Indirect Referencing

The name used to reference any local variable, PCL variable, CR variable, or table can be derived in many ways. The reference

V:XYZ

is the simplest, as it explicitly states the name of the variable. In this case, the reference is to the compute variable XYZ. The variable name can be stated indirectly by using a double prefix, as in the reference

V:N:ABC

This example refers to the compute variable whose name is the current value of the name variable ABC. If N:ABC has the value XYZ, the reference resolves to V:XYZ. Indirect addressing is also permitted in subroutines to reference names in the argument list. For example:

T:#1(A,B)

references the element in "A" row, "B" column of the numeric table whose name is the first argument in the list. Double prefixing in a subroutine such as:

V:N:#2

first resolves the reference N:#2, which is the name variable whose name is the second argument. Then, the current value of that name variable becomes the name of the compute variable. If the second argument has the value BCD, then the reference resolves to V:N:BCD. If the value of N:BCD is AXY, then the final resolution is V:AXY.

Of course, the calling program has immense flexibility in the construction of the values in the argument list. Any legal name expression can be used to produce the name-value argument. These examples illustrate the degree of indirect addressing allowed by DATAFORM—flexibility that was essential to the development of PLATOFORM.

3.4.3   DATAFORM Expressions

The DATAFORM syntax permits great flexibility in the use of expressions. The simplest expression consists of a "primary," which may be a single constant, variable, data-base reference, or function reference. A complex expression consists of two or more primaries combined with operators that define the operation to be performed, including possible structuring through the use of parentheses. DATAFORM allows three types of expressions:

- Compute expressions, which resolve to a numeric value.
- Name expressions, which resolve to an alphabetic value.
- Logic expressions, which determine truth values.

DATAFORM permits the use of a complex expression wherever possible. If the syntax of a statement specifies a numeric value, then a complex compute expression may be used to produce that value. If an alphabetic value is required, a complex name expression may be used. Complex logic expressions are permitted in a similar fashion. For example, one of the statements that is used to create a matrix intersection has the syntax

ROW   rowname, colname = value

In this example, rowname is the name of the constraint row, colname is the name of the activity column, and value is the coefficient at the intersection. In DATAFORM, rowname and colname can be any legal name expression, while value can be any legal compute expression. The statement

ROW   A, B = 1

references the constant values A, B, and 1, and produces the matrix coefficient value 1 at the intersection of the row named A and the column named B. The statement

ROW   N:A, N:B = V:A

uses local compute and name variables. The statement

ROW   TR&Z:A(0,1)&..&N:CONST, Z:A(0,1) = FV:SQRT(V:A)

uses much more complex expressions, including a call to the FORTRAN function SQRT to evaluate the coefficient.

Name Expressions. DATAFORM has a rich set of built-in character-manipulation functions that are used by PLATOFORM to implement the meta-table concepts. These name operator and name functions are summarized in Figure 3.9.

Expressions in Table References. Table references in DATAFORM are very flexible because of the language's ability to use expressions. The syntax for a table reference is:

T:name(r,c)

where: T = the table prefix, indicating a numeric table;
          name = the name of the table;

Facility	Sample Expression	Meaning
&	N:NAME = N:NAME & SF	Concatenate.
DATE	N:NAME = DATE(X)	Obtain current date.
FILL	N:NAME = FILL(N:ALPHA, N:BETA)	Replace blanks in ALPHA by corresponding character in BETA.
INCR	N:NAME = INCR(N:ALPHA)	Replace right most character of ALPHA by its successor.
MASK	N:NAME = MASK(N:A, XXXXOOXX)	Select characters in A corresponding to non−zeros in mask.
SHIFT	N:NAME = SHIFT(N:ALPHA,3)	Shift all eight characters of ALPHA three spaces to the left in circular fashion.
SQUOZE	N:NAME = SQUOZE(N:ALPHA)	Left justify, blank fill and delete internal blanks.
UNJUST	N:NAME = UNJUST(SHIFT(N:A,3))	Do not SQUOZE.
VTOD	N:NAME = VTOD(V:A,4)	Convert number to character string, specifying the number of decimal positions displayed.
VTON	N:NAME = VTON(V:A)	Convert number to character string.

Figure 3.9 Summary of name expression facilities.

r = either the row name or the row index;
c = either the column name or the column index.

Consider the numeric table:

$$
\begin{array}{lccccc}
\text{T:ATAB} & = A & B & C & D \\
\text{RX} & = 1 & 2 & 3 & 4 \\
\text{RY} & = 5 & 6 & 7 & 8 \\
\text{RZ} & = 9 & 10 & 11 & 12
\end{array}
$$

A simple reference by index is

T:ATAB(1,4)

which references the element at the intersection of the first row and the fourth column. In our example, this element has the value 4. Similarly, a simple reference by name is

T:ATAB(RY,C)

which references the element at the intersection of the row named RY and the column named C. In our example, this element has the value 7.

The DATAFORM language permits the use of an arithmetic or a name expression to indicate the row and column entries in a table. In fact, one table can be used to reference another table. Consider the table:

Z:LIST
RX
RY

This is an alphabetic table that consists of only row names; a degenerate stub table. The reference

T:ATAB(Z:LIST(1,0),A)

uses a table reference to produce the row name of the numeric table. The evaluation of the name expression Z:LIST(1,0) resolves to the value RX, which reduces the numeric table reference to

T:ATAB(RX,A)

which in our example has the value 1.

3.4.4 Automatic Indexing

The DATAFORM language contains special syntax for automatically indexing over the dimensions of a table. DATAFORM provides two forms of automatic indexing, index matching and name matching. Automatic indexing is invoked through the use of the special notation

$n or =n

where n is a one- to seven-digit integer value. The notation $n invokes index matching, while =n invokes name matching.

Index Matching. The automatic-indexing symbols are used to reference the row or column of a table. For example, the statement

CALC   T:A(1,$1) = 0

contains an automatic-indexing reference for the column index. The statement means vary $1 from 1 to the number of columns in table T:A. If T:A had five columns, the single statement above would be equivalent to the following five statements:

CALC   T:A(1,1) = 0
CALC   T:A(1,2) = 0
CALC   T:A(1,3) = 0
CALC   T:A(1,4) = 0
CALC   T:A(1,5) = 0

The statement sets all the elements in the first row of the table to zero. The statement operates over the current column dimension of the table. The statement

CALC   T:A($1,$1) = 0

will set to zero the elements on the diagonal. If automatic-indexing variables with different values for n occur in the same statement, they indicate nested loops. For example, the statement

CALC   T:A($1,$2) = 0

will set to zero all the table elements. The smallest n value identifies the outermost loop.

Name Matching. The name-matching variables also appear as a table row or column, but they reference name values instead of numeric indices. For example, the statement

CALC   T:T1(R1,=1) = T:T2(R3,=1)

means replace elements in row R1 of table T:T1 with elements from row R3 of table T:T2. The automatic-indexing variable =1 is assigned the alphanumeric values in the head of table T:T1 in sequence from the first to the last. The replacement takes place only when a matching name is encountered. If the tables had the following values:

T:T1	A	B	C	D		T:T2	G	E	D	C	A
R1	1	1	1	1		R1	10	10	10	10	10
R2	2	2	2	2		R2	20	20	20	20	20
R3	3	3	3	3		R3	30	30	30	30	30

then the single statement would expand into these four statements:

CALC   T:T1(R1,A) = T:T2(R3,A)
CALC   T:T1(R1,B) = T:T2(R3,B)
CALC   T:T1(R1,C) = T:T2(R3,C)
CALC   T:T1(R1,D) = T:T2(R3,D)

The references to the columns named A, C, and D have a match in the table T:T2, so the replacement takes place for those three references. The column named B does not have a match in the table T:T2 and, therefore, fails the name-matching criterion. The name-matching logic suppresses the replacement operation, and the element T:T1(R1,B) is not altered. The resulting table would be the following:

T:T1	A	B	C	D
R1	30	1	30	30
R2	2	2	2	2
R3	3	3	3	3

3.4.5 Putting It All Together

The next example combines indirect referencing and automatic indexing to illustrate the meta-table concepts that can be implemented using DATAFORM. Consider the tables

Z:CONTROL	T:REF1 $=$ A B C	T:REF2 $=$ D E F
REF1	R $=$ 1 2 3	U $=$ 4 5 6
REF2	S $=$ 4 5 6	V $=$ 7 8 9
	T $=$ 7 8 9	

In DATAFORM, the LOOP statement is used to control automatic indexing over multiple statements. All the statements in the range of the LOOP statement are executed for each value of the indexing variables. The statements

$$
\begin{array}{ll}
\text{LOOP} & \text{N:TN} = \text{Z:CONTROL(\$1,0), A} \\
\text{ROW} & \text{T:N:TN(\$2,0), T:N:TN(0,\$3)} = \text{T:N:TN(\$2,\$3)} \\
\text{A} \quad \text{CONTINUE} &
\end{array}
\left.\begin{array}{l} \\ \\ \\ \end{array}\right\} \begin{array}{l} \text{LOOP} \\ \text{range} \end{array}
$$

illustrate the LOOP statement. In this example, the LOOP statement retrieves stub entries from the table Z:CONTROL and places the retrieved value in N:TN. The LOOP statement and all the statements in its range are executed twice, based on the row dimension of Z:CONTROL. The ROW statement is executed over the dimensions of the table chosen in the LOOP. For the first execution of the LOOP, the ROW statement is executed as

ROW   T:REF1($2,0), T:REF1(0,$3) $=$ T:REF1($2,$3)

For the second execution of the LOOP, the ROW statement is executed as

ROW   T:REF2($2,0), T:REF2(0,$3) $=$ T:REF2($2,$3)

Notice that the execution of this set of statements is entirely dependent on the contents of the individual tables. The LOOP statement responds to the number of entries in the stub of Z:CONTROL. The ROW statement responds to the dimensions of the tables chosen in the LOOP statement. The user controls the submatrix generated by changing the contents of the tables, rather than by changing the program.

3.4.6 Subprograms in DATAFORM

The DATAFORM language is given added flexibility through its ability to reference subprograms written both in DATAFORM and other high-level languages.

DATAFORM Subroutines. Subroutines may be written in the DATAFORM language to accomplish repeated operations. The implementation of DATAFORM subroutine syntax continues the expression orientation of the language. Using DATAFORM subroutines is a two-step process:

- First, the user must define the subroutine (i.e., specify the operations to be performed). The name of the subroutine is placed in the label field of the initial statement. The statements in the body of the subroutine refer to the arguments by using the notation:
    #n
  where n is an integer constant that refers to the nth argument. Subroutines are terminated with the NEXT statement, which returns control to the statement following the subroutine call.

- Second, the user must refer to the subroutine in the appropriate place in the program with the desired arguments. The name of the subroutine is placed in the verb field as though it were a DATAFORM statement. This syntax was chosen to improve the readability of DATAFORM programs.

The following example illustrates the use of subroutines by showing an alternative programming scheme for the previous example. The example generates a submatrix from a control table and data tables, namely,

Z:CONTROL	T:REF1	= A	B	C		T:REF2	= D	E	F
REF1	R	= 1	2	3		U	= 4	5	6
REF2	S	= 4	5	6		V	= 7	8	9
	T	= 7	8	9					

The statements

```
              LOOP       N:TN = Z:CONTROL($1,0), A
              SUBMAT     N:TN
    A         CONTINUE
                 .
                 .
                 .
    SUBMAT    ROW        T:#1($2,0), T:#1(0,$3) = T:#1($2,$3)
              NEXT
```

illustrate the use of a DATAFORM subroutine. This example produces exactly the same submatrix as the previous sample coding, with precisely the same meta-data flexibility.

DATAFORM subroutines must be compiled with the calling program. The DATAFORM compiler will search a separate library to locate any subroutines not explicitly supplied.

External Subprograms. DATAFORM can also execute subprograms written in Assembler and any of the high-level languages such as COBOL, FORTRAN, or PL/I. This permits the use of computational procedures that already exist or that are more conveniently performed using a non-DATAFORM language. It also means that a user can provide subprograms that extend the functionality or improve the convenience of the DATAFORM language.

Using non-DATAFORM subprograms is also a two-step process:

- First, the user must define the subprogram (i.e., specify the operations to be performed). The subprogram then must be compiled and link-edited. For convenience, these subprogram load modules can be retained on a partitioned data set that serves as a subprogram library and contains all external routines.
- Second, the user must refer to the subprogram in the appropriate place in the program with the desired arguments.

DATAFORM provides access to external subprograms in two ways:

- Function subprograms that return a single-valued result and are admissible elements in DATAFORM expressions.
- Subroutine references that are executed by the DATAFORM CALL statement.

The choice between the use of a function or subroutine subprogram is determined by evaluating the differences in capability and implementation.

When a DATAFORM program begins execution, all the function subprograms referenced by that program are loaded into core and remain resident for the entire execution of the DATAFORM program. A subroutine subprogram is not loaded until the DATAFORM statement CALL is executed. At that time, the DATAFORM program is rolled out of core and the subroutine is loaded and executed. When the subroutine execution is complete, the subroutine is deleted, and the DATAFORM program is rolled back into core. Core storage is conserved because the subroutine is resident only while it is being executed, but this is at the expense of increased execution time for multiple loadings of the same subroutine. Figure 3.10 illustrates the core configuration during the execution of external functions and subroutines.

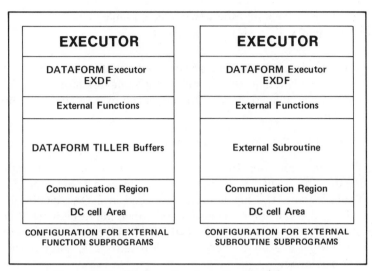

Figure 3.10   Configuration for external subprograms.

DATAFORM loads external subprograms during the actual execution of the DATAFORM program. This means that the writing and debugging of an external subprogram is independent of the DATAFORM program that calls it. If the external subprogram is changed, recompiled, and link-edited again, DATA-FORM will load and execute the altered version of the subprogram without requiring any recompilation or link-edit step for the DATAFORM program itself. This late-binding feature is unique to DATAFORM and provides great flexibility in the development and maintenance of external subprograms.

Access To The DATAFORM Data Base. External subprograms may use the DATAFORM data-management routine, TILLER. Special accommodations have been made so that the subprograms use the version of TILLER that is in the DATAFORM module. Function subprograms may have access to SYSACT, the DATAFORM working data base.

This feature means that external function subprograms may be written that extend the functionality, convenience, or efficiency of the DATAFORM language. For example, there is no single statement in DATAFORM for extending the number of rows in a table; to do this, one would use the following sequence of DATAFORM statements:

- Create a new table, X, with the proper row names.
- Copy the elements from the original table A to table X.
- Delete table A and recreate it with the proper rim from table X.

- Copy table X back into the newly created table A.
- Delete table X.

As tables become large, this becomes a very inefficient process, but a FORTRAN function can be written that will accomplish the same task more efficiently. The DATAFORM code outlined above is then replaced by a reference to the FORTRAN function:

```
IF      FS:APPEND(argument list), LABL1
```

The APPEND function uses TILLER to add row information to the table element directly. The prefix FS defines the function as a FORTRAN LOGICAL function subprogram. The function call is placed in a DATAFORM IF command so that if the function executes properly, then the result is set TRUE, and control passes to the statement at LABL1. If difficulties are encountered that prevent the proper execution, then the result is set FALSE, and control passes to the next statement.

3.4.7 TABULATE Statement

The DATAFORM language is designed to deal with a dynamic data base. The syntax contains constructs to process elements whose dimensions are continually changing. One of the major report writing statements, TABULATE, exemplifies these concepts. TABULATE automatically produces a multipaged report whenever the report dimensions are greater than a single page. TABULATE also provides for automatic translation of the stub and head into a more intelligible form. As a result, DATAFORM programs "float" on the associated data base, automatically responding to changes in the size of the data tables or lists. The TABULATE statement prints a table in the general format illustrated in Figure 3.11. The syntax for the TABULATE statement requires a reference to a FOR-

```
USER-PROVIDED HEADING INFORMATION

                    Column text    Column text    Column text  . . .

   Row text         xxx            xxx            xxx          . . .
   Row text         xxx            xxx            xxx          . . .
   Row text         xxx            xxx            xxx          . . .
     .               .              .              .             .
     .               .              .              .             .
     .               .              .              .             .

USER-PROVIDED TRAILER INFORMATION
```

Figure 3.11 General TABULATE format.

MAT statement that indicates how the report is to be formatted on the page and a reference to three tables:

- A BODY table that provides the information for the body of the report.
- An optional STUB table that provides the row text translation.
- An optional HEAD table that provides the column text translation.

The information to be printed is determined by matching:

- The head of the BODY table against the stub of the HEAD table.
- The stub of the BODY table against the stub of the STUB table.

Only those names appearing in both BODY and HEAD are printed as columns; only those names appearing in both BODY and STUB are printed as rows. The order in which the tables are stated determines the order in which the rows and columns are printed. If the optional tables are not provided, the row or column text is obtained from the BODY table itself. Given the table

T:A	TX	AZ	NJ	CA	MA
F3	25	21	24	22	23
F1	5	1	4	2	3
F2	15	11	14	12	13

then the simplest form of the TABULATE statement would be:

```
     TABULATE      FMT,BODY=T:A
FMT FORMAT         (1H , A2, 6F10.2)
```

which specifies a BODY table only. Table T:A will be printed according to the FORMAT statement labeled FMT, which is interpreted as follows:

- 1H is the carriage control specification. In this case, it means that the carriage will be advanced one line before printing.
- A2 is the row text specification. In this case, it means that the row text will occupy two columns.
- 6F10.2 is the numeric element specification. In this case, it means that six columns will be printed on a page and that each element will occupy ten columns including two decimal places.

The resulting report is

	TX	AZ	NJ	CA	MA
F3	25.00	21.00	24.00	22.00	23.00
F1	5.00	1.00	4.00	2.00	3.00
F2	15.00	11.00	14.00	12.00	13.00

which is just a pure tabulation of the table, duplicating the lack of order in the original data. Things get more interesting when we add the Z-tables:

Z:S	A1	A2	Z:H	A1
F1	POPULATI	ON	AZ	ARIZONA
F2	ENVIRONM	ENT	CA	CALIFRNA
F3	RESOURCE	S	MA	MAINE
F4	EMPLOYME	NT	MD	MARYLAND
F5	RECREATI	ON	NJ	NEW JRSY
F6	GOVERNME	NT	NY	NEW YORK
F7	ENERGY		TX	TEXAS

and extend the TABULATE statement to

          TABULATE FMT, STUB=Z:S(A1−A2), HEAD=Z:H(A1),
                   BODY=T:A
     FMT FORMAT   (1H , A10, 6F10.2)

The resulting report is

	ARIZONA	CALIFRNA	MAINE	NEW JRSY	TEXAS
POPULATION	1.00	2.00	3.00	4.00	5.00
ENVIRONMENT	11.00	12.00	13.00	14.00	15.00
RESOURCES	21.00	22.00	23.00	24.00	25.00

Notice that the HEAD and STUB tables have been used not only to translate the head and stub names in the BODY table, but also to order the tabulation. The ordering is based on the sequence of appearance of the HEAD, STUB, and BODY tables in the TABULATE statement. In our example, the STUB and HEAD phrases were placed before the BODY phrase, so the stub is printed in the STUB-table order and the head is printed in the HEAD-table order.

3.4.8 READTAB Statement

The DATAFORM language, as originally designed, does not contain a READ statement for accessing external files. It was considered that the access to external FORTRAN subprograms would provide sufficient capability in this area. However, DATAFORM does allow access to a specific type of external file via the READTAB statement. READTAB provides a method for establishing new tables and revising, contracting, or expanding existing tables in the working data base during the execution of a DATAFORM program. READTAB processes a symbolic input file, provided by the user, containing tabular data and control statements that indicate the action to be taken. The READTAB statement can be used to:

- Delete a table.
- Add a table.
- Replace an existing table.
- Add rows, columns, and intersections to a table.
- Delete rows and/or columns from a table.
- Insert rows, columns, and intersections in a table.
- Modify a table element.

In other words, READTAB permits the tabular data base to be modified in any way during the execution of the DATAFORM program. As we have already seen, DATAFORM programs automatically respond to the instructions in the meta-tables and the dimensions of the application data tables. Therefore, the execution of a given DATAFORM program can be altered by the data-base changes specified in a READTAB input file.

READTAB creates a Z-table whose name is specified in the argument list. This table contains the names, modes, and validity status of all the tables mentioned in the input file. The table names are in the stub of the table, the modes are in the column whose name is TYPE, and the validity status is in the column whose name is VALID. The validity status codes are:

MAJ	Major error encountered
MIN	Minor error encountered
blank	No errors_encountered

This provision gives the DATAFORM user the opportunity to determine if the READTAB execution was sufficiently valid to continue the execution of the program.

3.5   A FEW DESIGN ALTERNATIVES AND CHOICES

The purpose of this section is to explore the rationale for some of the basic decisions made in the design and construction of DATAFORM. As the contractor, Exxon set the priorities for the implementation and verified that the final product would be suitable for its needs. True to the original objectives, the highest priority was placed on those features of the system that provided capability and flexibility.

3.5.1   Eight-Character Name Limitation

The decision to limit the names of the matrix row and columns to eight characters was a difficult one. At that time, the EMPS design team had recently completed the LP/600 system that permitted twenty-four-character matrix names. The initial assumption was that the long name conventions would be carried over into EMPS. As a matter of fact, there is nothing in the design of the TILLER element processing, or the DATAFORM language, that would prevent such a design.

The final decision was based on the perceived difficulty of making the change throughout EMPS and on an evaluation of the benefits weighed against the required expenditure of time and money. EMPS was built on MPS/360, and, unfortunately, the eight-character name limitation was built into the basic MPS/360 PROBFILE and work file formats in a way that was difficult to change. The problem was compounded because MPS/360 did not have general routines that processed the PROBFILE and work file; each procedure included specific codes that read and unpacked the information on the file. This meant that every procedure in the entire system would have to be changed. The amount of effort involved was considerable, and the benefits were not justifiable. In addition, compatibility with the IBM system on the PROBFILE level would have been lost. Since most Exxon applications were on MPS/360, compatibility with the IBM system was important to minimize the effort of bridging applications to EMPS.

3.5.2   No Mixed-Mode Tables in DATAFORM

DATAFORM divides tables into two classes: those that contain numeric information (T) and those that contain alphabetic information (Z). Within DATAFORM, there is no way to define a table that contains some numeric elements and some alphabetic elements. The prime motivation behind this decision was a concern for the execution speed of DATAFORM. There can be no denying that the experience with MARVEL had a great impact on this decision. It was clear that DATAFORM had to execute in a reasonable amount of time or it would not be acceptable. It was judged that the overhead of keeping track of element type would degrade the DATAFORM executor and was therefore not worth the risk. The decision to keep all DATAFORM arithmetic data as double precision floating-point was based on the same consideration.

3.5.3 Execution Speed Versus Capability

Although everyone agreed that DATAFORM had to run in a reasonable amount of time, the maximization of CPU efficiency was never a priority issue. First, the initial tests showed that the execution speed would be within the limits of acceptability. Second, FORTRAN functions could be used to increase efficiency where necessary. When decisions had to be made between expending effort on increasing CPU efficiency or increasing functionality, the latter was almost always chosen. Some additional factors that contributed to this attitude included:

Advances in Computer Hardware. The computer resources needed to solve an LP model had been reduced significantly because of the increased hardware capability of the 360/165 system. The concern for the efficiency of arithmetic calculations had decreased. The emphasis began to shift away from solving the model to creating and maintaining the model, and reporting their results; in other words, to data-processing aspects.

New LP Techniques. A high incentive existed to solve many problems that were outside the capabilities of the current system. Emphasis was placed on developing functionality and flexibility so that new techniques could be employed to solve previously intractable problems.

Over time, the incredible rate of productivity improvement in computer hardware has borne out the original attitude. In fact, the role of the computer professional has changed to emphasize the saving of users' efforts and application-development time.

Kenneth H. Palmer

4. DATA AND FILE STRUCTURE

4.1 INTRODUCTION

One of the most important considerations in the design of a system to support mathematical programming applications concerns the data-handling aspects of the system. With powerful computing hardware capable of extremely rapid arithmetic operations, linear programming (LP) problems of several thousand rows can be solved economically. Such large models inevitably imply a large quantity of data (often over 100,000 data items), so that logical control of this data becomes an overriding consideration if an application is to become a working reality.

It is no longer conceivable that large LP applications be controlled at the model level, with data updates being made directly to the LP matrix. Not only do matrix coefficients frequently involve quite complex computations from more fundamental data, but the lack of control, or discipline, that such a procedure entails inevitably leads to ultimate breakdown as the model usage departs from the original base case. Moreover, control at the matrix level requires users to have a high level of skill and familiarity with LP fundamentals, distracting them from the more important tasks of problem-solving and decision-making.

Since the early 1960s, the need for some form of matrix generation facility to build the matrix from fundamental data has been recognized as an essential ingredient of an LP support system. But today a matrix generator alone is not sufficient to solve this problem. The matrix generator still has to get its data from somewhere. If data is buried in the program, the problem of controlling the data is not alleviated. Even if it is strung out on a separate sequential file, the problems associated with its maintenance are still formidable, particularly when one wants to carry out temporary side cases without disturbing the base case.

Today, some form of logically structured data base coupled with a data-management component becomes an almost essential feature of any comprehensive LP support system. It becomes mandatory if effective data sharing among a group of applications is to be a prime goal of the support system. Once established, it enables the users to interface with the system at the data level— a level naturally descriptive of the problem to be solved. The system can be given added generality by including in this data base not just the numbers from which the model is to be generated, but also logical instructions on "how" the model is to be built. In this way, a number of different models representing a wide range of company activities can be built using a single "generalized" matrix generator driven by logical data stored on the data base. One big advantage of such a data-driven system is the marked reduction in computer programs to be maintained by systems specialists. Maintenance is transferred from a specialized computer program to the data itself, which is more easily accessed.

One of the first tasks, then, in building a general support system for LP applications is to consider how the data is to be stored and manipulated. The concept of a data base with a data-management system component implies that the data must be subjected to certain rules or disciplines. What should these rules be and how rigid should they be made? These are fundamental questions that must be answered at the outset. They can be answered only by careful consideration of:

- The nature of the mathematical programming system (MPS) that is to be used to develop the system.
- The complete spectrum of applications that the system will be expected to support.

Once the MPS has been selected, the first of these considerations is simple to address. The second is far less easy to address, requiring careful judgment, foresight, and restraint against becoming overly ambitious.

To illustrate the thought processes involved in fixing the data-base rules, we trace the development of the PLATOFORM data base from the point at which the decision was made to use the Enhanced Mathematical Programming System (EMPS) and its associated language, DATAFORM, as the fundamental development tool. The initial goal was to produce a system capable of supporting all supply and refining applications throughout Europe, with the hope that the system would ultimately expand to support all of Exxon's foreseeable mathematical programming applications.

4.2 CONSTRAINTS IMPOSED BY THE CHOICE OF AN MPS

As discussed in Chapter 3, EMPS was developed for running on large IBM mainframes (360-50 and upward). As such, it is geared to an eight-character word length for matrix "identifiers" (row and column names). While prior studies of mnemonic requirements in the matrix had indicated that ten characters were probably optimal, there was no incentive to use multiples of eight (i.e., sixteen- or twenty-four-character names). These studies had also shown a strong incentive to equate the maximum character length allowed in the matrix to the maximum character length allowed in the data base, because optimal solutions, for instance, would need to be converted back to data tables in the reporting process.

DATAFORM, which combines the functions of data management, matrix generation, and report writing into a single language, is also geared to an eight-character storage system based on a table structure, as shown in Chapter 3. Here we should remind ourselves that the table names, row names ("stub"), and column names ("head") are each limited to a maximum character length of eight, and DATAFORM tables are of two distinct types, depending on their contents ("body"). These are referred to as:

T-tables. Body contains numeric values stored in double-precision floating-point form, undefined elements having a value of zero.

Z-tables. Body contains alphanumeric characters stored as an eight-character string, undefined elements being eight blanks.

DATAFORM does not allow mixed-mode tables, though this need not be a serious limitation in the design of a support system. The use of a simple DATA-FORM-callable FORTRAN (or PL/I) function to convert an alphanumeric digital string to its numeric equivalent easily removes this limitation. In general, however, a logically structured data base might well benefit from observing the segregation imposed here by the MPS. Numeric data, destined for the matrix, could form the contents of T-tables, while the logical (or control) data could be stored in Z-tables. There are, however, several instances of quasi-numeric data (e.g., calendar dates) that fall between the two and require intermodal conversion.

These aspects of the underlying MPS led to certain broad, basic design concepts for the PLATOFORM data-base structure. Although these were not mandatory, they were natural choices that, if acceptable in terms of the data requirements, would lead to simpler supporting programs:

1. The basic unit for data description would be an eight-character (maximum) string.
2. Data would be grouped and stored in DATAFORM-accessible tables.
3. Where possible, numeric data would be stored separately from control data.

These three design concepts arise from the nature of the DATAFORM language. To examine what further factors might influence the design of the PLATOFORM data base, it is necessary to consider the underlying file structure. This file structure is fundamental throughout EMPS and is critical to many of the key design features of PLATOFORM. It is described in Chapter 3 and is illustrated in Figures 3.6 and 3.7.

The key features of this file structure that influence the design of a data-base structure can be summarized as follows:

- Data is stored by ACTPROB. Each ACTPROB must have a unique name.
- T-tables are stored separately from Z-tables. Hence a T-table can have the same name as a Z-table on the same ACTPROB.
- All T-tables and all Z-tables must have unique names within an ACTPROB, but any one ACTPROB can contain tables of the same name and type as another ACTPROB.
- No conflict arises in naming the heads and stubs of tables, since these are stored separately in array segments.

These structural rules for storing tables on an ACTFILE offer an opportunity to solve one of the most important problems connected with data handling in a mathematical programming environment. This is the problem of carrying out side cases while leaving the base case untouched. In the simplest case, all the data from the base case—stored on a single ACTPROB—can be copied to a second ACTPROB, modified, and used to represent the side case. The original base case remains untouched on the first ACTPROB, and both cases are contained on the same ACTFILE. As will be explained later, the actual solution of this problem in PLATOFORM was much more elegant than this. It also avoided the inevitable duplication of data resulting from this rather naive approach (see Figure 4.1).

Having a superior ACTPROB level in the data structure hierarchy led to a fourth basic design concept. This resulted from consideration of the file structure as supported by the DATAFORM language:

4. ACTPROBs would be used to differentiate planning cases.

4.3 INFLUENCE OF APPLICATIONS TO BE SUPPORTED

Within the framework of these four design concepts, a more detailed data structure was needed to set the bounds on data-base design for any future PLATOFORM application. Here, it is important to distinguish the general data structure rules with which any future application would need to comply from a particular data structure selected for any one application. The emphasis here is on the thought processes that led to the establishment of the general rules. Chapter 12 gives an example of how these rules were applied to a particular problem.

Development of the general rules had to begin by considering the most complex model for which PLATOFORM was originally designed—the Esso Europe Supply/Refining model—and by considering its overall data require-

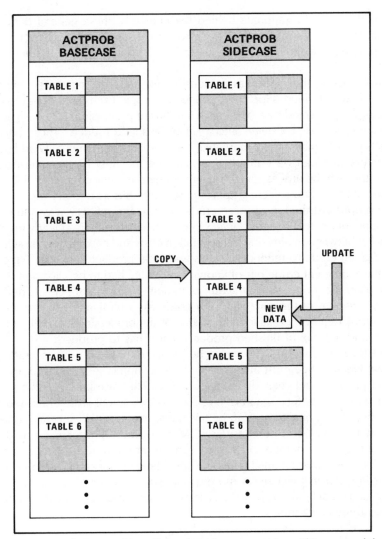

Figure 4.1 A simple approach to solving the side-case problem. This approach leads to much duplication of data.

ments. Any general rules, as a minimum, had to accommodate the data requirements of this model. Only experience would determine whether these rules would be robust enough to encompass all future applications. As it happens, the rules chosen for the first application have proved sufficient for almost all subsequent applications. Where these rules have fallen short, it has been possible to relax them without having to restructure the basic design.

To understand the rationale behind the choice of these general data-base rules, we should first look briefly at the Esso Europe supply/refining function and the data needed to build a representation of this function as a mathematical model.

Esso Europe at that time was responsible for coordinating the activities of some seventeen refineries located throughout Europe. Each refinery has its own technology represented by a number of interlinked processing units designed to refine the crude oils by stages into a set of finished products. These finished products have to match local marketing requirements both in quality and quantity, factors that can be controlled by altering the flow of the intermediate streams through the process units, changing their operating conditions, blending streams together, and selecting appropriate crude oils as primary input.

The proper distribution of these crude oils to each refinery was one of the prime functions of the Regional Head Office. The correct choice clearly involves a trade-off between minimizing transportation costs and maximizing the suitability of a given crude to refinery facilities and local product demand. Typically, some thirty different crude oils of known availability had to be allocated within a given planning period. While the transportation economics are clearly defined, the measurement of the suitability of a crude for a particular refinery in economic terms requires a detailed knowledge of the processing capability of that refinery and the mix of finished products that it has to produce.

Since local production seldom matches demand for every product grade, the Regional Head Office must also consider the economic implication of moving finished products between different countries. This distribution network can involve as many as fifty separate oil products. Variations in national requirements for product quality need to be taken into account, and several modes of transportation must also be considered. Many of these products are made by blending together a number of streams coming off the process units. A typical motor gasoline product might involve the blending together of some eight streams in order to meet six or more quality specifications.

Any such model must be based on a large amount of data—data that naturally falls into three categories:

- Region-specific data that is common to all locations, such as data that characterizes the crude oils, standard freight rates, or certain stream qualities.
- Location-specific data that would need to be shared between any Head Office model and the local model. Such data includes product quality specifications, process unit capacities and yields, plant shutdowns, and inventory capacity.
- Case-specific data relating to a particular time period or "case," such as market demands for products and various cost data.

Data in the first two categories—by far the majority of all data—remains relatively stable, and only the data in the final category requires revision for each new plan. This concept of data categorization was important in molding the design of a data-management system. Some mechanical means of transferring data between Head Office and the local affiliate (first category) and the reverse (second category) had to be included in the system design, which in turn would enable organizational responsibilities for data maintenance to be effectively implemented.

The Regional Model in the Head Office was used primarily to prepare medium-term plans every quarter for up to four time periods ranging from six to thirty months ahead. These reference plans form the basis for short-range plans prepared locally every month, plans that cover four or five immediate-future time periods. The Regional Model was also used to study the longer range outlook and to investigate investment opportunities. While the Regional Model looked at one time period at a time, some of the local models needed to examine several linked time periods in order to study inventory strategies.

Additionally, the models were used to study variations on the base planning situation. Thus they needed to be very responsive to answering the "what if?" type of question.

The system, then, had to be able to support many uses and accommodate many users. It had to support models varying in size from a single refinery to multirefineries, planning horizons from the present to five years ahead or more, single- or multi-time-period models, and routine or case study usage.

An initial estimate of the data requirements for such a model—or, better, family of models—showed that there could be up to 100,000 individual data items required to fully describe this planning situation. These included not just the basic numerical data, but also the "logical" data required to identify all the options available to the model (e.g., which crude oils are available for processing at each refinery). Clearly, some form of organized data base was necessary to accommodate this quantity of data. To illustrate the type of data involved, Figure 4.2 lists some of the numerical data by category. These categories begin to suggest a table structure to store the data.

One of the early decisions was whether to aim for a data base with just a few very large tables or with many small tables. This was not a difficult decision. Early tests with DATAFORM showed a clear execution efficiency for the smaller tables, an efficiency that was partly due to DATAFORM's storage mode in which all elements of a table are stored even if zero-valued. Large, sparse tables required much more storage space than equivalent small, compact tables. An example of the extreme case is shown in Figure 4.3.

Aside from storage space requirements, there are many other advantages for keeping tables small. An important one is the ability to print them compactly on the page. But perhaps more important than this was the concept of being able to share data among a group of applications. Here the "table" seemed the

CATEGORY	NUMBER OF DATA ITEMS
1. Crude Oil Yields/Qualities	37,132
2. Crude Oil Availabilities	233
3. Process Yields	3,912
4. Recipe Blends	732
5. Stream Qualities	7,578
6. Product Quality Specifications	2,823
7. Plant Capacities	673
8. Plant Shutdowns	183
9. Cost Data	1,419
10. Product Demands	2,602
11. Inventory Data	2,870
12. Blending Losses	8
13. Blending Tolerances	18

Figure 4.2　Categories of numeric data for the Esso Europe supply refining model. A total of 60,183 data items are stored in 1,103 data tables. Average data-table density is 83%.

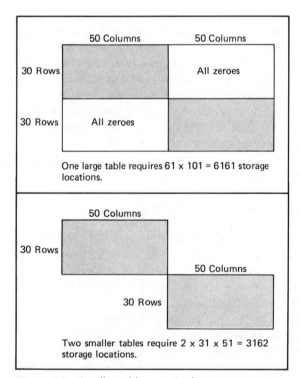

Figure 4.3　Smaller tables require less storage space.

natural choice to define a "unit" of data that could be shared. To share on an individual data-item level would involve excessively complex accounting routines, and to share by column or row within a table would require complex supporting programs that would almost certainly lead to inefficient execution. If the table was to become the unit of data within PLATOFORM, then effective sharing would be achieved only by allowing the data base to be subdivided into many relatively small tables. This, in turn, would require that the table names themselves contain several "fields" of information relating to the contents of the table.

In addition to the table names, any piece of data could be further identified by the name of the head and the name of the stub by which it was located within the table. This allowed a maximum of twenty-four characters (table, head, and stub names) to identify each piece of data. Would this be sufficient?

To answer this crucial question, it was necessary to establish:

- The maximum number of attributes required to qualify any particular data item.
- The number of characters required to identify each of these attributes.

To clarify these important concepts, consider the following example. One class of data that was important to the Esso Europe model concerns the quality specifications of any finished product. For instance, a particular heavy fuel oil made by blending a number of component streams may be required to have a sulfur content of not more than 3 wt% to be marketable. The number "3" is, then, a single data item in the total description of the business problem. How should "3" be qualified to ensure that it is properly identified? The attributes required are:

- It is the value of a particular quality, namely, "wt% sulfur."
- It belongs to a class of product specifications (as contrasted, for instance, to actual product qualities).
- It is a maximum specification (not a minimum).
- It applies to a certain finished product (heavy fuel oil).
- It applies only to a certain location (e.g., the United Kingdom).
- It applies to a certain time period (e.g., January 1982).

Thus six characteristics must be defined to qualify this data item: quality, quality class, max or min, product, location, and time period.

If we now assume that each of these parameters can be identified by a two-character code (Figure 4.4), then we see that we require twelve characters to describe this particular data item, which is well within our limit of twenty-four characters.

Quality	Wt% Sulfur	S U
Quality Class	Specifications for Blends	S B
Max or Min	Max	MX
Product	Fuel Oil	F L
Location	U.K.	U .
Time Period	January 1982	M 1

Figure 4.4   Example of attribute codes.

For the moment, we have begged the question of whether two characters is sufficient to describe, for instance, *all* products or *all* qualities that will need to be identified in our application. This clearly depends on the total number of such "products" and "qualities" to be described, but we have plenty of room for maneuvering. The key question is whether twenty-four characters will always be sufficient to describe every data item. In fact, the example chosen turns out to be one of the most demanding for qualifying description; hence we can confidently proceed with our chosen table structure.

A possible structure suggesting itself at this point is illustrated in Figure 4.5. Here, the table name is SBU., implying that all the data items contained in the table refer to product-quality specifications (SB) in the United Kingdom (U.). Each row of the table (e.g., FLM1) gives the specification data for a particular product (FL) in a particular time period (M1). The specification itself is identified by the head of the table (e.g., SUMX), which shows the particular quality (SU) and the specification type (MX).

Here, the data is grouped in terms of all product specifications by location. This, of course, is not the only possible grouping. We could have chosen to group product specification data by product, giving a row of data for each location/time-period combination. The choice of grouping depends on the application and is not relevant to the development of basic PLATOFORM data structure rules, provided these rules are flexible enough to permit such choices to be made.

Figure 4.5 also illustrates the choice between a few large tables or many smaller tables, as discussed previously. This example shows data for fuel oil (FL) and motor gasoline (GS) grouped in the same table. In practice, these two products do not share the same critical qualities. With motor gasolines we are

SBU.	SUMX	V1MX	RVMN	D1MX	ROMN
FLM1	3	500			
GSM1			10	60	99
FLM2	3.5	500			

Figure 4.5   Possible table structure to hold product quality-specification data using two-character identifier codes.

more concerned with vapor pressure (RV), distillation characteristics (D1), and octane number (RO); whereas with fuel oils, sulfur content (SU) and viscosity (V1) are of more interest. We could, therefore, save storage space by segregating data for all fuel oils and for all motor gasolines into separate tables. Note, however, that this would require introducing a seventh attribute—product class — that would need to be reflected in the table name in order to qualify any one data item and hence avoid duplicate table names. Figure 4.6 illustrates this segregation of data.

Here again, the choice of degree of segregation depends on the application, but the general PLATOFORM rules must allow for qualifying attributes beyond those that are obvious from a first inspection of the data. If product specifications are to be stored as shown in Figure 4.6, then PLATOFORM must be designed to allow at least seven qualifying parameters.

Before developing this table structure further, let us consider the advantages and disadvantages of using short (e.g., two-character) codes to identify the various attributes. If we try to utilize the full twenty-four characters available to us, we might come up with a table structure such as that shown in Figure 4.7.

It could be argued that such a table is more easily understood, when viewed on a printed page, than those shown in Figure 4.6, although even this argument is strained to carry conviction. At best, we have been able to expand certain qualifiers from two to four characters (FL to FUEL, SU to SULF, etc.), while product class (FL) and location (UK) have to remain as two-character codes. (UK

SBG.U.	RVMN	D1MX	ROMN
GSM1	10	60	99

SBF.U.	SUMX	V1MX
FLM1	3	500
FLM2	3.5	500

Figure 4.6   Specification data segregated by product class; G. = motor gasolines, F. = fuel oils. Storing data in this form saves 29% of the storage locations required compared to Figure 4.5.

SPECFLUK	SULFMAXM	VISCMAXM
FUELJA81	3	500
FUELFE81	3.5	500

Figure 4.7   Possible table structure using all available characters.

is a fortunate example of a location. Thessaloniki, in Greece, might not look so good this way.) In fact, in adopting a structure such as this, we fall between two goals. We have not used enough characters to make the reading of the table intuitively obvious, and we have increased the burden of having to memorize a particular set of characters chosen to represent a qualifier when the data is entered. This may not be too obvious in the simple example illustrated. However, a large application might need to reference sixty or seventy product qualities, and an arbitrary set using two-character codes is certainly easier to remember than an equally arbitrary set of four characters. The key factor is that when we read a table, we would like to see as many characters as possible printed, but when we put in the data, we would like to use as few characters as possible.

How can these two diametrically opposed requirements be satisfied? The answer is to use dictionaries to translate the codes when the table is printed.

4.3.1 The Concept of Dictionaries of Entity Classes

Consider the form that such a dictionary might take. To remain within the basic design concepts, it would need to be structured as a DATAFORM table, and because its purpose is to contain explanatory text, it would be a Z-table. It would seem logical to have the short (e.g., two-character) codes as the stubs of this table and the translation text arranged in eight-character fields in the body of the table. Figure 4.8 illustrates one possible arrangement for a dictionary table based on the previous examples.

Such a table could be used to translate the two-character codes that make up the table, stub, and head names whenever a table is printed. There are, however, a number of severe disadvantages to such an arrangement:

- Any such dictionary applied to a model the size of the Esso Europe supply/refining model would be an extremely large table, which could be expected to have a detrimental effect on execution efficiency.
- All the stubs in the dictionary would need to be unique. This imposes a severe strain on designing a code structure that would have any semblance of mnemonic significance.
- All codes throughout the model would have to be the same character length (two in the example), since table, stub, and head names would need to be decomposed into fixed blocks of characters. Thus any system would have to know that FLM1 is to be translated by first looking for FL in the dictionary and then M1—not F followed by LM1. This further restricts the ability to attach any mnemonic significance to the codes.
- The translation of a stub such as FLM1, using the dictionary of Figure 4.8, requires forty-eight characters of text. If spread across a line of print, this information will severely limit the number of columns of data that can fit

DICTNRY	D1	D2	D3
SB	SPECIFICATIONS - BLENDS		
G.	MOTOR GASOLINES		
F.	FUEL OILS		
GS	PREMIUM GASOLINE		
FL	HEAVY FUEL OIL		
RV	REID VAPOR PRESSURE -		
D1	VOL % DISTILLED AT 100C		
RO	RESEARCH OCTANE NUMBER		
SU	WT% SULFUR		
V1	VISCOSITY-CENTISTOKES		
MX	MAXIMUM		
MN	MINIMUM		
M1	JANUARY 1982		
M2	FEBRUARY 1982		
U.	U.K.		

Figure 4.8 Possible structure for a dictionary as a DATAFORM table.

in the rest of the page. The problem is not as severe with heads, since the translation could be "tiered" in six fields of eight characters above the columns.

How can these limitations be overcome? One obvious approach would be to allow multiple dictionaries instead of one large dictionary. The single large dictionary (as illustrated in Figure 4.8) contains in its stub a mixture of entities that together describe all the attributes of the problem. Thus, SU and RO describe stream properties (qualities), FL and GS describe the nature of the streams (products of distilling and further processing crude oils), M1 and M2 describe time periods, and so on. In concept, each of these entities belongs to an "entity class." Since there may be large entity classes (products) and small entity classes (time periods associated with a plan), it seems sensible to subdivide the dictionary by entity class. This way, the small classes could be characterized by a one-character code, while the large classes might require three-character codes. To gain full advantage from this subdivision, each dictionary would still need to contain entities having the same number of characters in each code (the system reasons for this become more evident later). In this way, we can talk of a three-character PRODUCT dictionary or a two-character QUALITY dictionary. The choice of how to group entities into entity classes becomes one of the fundamental design considerations in setting up a new application. However, once again, this choice is application dependent and is not determined by the basic structural rules.

This multidictionary approach was adopted in PLATOFORM. Figure 4.9 shows how the "large" dictionary of Figure 4.8 could be represented by six smaller PLATOFORM dictionaries. (The seventh attribute, quality class, is handled in a rather different way; see Section 4.3.2.) Besides allowing more flexibility in selecting the number of characters to represent a class of entities, this

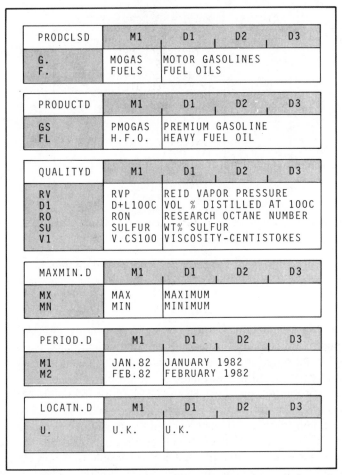

PRODCLSD	M1	D1	D2	D3
G. F.	MOGAS FUELS	MOTOR GASOLINES FUEL OILS		

PRODUCTD	M1	D1	D2	D3
GS FL	PMOGAS H.F.O.	PREMIUM GASOLINE HEAVY FUEL OIL		

QUALITYD	M1	D1	D2	D3
RV D1 RO SU V1	RVP D+L100C RON SULFUR V.CS100	REID VAPOR PRESSURE VOL % DISTILLED AT 100C RESEARCH OCTANE NUMBER WT% SULFUR VISCOSITY-CENTISTOKES		

MAXMIN.D	M1	D1	D2	D3
MX MN	MAX MIN	MAXIMUM MINIMUM		

PERIOD.D	M1	D1	D2	D3
M1 M2	JAN.82 FEB.82	JANUARY 1982 FEBRUARY 1982		

LOCATN.D	M1	D1	D2	D3
U.	U.K.	U.K.		

Figure 4.9   Subdivision of dictionary into six PLATOFORM dictionaries. In this example, each is a two-character dictionary.

arrangement permits the same code to be used in different contexts. An example is the code BA, which might be used to represent:

- Basrah Crude in the CRUDES dictionary.
- Belgium-Antwerp in the LOCATION dictionary.
- A vacuum distillate stream in the PRODUCT dictionary.

Later we see that no ambiguity arises from the multiple use of the same code, provided each occurrence is in a separate dictionary. This allows much more opportunity for giving some mnemonic significance to the chosen codes; duplication need be avoided only within a single entity class.

Notice that the dictionaries illustrated in Figure 4.9 have an additional column (M1). This contains a short (eight-character maximum) version of the translation text that can be used to translate the stub or head names of tables containing more than one entity, hence avoiding long strings of text.

To give some idea of the number of entity classes in a typical large problem, Figure 4.10 shows the dictionaries used by the Esso Europe Supply/Refining model with some details of their contents. Notice that the names of these dictionaries are all eight characters long and that the terminal character is always D (for dictionary). This is part of a standard table-naming convention adopted throughout PLATOFORM; the last character identifies the table class. Apart from this terminal D, PLATOFORM in no way preempts the name of a dictionary suitable to the application. Thus, product qualities do not have to be stored in a dictionary called QUALITYD. They could just as easily be contained in one called PROPRTYD. It is, however, important to tell PLATOFORM the names of all dictionaries used by an application. This is done through a dictionary catalog.

Entity Class	Dictionary Name	Number of Entities	Character Length
Blend Actions	BLEND..D	7	2
Capacity Investment Type	CAPINVTD	9	2
Capacity Types	CAPTYPED	14	2
Chemical Products	CHEMPROD	84	2
Countries	COUNTRYD	21	1
Crude Oils	CRUDES.D	107	2
Cost Basis	CSTBASED	2	1
Cost Types	CSTYPE.D	22	2
Time Period Definitions	DATE...D	9	2
Distillation/Reforming	DISREF.D	68	2
Demand Types	DMTYPE.D	2	2
Locations	LOCATN.D	128	2
Max or Min	MAXMIN.D	11	2
Blending Codes	MBLEND.D	17	2
Miscellaneous	MISC...D	19	2
2-Char. Time Periods	PERNAM2D	18	2
Cut-Points	PIE....D	3	1
Pipelines	PIPELIND	5	1
Product Classes	PRODCLSD	30	2
Products	PRODUCTD	337	2
Qualities	QUALITYD	67	2
Recipes	RECIPE.D	10	2
Residua Type	RESID..D	8	1
Special Structures	SPECSTRD	9	2
Sulfur Levels	SULEVELD	3	1
Data Source	TABSORCD	3	1
Tax Basis	TAXBASED	2	1
Transportation Types	TRTYPE.D	4	1
Process Units	UNITS..D	131	2

Figure 4.10   Example of entity classes in a large European application. A total of 1,150 entities are grouped into 29 dictionaries.

Whereas dictionaries describe the elements of a problem to be modeled, catalogs are required to describe the elements of the model itself. Several examples of catalogs are described in this chapter. Basically, they define the names of all the "pieces" that together form a complete description of the problem. Such pieces include the names of dictionaries, data tables, ACTPROBs, and solution cases. We see that some of these catalogs are maintained entirely by the system, whereas others require user maintenance.

Without catalogs, the underlying supporting programs become much more complex. In many cases, these catalogs are a key to providing powerful supporting features transparent to the user.

Unlike dictionaries (and data tables), the names of catalogs need to be preempted and they form an integral part of the system. In PLATOFORM, the catalog used to record the names of all dictionaries is called DICT . . . C and is illustrated in Figure 4.11 for the dictionaries shown in Figure 4.9.

Notice that the name of this catalog table is again eight characters long. However, the terminal character is now C to indicate a catalog. Columns D1 to D3 contain twenty-four characters of descriptive text to identify the entity class contained in the dictionary.

Column C1 contains a code that provides a "generic" reference to the dictionary. These codes must have the same number of characters as the codes in the dictionaries themselves—all two-character codes in this example. These "dictionary codes" play a vital role in tying together all the elements of the system, and we see a number of examples of how they are used. The codes are selected by the user (DICT . . . C is an example of a user-maintained catalog) and need not be mutually exclusive with the codes used in the dictionary to which they refer. Thus while PR refers to the class of products contained in dictionary PRODUCTD, there could still be an individual product PR contained in the stub of this dictionary.

At this point, let us summarize the structural rules connected with PLATO-FORM dictionaries:

- All descriptive elements of a business problem to be modeled are first grouped into a set of "entity classes." Each entity class requires a PLATO-FORM dictionary.

DICT...C	D1 D2 D3	C1
PRODCLSD	PRODUCT CLASSES	PC
PRODUCTD	PRODUCTS	PR
QUALITYD	QUALITIES	QU
MAXMIN.D	MAX-MIN	MM
PERIOD.D	TIME PERIODS	N2
LOCATN.D	LOCATIONS	LO

Figure 4.11   PLATOFORM catalog of dictionaries.

- Each entity within an entity class is then given three levels of description:

Example.   Entity.   Heavy Arabian Crude Oil (belongs to the entity class of CRUDE OILS)

Code	(C)	AH	Can be one, two, three, or more (up to six) characters; must be same length for all members of class
Mnemonic	(M)	ARABHY	Short descriptive text (eight-character maximum)
Descriptor	(D)	ARABIAN HEAVY	Unambiguous text of up to twenty-four characters

- Each dictionary must be given an eight-character name whose last character is D.
- The name of each dictionary must be entered into the Dictionary Catalog (DICT . . . C) and allocated a unique generic code having the same number of characters as the entity codes in the dictionary itself.

The format in which this information is stored in tables has been illustrated in previous figures. Notice that the head names of these tables are all two characters long. Chapter 7 describes how this is significant in setting a syntax for updating these tables. The head names cannot be varied, but Chapter 11 shows how additional "tag" columns can be added to a dictionary to aid report writing.

4.3.2   The Structure of Data Table Names

The decision to use multiple dictionaries helps users to develop more meaningful entity codes, but it adds complexity to the problem of translating these codes when tables are printed. In translating a stub such as FLM1, the system must know, first, how the character string breaks down (if at all) and, second, in which dictionaries to find translations of the component parts, since codes need no longer be globally unique or of uniform character length. Clearly, some form of data-table catalog is required to contain this type of information.

In any effective modeling system, the data base will inevitably be highly dynamic. For instance, new tables will be required whenever a new crude oil is to be studied, a new product is to be made, or a new location is introduced. The supporting system must allow these additions without requiring a complex entry in the table catalog for every new table—and yet only the user would know the structure of the new table! Aside from this, a typical application, with emphasis on many small tables, might be concerned with as many as 3,000 tables at any one time, making the catalog virtually impossible to control. How could such a catalog be maintained?

A way around this impasse had to be found in order to retain the obvious advantages of adopting multiple dictionaries. A clue to the solution can be found by referring to Figure 4.6. Here, a large table was broken down into two smaller ones, but the structure of these two smaller tables is identical. The table names are composed of three attributes: quality class, product class, and location; the stub names consist of product and time period; and the head names consist of quality and specification type. In other words, one catalog entry would serve for both these tables and could contain enough information to allow each to be fully translated. In fact, both of these tables contain the same "type" of data—quality specifications for blends.

In a large model, there could be many such tables, all containing the same type of specification data. There is no reason why they should differ in structure. A typical Esso Europe model has sixty such tables, so that one catalog entry covers sixty of the 3,000 tables to be cataloged. Instead of cataloging individual tables, therefore, we can catalog a "set" of tables characterized in this instance as containing quality specifications for blends (code SB). This is the seventh parameter that was not included in a dictionary. It is used to play a slightly different role— that of characterizing the SB set of tables. Of course, this same argument applies to other categories of data; instead of cataloging some 3,000 individual tables, we might typically catalog some fifty or sixty sets of tables. Additional members of a set of tables can then be added without cataloging by the user.

But how stable are the "sets" of tables? To answer this, we need to consider the context of the support system. The primary purpose of the structured data base is to provide input data (numerical and logical) to the "matrix generator" that converts this data into a matrix. Before setting up a new application, both the structure of the data and the required structure of the matrix must be designed. A matrix generator is then written to convert the one into the other. (If an existing matrix generator is to be used, both the data structure and matrix structure are already predefined, at least in a broad sense.) The matrix generator, then, must know the exact basic structure of the data base. It must know, for instance, to look in SB tables to find quality specifications, and it must know where to put these into the matrix. This does not mean that matrix generators cannot be written to handle a wide range of problems; they must be geared to a certain predefined basic data structure. In Chapter 9, we see how this generality is introduced. The fact is that once the matrix generator is written, adding further table sets (i.e., categories of data) into the data base is not going to achieve anything. The matrix generator will not even know that they are there.

Of course, there will be occasions when data is required for other purposes (e.g., reporting) or changes are made to the matrix generator. In general, however, table sets are particularly stable for any one application, and the need to expand the catalog becomes a rare event.

What is meant by a set of tables, and how can these sets be efficiently cataloged? To answer this, let us first consider the names of the tables themselves.

Here, there are eight characters to work with. First, data tables must be distinguished from catalogs and dictionaries. This can be done by using an appropriate terminal character. Since DATAFORM will be the supporting language, there are advantages in further distinguishing numeric (T) tables from control, or text, (Z) tables. Therefore, these two characters (T and Z) appear to be appropriate choices for eighth characters.

Next, the set to which a table belongs must be identified. In the previous example, the code SB was used for this purpose. Is two characters the right length for such a code? This is a crucially important decision. Clearly, one character is insufficient, since at least fifty sets are required for the prototype application. Choosing between two or three is harder, since table sets benefit from having some mnemonic significance. The original PLATOFORM design was based on three characters, but it soon became clear that this left insufficient characters (four) to qualify a table's contents. This, in turn, implied fewer larger tables (compare table names in Figures 4.5 and 4.6). Therefore, two characters was adopted as the standard for table set codes, and this became a basic PLATOFORM rule for all applications.

Assuming characters 1 and 2 are used to identify the set, five internal character positions (3 through 7) are left to qualify the table's contents. The way in which these five characters are used can be left to the data-base designer, and will depend on how many qualifying attributes are needed and how many characters are needed for each attribute. They need not be the same for each table set.

Let us see how we might structure the two quality specification tables shown in Figure 4.6 according to these rules. One such structure might be

SB.G.U.T
SB. F.U.T

The internal character positions (3 through 7) have been arbitrarily split into three fields. The total structure now consists of five fields as follows:

SB.pcloT	Content
SB	Quality specifications; defines the set
.	Invariant; used as fill character
pc	Product classes (e.g., G., F., etc.)
lo	Locations (U., etc.)
T	Invariant; numeric data table

The variable fields are character positions 4, 5 and 6, 7, which may contain *any* product class and *any* location, respectively. To convey this information, this set must be cataloged as a single entry in the table catalog. Since positions 4 and 5 always contain members of the entity class "product classes" (dictionary PRODCLSD), one way to do this would be to substitute the generic code for this dictionary as given in column C1 of the dictionary catalog (i.e., PC in Figure 4.11).

Since these codes have the requisite number of characters, they will always fit. Similarly, we could substitute LO in positions 6, 7 of the table name to indicate the entity class "locations"—dictionary LOCATN.D. Therefore, the set of SB tables can be written generically in the form

SB.PCLOT

But this does not entirely solve the problem, since the internal fields have not been broken down. The generic description of the table set could, for instance, be interpreted as:

SB.PCLOT	Content
SB	
.	
P	Entities from the one-character entity class P
CL	Entities from the two-character entity class CL
O	Entities from the one-character entity class O
T	

A "mask" must be provided to enable the field structure to be interpreted. In PLATOFORM, the adopted mask notation is in the form:

XXXAABBX

where the non-X characters indicate the variable parts of the table name and like characters, appearing consecutively, define individual field widths.

Without this mask, the field structure would need to be fixed for all table sets (e.g., 2-1-2-2-1). However, with only eight characters available, such a fixed structure would be prohibitively restrictive; it would, for instance, rule out dictionaries of more than two-character entity codes.

The table name structure of the SB table set is now fully defined by the generic name SB.PCLOT and the mask XXXAABBX. Next, we consider what further information is required to catalog the set.

4.3.3 Cataloging Table Sets

In designing a structure for naming data tables and grouping these into sets, notice that we have allowed five character positions within the table name to qualify the contents of a particular member of the set. In the example of the SB table set (quality specifications for blends), two attributes (product class and location) are required to do this. Since we have also decided to characterize each of these attributes by two-character entity codes, we need use only four of the available five positions. In the extreme, a member of a set could be qualified by five attributes, provided each could be represented by a single-character entity code. But, of course, not many entity classes could be characterized by one-character codes, and it would be more reasonable to assume a practical maximum of three attributes (two of two characters and one of one character). This means that room must be found for three or four additional qualifying attributes, in the stub and head of the table, to allow for the maximum foreseen requirements of six or seven attributes. Remember that one of these has now taken on the role of the table set name (SB in this example).

The problem is to determine the maximum number of attributes that should be permitted in the stub or head of a table, or to determine if any such maximum should be imposed at all. The second part of the problem is quickly disposed of, since in the extreme case of a stub representing eight different one-character entities, not even the short eight-character mnemonic description of each entity (column M1 of the dictionary) would leave sufficient space to print effectively the contents of the table. Some maximum has to be imposed.

Careful consideration of the data requirements for the large Esso Europe model showed that a maximum of two in both the stub and the head would suffice for this application—but just barely. Should the maximum be three? To answer this, the complexity of the supporting print routines to be written in DATAFORM had to be considered. At least initially, a maximum of two was selected. In practice, this maximum has never had to be relaxed.

These rules allow any one data item to be qualified by a maximum of seven (or nine in the extreme) qualifying parameters: two in the stub, two in the head, and three (or five) in the table name. This has proved sufficient for all of Exxon's mathematical programming applications, but may prove restrictive in a pure data-base environment, such as a management information system.

Now that the structural rules for sets of tables have been fixed, we can construct a catalog using the generic table set names developed in Section 4.3.2. An example of an entry for the SB-table set in this catalog is shown in Figure 4.12. This Generic Table Catalog, named GENTAB.C, is one of the key catalogs in the PLATOFORM system. It ties together the data stored in tables with the dictionaries that describe the data. It enables any data table, even if it has only just been included in the data base, to be fully translated when printed; that is,

GENTAB.C	S1	S2	H1	H2	M1	F1	D1 - D3
SB.PCLOT	PRODUCTD	PERIOD.D	QUALITYD	MAXMIN.D	XXXAABBX	D3	BLEND SPECIFICATIONS

Stubs - Generic description of table set, first two characters identify the set and must be unique.

Heads - S1   Name of dictionary identifying entities in the first segment of the stub names
        S2   Name of dictionary identifying entities in the second segment of the stub names
        H1   Name of dictionary identifying entities in the first segment of the head names
        H2   Name of dictionary identifying entities in the second segment of the head names
        M1   Mask, identifying variable fields within the table name
        F1   Format for printing values in the table (D3 means three decimal places)
        D1-D3 24-character description of table set (i.e., translation of Code SB)

Figure 4.12   The PLATOFORM Generic Table Catalog.

the table name as well as the stub and head names will be translated. Examples of translated tables are given in Chapter 7.

More importantly, perhaps, this catalog serves to define the entire data structure that is valid for a particular application. No data is allowed into the system unless it belongs to a set cataloged in GENTAB.C. In general, it is set up once, when the application is first designed, and rarely needs modification. In a sense, it is geared to the matrix generator that the application will use, and describes the data requirements of that generator (though this is somewhat of a "chicken-and-egg" argument). GENTAB.C is another example of a user-maintained catalog as typified by the two-character head names (as in DICT. . .C).

4.3.4 Summary of Rules for Structuring Data Tables

The rules that the system imposes on the user in designing a data structure can now be summarized as follows:

1. All data relating to the business problem must first be grouped into sets. In the extreme, a set may consist of one member, but this is not an efficient way to use the system.
2. Each set must be identified by a two-character code that becomes the first two characters of the table names making up the set.
3. The eighth character of the table names must be T for numeric data and Z for nonnumeric data.
4. Each set of data is then subdivided into tables; attributes describing this grouping are identified in positions 3 through 7 of the table names using generic dictionary codes. This table name structure must be consistent for each member of the set.
5. Further attributes qualifying the description of individual data items within a table can be given in the stub and head names, with a maximum of two attributes for each. Each stub (or head) in a table must have the same ordered combination of attributes, and this consistency must extend to all members of the set.
6. Finally, each set must first be cataloged in the Generic Table Catalog before any member of the set can be added to the data base.

Clearly, such rules permit a great diversity of data structures to be designed to match different applications. But are these rules too stringent?

The answer is unequivocal: yes. There will always be occasions when users find that unless they can break a rule, they cannot or will not use the system—and the system has to respond to this situation. In PLATOFORM, regarding these rules as "ideal rules" resolves the problem. If users choose to break them, they obtain less from the system, but the system will not collapse. This has proved to be a very important decision in determining the success with which PLATO-FORM has been adopted as a general support tool for all mathematical programming applications. Of the six given rules, only rules 3 and 6 cannot be broken. Examples of the effects of breaking the rules are given in Chapter 7, but, in general, our experience has shown that users will choose to break the rules only when there is little option but to do so. In some cases, this necessity is deliberately imposed by the design of a particular matrix generator.

4.4 DATA SHARING

Now that we have seen how the total data required to describe a business problem can be structured in sets of data tables and dictionaries of descriptive text, a somewhat different problem requires attention. As discussed earlier in this chapter, a DATAFORM ACTFILE, on which all data will be stored, can optionally be subdivided into ACTPROBs. How can we take advantage of this further subdivision? Figure 4.1 has already suggested one way in which the use of more than one ACTPROB could be advantageous in helping to solve the side-case problem, but first it is necessary to look a little deeper into the issues involved here. These issues basically concern the ways in which a mathematical programming model is used in practice, ways that themselves vary with each application.

A typical application is one in which a model is used to produce regular plans, perhaps once a month or once every quarter. As long as the model is used exclusively for this purpose, no real problems arise. A full set of data is loaded the first time the model is used. At the next planning cycle, the variable case-specific data is changed, and the model is used as before. Problems begin when, between planning cycles, the model is required to investigate a side case: for example, what would be the effect of a further $10 increase in crude oil prices? Now, the crude oil prices must be changed in order to carry out the side case, but must later be restored before the next regular planning cycle. The incidence of several such side cases quickly leads to confusion and data errors.

In another scenario, a single model is required to compute both short-range plans every month and longer range plans every quarter. The planning cycles inevitably overlap. Ninety percent of the data is common to both sets of plans. If the model is duplicated into two models, then any single change to the common data requires that both models be updated. If not rigorously applied, this process can easily lead to inconsistent data and, in turn, to inconsistent results.

One of the prime goals of Exxon's PLATOFORM system was to try to solve this data-sharing problem. Early attempts had already shown that the only effective solution lay in a "one-file/one-application" approach. Previous systems had utilized as many as a hundred computer tapes, each representing a minor variation of the data base, in a desperate but highly unsuccessful attempt to overcome these problems.

The advent of DATAFORM, with its ability to subdivide a single ACTFILE into any number of ACTPROBs, clearly offered a prospect of being able to solve this problem, provided the file did not rapidly increase in size as new cases were added. The solution proposed in Figure 4.1 is not too different from the hundred-tape solution that had already been tried, except that it reduces the interaction with the Computer Center by requiring only one physical file. The problem of having to update "common" data in more than one place is in no way alleviated by this solution, and any such file would grow too fast.

How, then, is it possible to take proper advantage of the ACTPROB file structure to:

- Allow "common" data to be updated in one place only?
- Allow "common" data to be shared without replication?

The solution is illustrated conceptually in Figure 4.13, showing a single application with two users. User 1 is doing short-range plans every month; User 2 is doing long-range plans every quarter. Additionally, User 2 is doing a side case to study the effect of new crude oil prices in the long range. A block of data (e.g., process unit yields and product qualities) is common to all three studies and can be shared (area ABC in the diagram). The short range differs from the long range, however, in data such as product demands, crude prices, and so on. Such data must be duplicated—one block for User 1 (area ADB) and one for User 2's base case (area AEC). Finally, in order to do the side case, the crude prices that are

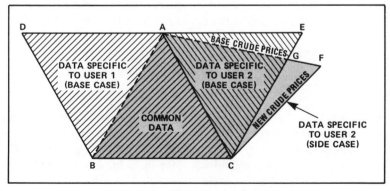

Figure 4.13 Concept of data sharing.

already specific to User 2 must be duplicated with new data (area CGF). In order to build a full model, each study requires a full set of data made up as follows:

BASE1		ADB + ABC
BASE2	AEC	+ ABC
SIDE2	CGF + (AEC − AEG)	+ ABC

Notice that this arrangement depicts the minimum amount of data required to represent all three studies. There is no unnecessary duplication. Furthermore, any change to a piece of common data in ABC will be immediately reflected in each study, as it should be, whereas continual alteration of crude prices in study SIDE2 (area CGF) will in no way affect what is happening in BASE1 or BASE2, as it should not. Each study can continue independently and yet share common data. Note, here, that if a table is used as a unit of data, the argument for which has already been presented, then there may be some minor duplication of data, since not all the values within a table may be different from case to case. However, the logical grouping of data into sets will tend to reduce this redundancy.

The solid lines in Figure 4.13 suggest how this concept can be translated into an ACTPROB structure on an ACTFILE. Suppose three ACTPROBs named BASE1, BASE2, and SIDE2 are set up. On BASE1 is a full set of tables, including those that contain common data and those that contain data specific to User 1 (area ACBD in the diagram). On BASE2 are only those tables specific to User 2's base case (area AEC). They will, of course, have the same names as the corresponding tables on BASE1, but with different values. On SIDE2 are the crude price tables with the side-case values in them (area CGF).

While this effectively segregates the data, it does not solve the problem. How does ACTPROB SIDE2 know, for instance, that it is to use demand data from BASE2 and not from BASE1? There are several approaches to solving this problem; one is the "parent/child" approach. If ACTPROBs have to be set up in sequence, then the three ACTPROBs could have evolved as:

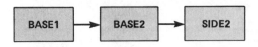

Here, BASE2 is the "child" of BASE1 and the "parent" of SIDE2. Now, if SIDE2 requires data it does not "own," then it will seek this data from its parent. If the parent does not have it, SIDE2 will seek it from the grandparent, and so on. This is a simple concept and not difficult to implement, but it suffers from several drawbacks. If the concept is extended to allow multiple children from one parent, then no intersibling data sharing is possible. More importantly, it prohibits the presence or absence of a table from indicating to the matrix generator how to build the matrix. Of course, it is possible to write matrix generator programs that do not rely on this type of information, but then they require more control tables to guide them. In PLATOFORM, we preferred an approach that allowed flexibility in how the matrix generators would be designed and that permitted data sharing between any ACTPROBs, regardless of their ancestry.

The key to this PLATOFORM approach lies in another table catalog, but one that is maintained wholly by the system. Its name is CONT. . .C (Contents Catalog), and every ACTPROB on the ACTFILE has its own CONT. . .C. In essence, these catalogs contain a list, as their stub, of all the individual tables (not sets of tables) that the ACTPROB requires. A column of the catalog indicates on which ACTPROB each table is physically resident; it acts as a "pointer" to the required ACTPROB.

To see how this important concept functions, take a much simplified example based on the data-sharing environment discussed for Figure 4.13. Assume that only four tables are involved, namely,

Y P . . . . . T	Process yields	Common data
QB . . . . . T	Product qualities	Common data
B S . . . . . T	Demand data	Specific to User 1 or 2
F C . . . . . T	Crude prices	Specific to all three studies

As a first step, set up a single ACTPROB, BASE1, and load in all four tables. The BS and FC tables will contain values specific to User 1. During this process, the system will compose and file a fifth table on this ACTPROB—the Contents Catalog—that will record the presence of the four data tables. At this point, ACTPROB BASE1 will contain five tables, as shown in Figure 4.14.

Here, the Contents Catalog shows that ACTPROB BASE1 has access to four data tables and that all four are physically resident on this ACTPROB (the "pointers" are self-pointing).

To set up a second ACTPROB, it is necessary to nominate an existing ACT-PROB from which to create it. So, ACTPROB BASE2 is now created from BASE1. The system action is to set up a second ACTPROB (BASE2) and file on it a physical copy of the CONT. . .C from BASE1. No other tables need to be copied across. Figure 4.15 shows the state of the ACTFILE at this point.

Notice that BASE2 at this point also has access to four data tables, but that none of them are physically resident on ACTPROB BASE2. The pointers all refer to ACTPROB BASE1. BASE2 is an exact logical copy of BASE1.

Now, User 2 must update or replace the BS and FC tables to reflect his or her own data. Any such updating results in a copy of the table being moved across to the new ACTPROB, in the data values being modified, and in the

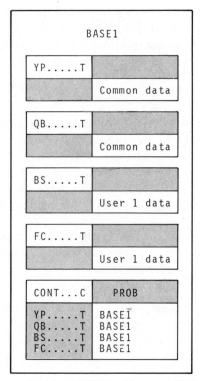

Figure 4.14   An ACTFILE with one ACTPROB.

CONT. . .C pointer being changed to reflect the new residency. Figure 4.16 shows the state of the ACTFILE after the updating.

Finally, User 2 wants to do the side case. Therefore, User 2 creates a third ACTPROB SIDE2 from BASE2 and updates the contents of the FC table, giving the final ACTFILE structure shown in Figure 4.17.

Notice the economy of data storage for study SIDE2. In practice, such a study may require access to some 300 data tables, but may involve the physical storage of only two tables—a Contents Catalog and a data table.

This approach, of course, parallels very closely the parent/child approach. But, by use of a Contents Catalog, it provides the additional flexibility required to overcome the latter's shortcomings. A data-base management system component can now provide capabilities to indirectly edit these catalogs, thus permitting unwanted tables to be deleted or cross-references to any sibling ACTPROBs to be made by changing pointers. Furthermore, a second column of these catalogs can be used to "date stamp" each table whenever it is modified (or created), thus providing an effective audit trail for use by other system components. These features are discussed more fully in Chapter 7.

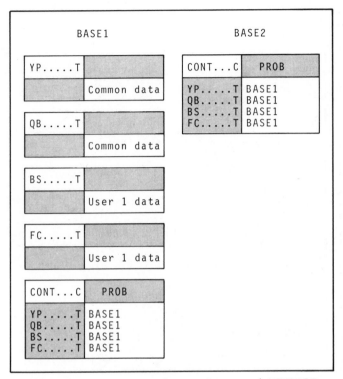

Figure 4.15   An ACTFILE after creating second ACTPROB.

4.5 ANNOTATING DATA TABLES

The automatic date stamping referred to at the end of the last section provides one way of monitoring data changes. While this identifies exactly when a data change was made, it does not indicate why the change was made. Monitoring data changes can be an important function, particularly where a model is being used by different analysts. Also, one of the original design goals of PLATOFORM was to enable an analyst to annotate the data in any table. These data table "comments" would become an integrated part of the data base, to be printed whenever the "host" table to which they referred was printed.

To do this, it was necessary to introduce a fifth class of tables, known as "Comments Tables." An eighth character of "X" is used to distinguish these comments tables (see Figure 4.18). The comment and host table are linked by having the first seven characters of their table names identical.

Any number of lines of fifty-six character (maximum) text can be included in a comments table, each line being identified by a unique line number that forms the stub of the table. An example is shown in Figure 4.19.

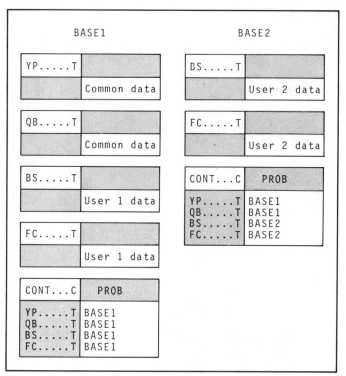

Figure 4.16   An ACTFILE after updates to tables on BASE2.

The type of syntax required to enter and edit such comments is discussed in Chapter 7. Note, however, that the adoption of this type of data structure adds significantly to the complexity of the supporting programs. These programs, in order to retain the integrity of the data base, need safeguards to prevent comments tables from becoming physically separated from their hosts as a result of Contents Catalog cross-referencing.

Notice also that this way of handling comments effectively rules out the possibility of having T-tables and Z-tables of the same name (apart from the eighth character), since the comment table's host is then ambiguous. Although DATAFORM allows identically named T- and Z-tables, this was a degree of flexibility deliberately forsaken in the design of the PLATOFORM data base.

4.6 CATALOGING ACTPROBS

The introduction of a multi-ACTPROB environment implies the need for a catalog that will record the names of all ACTPROBs created on the ACTFILE. By this

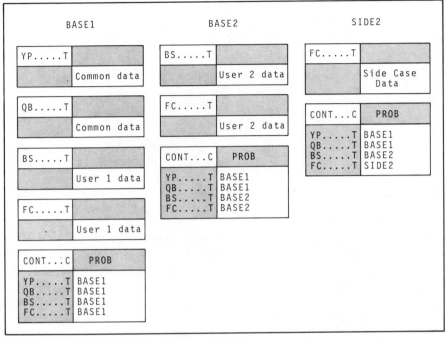

Figure 4.17   An ACTFILE with three ACTPROBs.

means, requests to create ACTPROBs that already exist can be quickly trapped; references to ACTPROBs assumed to exist can be checked; and a number of global system routines can be developed to carry out functions on the entire ACTFILE. A typical ACTFILE in routine use might contain as many as thirty different ACTPROBs at any one time; their control is carried out through the problem catalog named in PLATOFORM PROB. . .C. This catalog is wholly maintained by the system and can in no way be updated by the system user. An example of a PROB. . .C is shown in Figure 4.20 for the ACTPROB environment discussed in Section 4.4.

Whenever a new ACTPROB is created by the user (e.g., ACTPROB SIDE2), the system will automatically record this in PROB. . .C. Utility routines can be made available to remove ACTPROBs from the system, and they will modify PROB. . .C accordingly.

Figure 4.20 refers to two ACTPROBs not previously mentioned—DATABASE and DUMMY. These are two system ACTPROBs required for any PLATOFORM application and are set up when an ACTFILE is initialized. DATABASE is used primarily to store the application dictionaries and system catalogs, since these apply globally across all ACTPROBs. PROB. . .C is stored on ACTPROB DATA-

SUMMARY OF PLATOFORM TABLE CLASSES	
8th Character of Name	Class
C	Catalog
D	Dictionary
T	Numeric data table
Z	Control data table
X	Comments table

Figure 4.18   PLATOFORM table classes. All classes except T are DATAFORM Z-tables.

BASE. The concept, here, is that the dictionaries and catalogs describe an entire application, while the ACTPROBs define different cases within that application. Contents Catalogs do not "point" to dictionaries or catalogs; they point only to data tables (classes T, Z, or X).

DUMMY is an empty ACTPROB with an empty Contents Catalog. Its sole purpose is to provide a starting point for the first (or any subsequent) "case" ACTPROB, since all ACTPROBs must be created from some predecessor.

SB.F.U.T	SUMX	V1MX
FLM1	3	500
FLM2	3.5	500

SB.F.U.X	
L01	DATA TAKEN FROM CABLE 10 DEC 1981 FROM FAWLEY
L02	SULFUR RELAXATION OBTAINED FOR FEBRUARY

Figure 4.19   An annotated table. When the host table SB.F.U.T is printed, the textual contents of the comments table are added as footnotes.

PROB...C	DATE	MODEL
DATABASE	80100601	NO
DUMMY	80100601	NO
BASE1	80121002	YES
BASE2	80121101	YES
SIDE2	80121103	YES

Figure 4.20   A Problem Catalog.

4.7 SUMMARY

This chapter discusses some of the problems involved in structuring a data base within a mathematical programming environment and how these problems were resolved in Exxon's PLATOFORM system.

We have seen how a multidictionary approach, based on entity classes, can be used to enable short codes to retain some mnemonic significance. These short codes are necessary not only to pack the maximum of information into a two-dimensional table structure, but also to build the eight-character row and column names that will eventually form the LP matrix identifiers.

This use of multidictionaries leads to a need to catalog data tables in some detail, a task that can be greatly reduced by grouping data items into structurally consistent sets of tables.

These concepts lead to a set of structural rules that apply to any PLATOFORM application and result in a degree of mnemonic discipline necessary for the development of powerful system components. Adherence to these rules allows any PLATOFORM data table to be fully and automatically translated when printed. Routines can also be developed that allow individual entity codes to be changed wherever they occur throughout an entire data base (Chapter 7). Similar catalogs can be used in conjunction with the dictionaries to allow automatic translation of matrix row and column names when solutions are printed (Chapter 10). If the rules are broken, users lose some of these advantages, but the system will still function.

This chapter also shows how data can be shared without unnecessary duplication by using a higher level subdivision of the data file and introducing a cross-referencing concept based on Contents Catalogs. Chapter 9 presents further ways of reducing data redundancy by introducing the concept of "data overlay."

Before going on to see how users can interface with this data base, we must first, in Chapter 5, address the question of how to structure the system components (program modules) in the same way as data components.

Kenneth H. Palmer

5. SYSTEM STRUCTURE

5.1 INTRODUCTION

In Chapter 4, we saw how the design of a data structure is, to some extent, dependent on the nature of the underlying language that will be used to develop the supporting system. The chosen structure must be consistent with the structure supported by the language. For DATAFORM, the data structure had to be based on two-dimensional tables conforming to certain DATAFORM rules. Given such a structure, program modules can be written in DATAFORM to access and manipulate these tables. Now, we want to consider how these program modules can themselves be structured and linked so as to form a single integrated system. Clearly, the choice of structural options here is more heavily dependent on the nature of the underlying mathematical programming system (MPS) than in the case of data structure.

In following the development of the PLATOFORM system structure, it will be necessary to discuss some of the fundamental features of the Enhanced Mathematical Programming System (EMPS) that have influenced this design. The basic design issues, however, are independent of the underlying MPS, which will influence only their resolution and implementation.

Before considering these design issues, we must first clarify what we are trying to achieve. What is meant by a "single integrated system," and why is it important? Before the development of PLATOFORM in the early 1970s, linear programming (LP) models (at least within Exxon) tended to be supported by a heterogeneous collection of programs; each performed a particular function and frequently ran quite independently. There might be one tool for updating data, a quite different tool for generating a matrix, a third tool—usually the MPS itself —for finding an optimal solution, and a fourth tool for translating the solution values into readable reports. Often, several machine runs, or at least machine steps, were required to solve any one problem. Each tool would be written in the computer language best suited to its task.

Such systems had obvious disadvantages for the users. Elapsed times to complete a job were long; setting up the necessary series of jobs required systems skills; communication among the separate phases was difficult and provided little flexibility to control the course of a study.

One of our prime objectives in PLATOFORM was to provide users with just one system capable of carrying out any or all of the above tasks in a single machine run. This, in turn, implied providing users with a simple syntactic language with which they could specify a series of actions required to carry out a task. Such a task (for example, a side case) might typically involve the following actions:

- Set up a new ACTPROB as a logical copy of the base case.
- Modify some data.
- Regenerate those parts of the matrix affected by the change in data.
- Reoptimize, starting from the optimal basis of the base case.
- Print a solution.
- Run a set of output reports.
- Delete the new ACTPROB.

A single integrated system would allow the user to define these seven action steps (or any other combination of actions) through a single input stream using the syntactic language provided.

This chapter describes the system structure required to support such a capability. Later, in Chapter 6, we demonstrate the syntactic language that can be developed to provide user interface. We see that these are two quite independent issues that can be reconciled by means of a simple translation module. This vitally important concept means that the system structure can be designed without regard to the ease of user interface.

5.2   MODULARITY

In most MPSs, the optimization algorithms are called from the basic language of the MPS. This will be referred to as the Program Control Language (PCL). These algorithms are accessed via PCL calls to procedures such as PRIMAL, WHIZARD (an EMPS optimizer), PARARHS (for parametrics), and SOLUTION. In addition, the PCL will normally support calls to precompiled programs written in computer languages supported by the MPS (e.g., DATAFORM, FORTRAN, PL/I, etc.; see Figure 5.1). Since the optimization phase of a typical job occurs somewhere in the middle (i.e., after matrix generation and before report writing), the basic design of the PCL already imposes some form of modularity on the system design. A single large FORTRAN program, for instance, containing all the functions of data management, matrix generation, and report writing, would not suffice, since we would need to pause in the middle of the program to access the optimization algorithms.

It is true that recent trends in the design of MPSs have brought the concept of a single, all-embracing program module closer to reality by permitting access

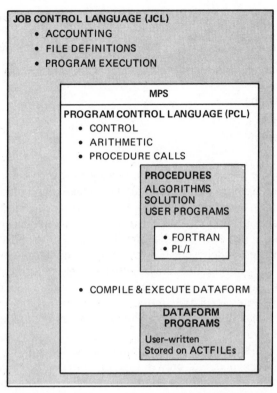

Figure 5.1 System environment for running an MPS (based on EMPS).

to the algorithms from a host language other than the PCL (e.g., IBM's MPSX370). In the future, PCLs may be gradually phased out, with all of their capabilities transferred to a host language, or vice versa.

The real question is: Even if the capability existed, should we aim for one large program module or for many smaller modules? Let us look, first, at some of the advantages of a multimodule approach and then examine the consequences.

First, the functions involved in using a mathematical programming model can be naturally broken down into a series of well-defined activities. For example:

- Data management.
- Matrix generation.
- Optimization.
- Report writing.

These already suggest a modular approach that could be matched by corresponding program modules. Such a matching would provide flexibility to vary the sequence in which the activities are carried out. For example, a user may wish to start a run by getting a report from an optimal solution obtained the previous day, and may then go on to optimize a second case (from the same matrix) and to run some reports on this case—implying the following sequence of activities:

- Report writing.
- Optimization.
- Report writing.

A modular program structure could more easily handle this resequencing.

A second, and perhaps more important, advantage of the modular approach concerns long-term system evolution. If the system is designed to have a long life and to be able to support a wide spectrum of mathematical programming applications, then it can evolve from a small beginning by increasing the number of program modules as needs arise. Each of these modules can be developed independently. It is, for instance, unlikely that a single matrix generator will be capable of serving the needs of all applications. Generators for other matrices can be developed later as separate modules, but the prototype system need not consider them. Once the basic system structure is established, the system is free to evolve naturally.

Splitting the total system into discrete modules also has important implications for program maintenance. Not only is it easier to maintain a smaller program, but the maintenance load can also be more readily spread among a group of system analysts, each responsible for one or more entire modules. If the system is to become large and comprehensive, this is an important consideration.

Thus the obvious advantages of a modular approach are considerable, but what does this mean in terms of overall system design? At first, it might appear that such an approach is regressing to designs used in earlier systems, with all of their associated problems. How can a multiprogram design be reconciled with our goal of a single integrated system? Clearly, some form of system control structure needs to be developed to tie together and sequence the execution of the various modules, just as the use of multidictionaries led to a Generic Table Catalog in the design of a data structure.

The question, then, is what language should be used to develop this control structure. It turns out that if the MPS control language (PCL) is used, we can resolve the conflict by creating a single global program in PCL while retaining full modularity at a lower level (DATAFORM, FORTRAN, procedural algorithms, etc.). This concept, of course, depends on the MPS used, and not all commercially available MPSs will support the development of such a system structure. Communication between the PCL and its subordinate languages/procedures is all-important here. In this respect, EMPS and its associated language DATA-FORM are particularly strong.

5.2.1 Communicating with PCL

From this point, we follow the development of the PLATOFORM system structure with respect to EMPS/DATAFORM. The essential ingredients for the development of such a structure become apparent, and many of these ingredients can be found in other MPSs. The precise form a system structure takes, of course, will be dictated by the available ingredients.

A key feature of EMPS is an area of memory referred to as the Communication Region (CR), which is set aside throughout the execution of a PCL program. This region is divided into a large number of "cells" (about 380). Their purpose is to pass information to and from the subsidiary procedures. Most cells have a particular prespecified function to perform, and each can be referred to by name. Thus the cell XOBJ is used to tell the optimizing algorithm the name of the matrix row to be used as the objective function. Other cells are set aside for use by the system designer. The one feature that has contributed most to the construction of the PLATOFORM system structure is the ability of the subsidiary language DATAFORM to both read from and write to the CR. Without this capability, the twin goals of a single integrated system combined with lower level modularity would have been difficult to achieve.

To see how this capability helps to design a PCL structure that provides all task capabilities within a single program and yet retains flexibility to vary the task sequence, consider the chain of PCL statements illustrated in Figure 5.2. This example employs one of the CR cells (XCHAR10) to direct the program flow to an appropriate activity represented either by a DATAFORM program or an algorithmic procedure. Thus, if XCHAR10 contains the name

```
L1          •
            •
            •
            IF        (XCHAR10.EQ.'DBM',ZDBM)
            IF        (XCHAR10.EQ.'GENER',ZGENER)
            IF        (XCHAR10.EQ.'OPT',ZOPT)
            IF        (XCHAR10.EQ.'REPORT',ZREPORT)
            IF        (XCHAR10.EQ.'STOP',ZSTOP)
            GOTO      (LERROR)
ZDBM        EXDF      ('NAME','DBM')
            GOTO      (L1)
ZGENER      EXDF      ('NAME','GENER')
            GOTO      (L1)
ZREPORT     EXDF      ('NAME','REPORT')
            GOTO      (L1)
ZOPT        READY     ( - - - - )
            WHIZARD
            SOLUTION
            GOTO      (L1)
ZSTOP       EXIT
```

Here the EXDF statements are calls to named, precompiled, DATAFORM programs.

XCHAR10 is a reference to one of the CR cells available to the system designer.

READY, WHIZARD, and SOLUTION are procedural calls to the optimizing algorithms and peripheral functions.

Figure 5.2 Part of a PCL program.

"GENER," the program will invoke the execution of the DATAFORM program named GENER. If XCHAR10 contains "OPT," the program will invoke the procedures READY/WHIZARD/SOLUTION in that sequence. These will prepare a matrix for optimization, find the optimal solution (if one exists), and print a solution.

In Figure 5.2, the GOTO (L1) statements are included to imply some recycling through the program so that a number of different activities can be executed in any user-requested sequence. The question now is: What should happen at L1? How should the contents of cell XCHAR10 be set successively in response to an activity sequence requested by a user?

Suppose that at L1 we execute another DATAFORM program whose task is to read in from an external file a required list of activities. Then, since DATAFORM can "talk" to the CR, this same program can store in the cell XCHAR10 the first name on the list before EXITing. At the next entry into this program, the second name on the list will be stored in XCHAR10, and so on until the list is exhausted. DATAFORM accesses an external sequential file by means of the READTAB verb (see Chapter 3). This verb will read in data arranged as DATAFORM tables contained in a single data set identified by surrounding NAME and ENDATA records (see Figure 5.3).

The basic functions, then, of this controlling DATAFORM program, which is called PLEX (PLATOFORM EXECUTOR) in PLATOFORM, are:

- READTAB—data set EXCTL.
- Set XCHAR10 with next name in column ACTION of table EXCTL.

Several technical problems remain to be solved before such an approach can be implemented. One is to provide a mechanism whereby PLEX can determine how many prior entries to the program have been made in order to select the correct action from the control table. This problem can be solved by utilizing another CR cell as a counter. The need to READTAB the data set each time can be avoided by filing the control table (EXCTL) in the data-base ACTFILE and

NAME	EXCTL	
EXCTL	**ACTION**	
001	DBM	
002	GENER	
003	OPT	
004	REPORTS	
005	DBM	
006	STOP	
ENDATA		

Figure 5.3 A data set containing a control table readable by DATAFORM. Several possible syntax options exist to express the table that is pictured here. Actions are listed in an internal column of the table to accommodate possible duplicates. Stub names are arbitrary, but unique.

removing it at the final entry. Storing the table here enables other program modules to have access to it.

Suppose a PCL statement (see Figure 5.2) is included at label L1 as follows:

L1   EXDF('NAME', 'PLEX')

This, then, affords the germ of a system structure in which a single PCL program contains all the capabilities required to complete a mathematical programming task and the flexibility to execute individual activities within that task in any sequence (see Figure 5.4). This structure is made possible by intercommunication through the CR region. But, of course, there is still a long way to go. We must address such questions as:

- How can we communicate to each broad activity (DBM, GENER, etc.) what details we want that activity to include?
- How can we implement our evolutionary plan for system growth (by adding new modules) without continually expanding the underlying PCL program?

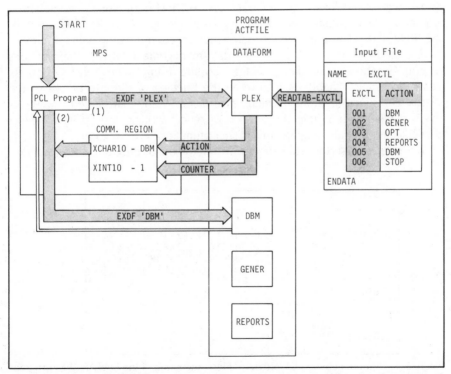

Figure 5.4   An embryo system structure. The PCL program provides a single integrated system. DATAFORM programs are separate modules callable in sequence from PCL.

An answer to the second question is suggested by the PCL program shown in Figure 5.5. In this program, only the key words OPT and STOP require special action. The program need not know the names of other DATAFORM program modules. This independence permits these modules to develop without expanding the PCL program. This, of course, is still a simple approach, but it illustrates the essential ingredients from which a full-scale PCL program can be built. Most of the problems center around accessing the optimizing algorithms, which can include nonlinear procedures and require exhaustive off-normal processing actions. Some of these problems are pursued further in Chapter 10. It took about five years before a stable PCL program suited to all applications evolved. This program eventually comprised some 1,500 statements.

Let us now return to the first question to see how we can communicate more detailed instructions to each major activity.

5.2.2 Modules and Submodules

So far, we have talked about separate DATAFORM programs to handle entire activities—data management (DBM), matrix generation (GENER), or report writing (REPORTS). These are very broad activities, each encompassing a number of discrete functions. Thus within the data-management activity, the functions of printing tables and updating tables are quite distinct. The activity of matrix generation might enable us to foresee the building of discrete blocks of the matrix as distinct functions, particularly if pieces of the matrix are to be selectively regenerated. Within the activity of report writing, the ability to select particular reports also implies that distinct functions are necessary.

This suggests a further substructure within the major program modules them-

```
                    .
                    .
                    .
ZPLEX     EXDF      ('NAME','PLEX')
          IF        (XCHAR10.EQ.'OPT',ZOPT)
          IF        (XCHAR10.EQ.'STOP',ZSTOP)
          EXDF      ('NAME',XCHAR10)
          GOTO      (ZPLEX)
ZOPT      READY     ( - - - - )
          WHIZARD
          SOLUTION
          GOTO      (ZPLEX)
ZSTOP     EXIT
```

Figure 5.5   A more general PCL program. Here, the name of the DATAFORM program to be executed is referred to indirectly via cell XCHAR10. The more descriptive label ZPLEX is used in place of L1.

selves. Just as the overall system design demonstrates how a control module (PLEX) can invoke (via the main PCL program) a sequence of DATAFORM programs (or PCL procedures), it is possible to envisage each of the DATAFORM programs containing a control segment that itself invokes, in turn, a sequence of subsegments. In DATAFORM, as in most such languages, this type of structure can conveniently be achieved by the use of subroutines. Figure 5.6 illustrates this concept of second-level modularity.

Now, we must determine how to pass instructions to these modules and submodules so that they can carry out their tasks in a predefined sequence. To

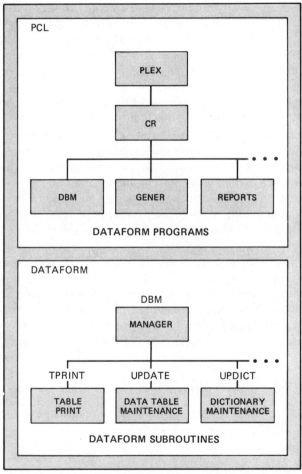

Figure 5.6   Concepts of modularity (top) and submodularity (bottom). Each of the major modules (DBM, GENER, etc.) is a single, complete, DATAFORM program that is itself substructured into a number of functional submodules (subroutines) under the control of a "manager" component.

help understand this process, consider an example in which a user wants to start a run by using the DBM to perform two tasks:

- Change a fuel oil specification in the U.K. in time period M2.
- Print all the U.K. product specification tables.

Previous discussion showed how the DBM module (Figure 5.4) could be accessed by including this request in the data set EXCTL to be read by the PLATO-FORM control program PLEX. Now, within the DBM two functional submodules need to be accessed—first, UPDATE to change the data in one of the tables and, second, TPRINT to print a set of tables. The manager component of DBM requires this information to control the ordered selection of the appropriate subroutines. One way to do this would be to add a second data set to the input file to be read (via READTAB) by the DBM. Such a data set might have the form displayed in Figure 5.7.

In this example, it would be the function of the DBM (manager component) to READTAB the data set named DBM and hence the table DBM. A simple loop through the column ACTION of this table would then control the ordered calling of the appropriate subroutines. But this simple strategy will not suffice. Suppose DBM is used more than once in the same run. Figure 5.3 gave an example where DBM was required twice: once at the start of the run and again after REPORTS. Which input data set should DBM read when it is called the second time? Since READTAB needs to know the name of the data set to be read, the second data set cannot also be named DBM, since the system may select the first data set again.

There are a number of ways around this problem. The preferred solution, which retains maximum flexibility and avoids the problems of precise sequencing of data sets on the input file, involves passing an additional piece of information via the CR. If the control information to PLEX is expanded by including the name of the data set required by each major module, then this data set name can be passed to the module through another CR cell (see Figure 5.8). This structure enables the user to call any of the major modules several times during a single run.

However, we still have not conveyed all of the information required. The

Figure 5.7 Input data set used to pass required submodule names to DBM.

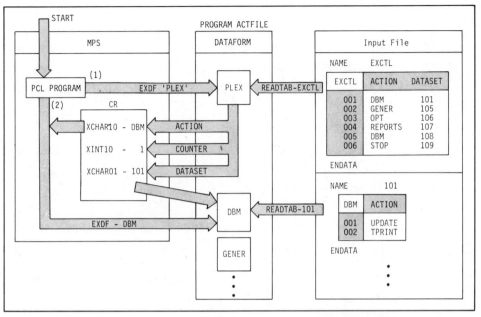

Figure 5.8   System structure passing input data set names to major modules. Numeric characters are used for data set names to ease system development. CR cell XCHAR01 is used to communicate the name of the data set to be read by the major module.

submodule UPDATE needs to know what data changes to make to which tables, and TPRINT will need to know which tables to print. Instructions to these steps can be handled by a simple extension to the hierarchical structure of the input file (illustrated in Figure 5.9). The CR does not need to be used to pass information this time, since control of the submodules is all within a single DATAFORM program.

In the example just discussed, several instances of detailed syntax have been illustrated (e.g., the control information for TPRINT in table TPCTL). This syntax and the reasons why UPDATE requires two separate data sets are discussed more fully in Chapter 7. The purposes here are to illustrate the mechanism that holds the system structure together and to show how a single integrated system can be both modular and submodular.

It is important to notice the nature of the sequential (as compared to random-access) input file with its strongly hierarchical structure, linked by cross-references to data set names. This structure is necessary to drive what is essentially a hierarchically structured system. It is not an input structure that naturally lends itself to a simple user interface. For reasons mentioned at the end of Section 5.1, discussion of the ease-of-use issue has been deliberately avoided. As we see in

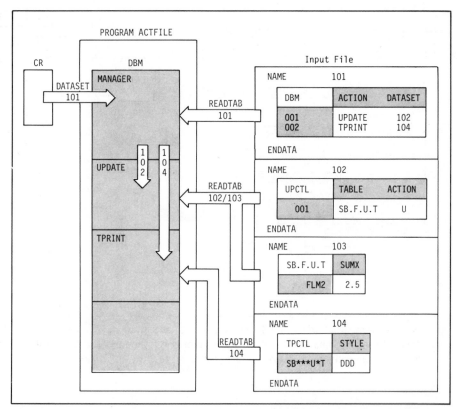

Figure 5.9   System structure (submodules). Manager reads data set whose name is in CR cell XCHAR01 (101); manager calls UPDATE passing data set name 102; UPDATE reads data set 102 (and later 103—the increment of 102) to make the required updates, returns control to manager; manager calls TPRINT passing data set name 104; TPRINT reads data set 104 and prints the requested tables; DBM then returns control to PCL, which calls PLEX to obtain the name of the next major module.

the next chapter, the actual user input to PLATOFORM is totally different from the input file described here. The user input file is built by a system component (a PCL-callable FORTRAN program called PLATORD) from a very simple (sequentially ordered) user input syntax. Thus all of the intricacies described above are completely hidden from the user, and by generating the input mechanically, the system can be made very robust. There is no need for the individual system components to check the validity of the data set cross-references. This concept of divorcing the system needs from the user needs leads to much simpler supporting programs—by having both an input format that already matches the system structure and one that is known to be free of inconsistencies. To complete this section, Figure 5.10 shows a PCL program that includes this input conversion step.

5.3 OPEN-ENDEDNESS

We have now seen how a basically simple system structure can be used to support a growing number of system components. Any number of discrete DATAFORM programs, each performing a specific activity, can be included on the PROGRAM ACTFILE with no change to the basic system. We could, for instance, include a data preparation module (called, say, DPREP) to convert data tables from one format to a format required by a matrix generator. We would need only to compile this new DATAFORM program and add it to the PROGRAM ACTFILE for the new member to be immediately recognized by the system.

The program itself would have to include a READTAB of the data set whose name is in XCHAR01 to link into the system framework. The contents of this data set would reflect the input requirements of the program as set by the design specifications for the module. PCL, PLEX, and even PLATORD would require no changes (the general design of PLATORD is described in Chapter 6). Nor would the surrounding Job Control Language (JCL) require changes, since a single physical file would store all the DATAFORM programs.

With the gates wide open for system evolution to begin, we must now turn our attention to how this evolution can, or should, be controlled. To do so, we need to reflect on the ultimate purpose of the system. The initial purpose of Exxon's PLATOFORM system was to support supply/refining models throughout Europe. This goal itself implied many users with models of varying complexity. Moreover, it was hoped that the system would ultimately be used for any mathematical programming model throughout Exxon, including models not necessarily connected with supply/refining functions.

```
          PLATORD
ZPLEX     EXDF        ('NAME','PLEX')
          IF          (XCHAR10.EQ.'OPT',ZOPT)
          IF          (XCHAR10.EQ.'STOP',ZSTOP)
          EXDF        ('NAME',XCHAR10)
          GOTO        (ZPLEX)
ZOPT      READY       ( - - - - )
          WHIZARD
          SOLUTION
          GOTO        (ZPLEX)
ZSTOP     EXIT
```

Figure 5.10   PCL program including input translation. FORTRAN module PLATORD takes a simple, sequentially ordered user input stream and creates the hierarchically structured input file required by the rest of the system.

The initial system, developed in 1973, contained just seven DATAFORM components, specifically tailored to handle the large European Regional model used by Regional Headquarters in London. These were:

PLEX   Basic control module
DBM   Data-base management
UTILITY   A set of file-management routines
GENER   A matrix generator
OPT   A module controlling selection of appropriate PCL procedures
FINSBS   A financial side-by-side report writer
REPORTS   A collection of solution reports

The initial system also contained the non-DATAFORM PLATORD and PCL components.

When the system was introduced to the European affiliates for use with their local models, it became apparent that certain of these modules (notably PLEX, DBM, and UTILITY) required no modification, others could be used with minor enhancement (e.g., GENER, OPT), while others (REPORTS) often required complete replacement. This led to the concept of classifying program modules according to their degree of generality. PLEX, DBM, and UTILITY were classified as Type-1 modules, GENER and OPT as Type-2, and REPORTS as Type-3. This classification was important in determining an organizational structure for maintaining these modules. Maintenance of Type-1 modules was centralized in London, but locally developed Type-3 modules needed to be maintained locally.

It was no surprise that modules such as DBM turned out to be completely general in purpose. They had been carefully designed that way, and, since they were geared to an underlying data-base structure to which all applications had to adhere, this was not difficult to achieve. With the matrix generator, the situation was different. Although it obtained its input from a data base of known structure, it had to build a matrix that was a direct representation of the planning environment being modeled. This environment differed in detail from application to application.

While the Type-1 and Type-3 program modules present no problems with regard to maintenance responsibilities, maintenance of the Type-2 program modules is less easily defined. The matrix generator is a typical example. These modules also tend to be among the largest of the system components; the original PLATOFORM generator comprised about 10,000 DATAFORM statements. It is not practical to allow each application group to alter the basic 10,000 statements any way it wants in order to meet local requirements, since all hope of central control and subsequent sharing of matrix generation techniques would then be lost.

Modules such as these require special design considerations. We have already seen how PLATOFORM is open-ended with respect to the addition of entire modules (DATAFORM programs). The problem, now, is to structure any one of these modules to be open-ended itself with respect to the addition of new submodules. If this can be achieved, the maintenance problem can be resolved by providing, centrally, a standard module consisting of a manager segment and a set of standard submodules. Each application group can then substitute any of the standard submodules or include additional submodules of its own. These locally written submodules would also be locally maintained. For this arrangement to work successfully, any substitutions or additions would need to be made without alteration to the centrally distributed base program. Even alterations to the manager segment should not be made.

Understanding of how this was implemented in PLATOFORM requires detailed knowledge of the way DATAFORM handles and compiles subroutines. Although a detailed explanation along these lines is outside the scope of this book, it is appropriate at this point to briefly review the main problem and to point to the type of solution adopted in PLATOFORM.

Basically, the problem concerns the design of the manager segment. Broadly, the functions of this segment are to READTAB the appropriate input instructions from the input file and then to pass control to each of the submodules requested in sequence. If the module is open-ended, this may involve calling a subroutine whose name is not known to the "manager." Unless the language supports indirect references to subroutine names (and most, including DATAFORM, do not), this can present a problem to the programmer.

One way out of this dilemma—the one adopted in PLATOFORM—is for the manager, whenever it encounters an unknown subroutine, to call instead another subroutine, one that is external to the base source program. This subroutine will, in turn, pass control to the required application subroutine. The external subroutine, whose name is fixed (see GNMISC in Figure 5.11), becomes the property of the application group and contains within it all the submodules specific to that application.

The compiled program consists of the combination of the two source codes that effectively segregate standard coding from application coding. Therefore, these programs can be maintained separately. Any number of application submodules can be embedded within the external subroutine without the need to change a single program statement in the standard block. Each application group will have its own GNMISC subroutine (see Figure 5.11).

It is important to note that this type of open-ended structure is not necessary in modules wholly maintained by one systems group. If a new submodule is added to one of these, then the manager can be made aware of the new addition at the same time.

Chapter 9 returns to the design of matrix generators and shows that there are other areas where open-endedness is important. In each case, this can be easily supplied by the provision of a "hook" to an application subroutine of predefined name.

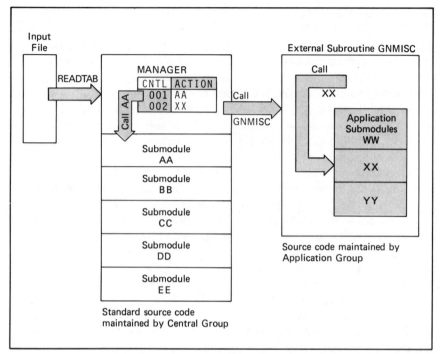

Figure 5.11   An open-ended module. When manager encounters an unknown sub-module (XX), it calls instead subroutine GNMISC, passing the name XX. GNMISC then calls subroutine XX. Application subroutines are nested one level deeper than standard subroutines.

The preceding paragraphs have demonstrated the influence of the total environment on the structural design of system components. These factors have to be recognized at the outset. If a system is being built to support a single application and will be maintained by a single group of systems analysts, there is no need to build open-ended components; the submodular nature of these components already provides sufficient scope for rational expansion. An important point is that the subsequent maintenance of a system is an issue that must be addressed before, and *not* after, the system is designed.

Of course, it is not always possible to foresee the full impact of future system evolution. Exxon could not have predicted, for instance, that six years after the first introduction of PLATOFORM, the initial seven DATAFORM modules would have grown to more than 200. Furthermore, over two-thirds of these modules were Type-3 (i.e., maintained by local application groups), and PLATOFORM was supporting more than a hundred different applications in all regions worldwide throughout the Corporation. Central support could no longer be truly central, but needed to be divided between two supporting locations separated by 3,000 miles.

In the remainder of this chapter, we consider how such an explosion in system usage influences further enhancements to the system's design structure. These are necessary to ensure proper control and maintenance of a major investment in software that quickly becomes an indispensable part of a corporation's business.

5.4 SYSTEM UNIFICATION

Once a system has evolved to the point at which it is a recognized tool used throughout a large corporation, then a highly professional approach is necessary to both system maintenance and system documentation. Enforcement of this professional approach can be achieved only if as much as possible of the system comes under control of a highly skilled professional group of systems analysts.

The initial open-ended structure of the system design rightly encourages diversification. This is necessary, in a way, to establish the system and to promote its acceptance. It is also an effective way to determine the true needs of a wide cross section of different applications. Later, this diversification must be discouraged by reducing the scope for local extensions, but this must be done in a way that will not deprive users of existing capabilities.

In order to see what problems this goal of system unification presents, we must examine the chief areas in which diversification can occur. With a large number of separately maintained program modules and submodules, some degree of functionality overlap is inevitable. One of the goals of a unification strategy must be to seek out this overlap and to merge small, specific modules into larger, more general-purpose modules. This is an ongoing process that must encourage a slow migration from locally supported Type-3 programs to centrally supported Type-1 programs. This process does not present any fundamental problems, but it should not be expected to remove entirely the need for modules that are locally supported, or at least application-specific.

Reducing diversity at the level of the surrounding system environment can influence the basic system design. We are talking, here, about the central PCL programs and the surrounding JCL.

5.4.1 Unifying PCL

The design of an appropriate PCL program, which forms the integrating component of any PLATOFORM system, was deliberately left open when PLATOFORM was first introduced. Two reasons for this were:

- The lack of sufficient information on the required capabilities.
- The apparent difficulty in designing a standard PCL program that encompasses a wide range of model requirements.

The PCL program discussed previously in this chapter and illustrated in Figure 5.10 looks deceptively simple. To support an operational application, this "simple" program must be enhanced to address such areas as file security, output selection, the processing of conditions in which an optimization run does not achieve an optimal solution, extending the range of algorithms (e.g., parametrics or mixed integer), and matrix displays, to name a few. As noted earlier, the standard Exxon PCL program eventually consisted of some 1,500 statements.

If we look back at the simple PCL program of Figure 5.10, we notice that there are two types of branch shown. One is to a general call to any DATAFORM program (the name of the program being identified via cell XCHAR10), and the other is a call directly to a PCL procedure (examples are OPT, STOP). In practice, there is a need for a third type of branch that combines the previous two. In this type of branch, a very simple DATAFORM program is executed before passing control to a PCL procedure. This enables the user to communicate certain parameters (via the CR) to that procedure. The optimization function (OPT) requires such a DATAFORM precursor in order to establish such things as case name, off-normal action, and parametric definition. This particular DATAFORM precursor is far from being "very simple" (see Chapter 10).

Any general PCL program, then, must include specific branches whenever a PCL procedure is involved, either directly or via a preceding DATAFORM program. Based on the type of structure shown in Figure 5.10, this involves an additional IF statement and a set of procedural instructions for each such branch. Figure 5.12 illustrates how the PCL can be expanded to include a function to copy an ACTFILE from one physical file to another.

We see, later, how we can eliminate the need for the additional IF statement whenever additional functionality is built in. However, a generalized PCL program can be expected to increase significantly the number of Type-1 DATAFORM modules required to pass parameters to PCL procedures. These parameters can often be fixed for a single, local application, but need to be variable in a general program.

It is worth noting the nature of these usually simple, parameter-setting modules. Unlike the major modules that support a mathematical programming activity, they do not require substructure. Their function is only to set cells in the CR whose values will be picked up by the PCL procedures that follow. The type of data set that forms their input is illustrated in Figure 5.13. Unlike the data sets that drive the major modules such as DBM (see Figure 5.9), this input is not

```
              PLATORD
ZPLEX         EXDF        ('NAME','PLEX')
              IF          (XCHAR10.EQ.'OPT',ZOPT)
              IF          (XCHAR10.EQ.'STOP',ZSTOP)
              IF          (XCHAR10.EQ.'UNLOAD',ZUNLOAD)
              EXDF        ('NAME',XCHAR10)
              GOTO        (ZPLEX)
ZOPT          EXDF        ('NAME',XCHAR10)
              READY       ( - - - - )
              WHIZARD
              SOLUTION
              GOTO        (ZPLEX)
ZUNLOAD       EXDF        ('NAME',XCHAR10)
              UNLOAD      ('ACT','FROM',XACTDD,'TO',XOLDNAME)
              GOTO        (ZPLEX)
ZSTOP         EXIT
```

Figure 5.12   PCL program with branch to new function, UNLOAD. Extensions include one more IF statement and procedural statements at label ZUNLOAD. The DATAFORM program sets the CR cells XACTDD, XOLDNAME to "file names" specified at input time. This example also shows a DATAFORM precursor for the optimization function.

a list of submodules to be executed in sequence, but a simple statement of parameter requirements.

Illustrating the type of functionality that can be provided by such minor modules within the framework of a single, unified PCL program, Figure 5.14 shows a typical set as used in the standard Exxon system.

If a single, universal PCL program is to be produced to meet the needs of all applications, a full range of functional capabilities essentially matching those of the MPS must be provided. The variables required to drive the PCL procedures can be supplied through basically simple, minor DATAFORM modules that can read user input requirements and pass them to the procedures through the CR. This approach leads to an increase in the number of Type-1 program modules required. By these means, the PCL can be effectively removed as a candidate for local extension, yet retain all the flexibility it offers the user.

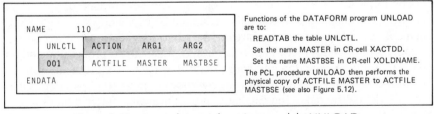

Figure 5.13   Input data set for minor module UNLOAD.

Module	Function	DF Precursor	No. of DF Statements
ABORT	Terminate on major error	NO	
ABSNORM	Normal print routing	NO	
ABSOFF	Second print file switched off	NO	
ABSON	Full printout to second file	NO	
COMPILE	Compile temporary DF program	YES	99
COMPRESS	Free disk space on ACTFILE	YES	22
CONTROL	Define input/output files	YES	169
DETAIL	Error message level control	YES	17
FREE	Print available free core space	NO	
LIST	Display all elements on ACTFILE	YES	17
MATRIX	Various matrix displays	YES	334
NOABORT	No termination on major error	NO	
NOBACKUP	Suppress back-up to tape	NO	
OPT	Optimization	YES	2166
PASSWORD	Password protection of ACTPROBs	YES	49
PRINTMIN	Minimize printout	NO	
PRINTOFF	Switch off primary print file	NO	
PRINTON	Switch on primary print file	NO	
TITLE	Define page title	YES	83
UNLOAD	Copy ACTFILE	YES	41

Figure 5.14   Functions provided by standard PCL procedures with or without a DATA-FORM precursor. In addition there are about thirty stand-alone DATAFORM Type-1 modules (such as DBM).

5.4.2   Unifying JCL

While it has proved possible, although not easy, to develop a single PCL program that can be maintained centrally, precompiled, and used by any application, this same goal is rather beyond reach in the area of JCL. This stems from the fact that each computer center is likely to have its own set of standards for naming of data sets, accounting procedures, and so on. JCL is, of course, dependent on the hardware and basic operating system in use at the computing center.

Since PLATOFORM (and EMPS) were designed to run on large IBM machines, discussion is confined to the type of standardization possible within the JCL supported by these computers. In particular, we are concerned with the data definition statements that define the files and data sets required for the execution of a computer job. A complete JCL "program" can be wrapped up into what is known as a cataloged procedure that can then be executed via a single EXEC statement. By this means, input streams can be reduced to a few records— basically, a job control record, an execution of a cataloged procedure, and an inline data set containing the job input stream. The cataloged procedure can be generalized to a wide extent by allowing overrides to numerous parameters that indirectly define the data. The cataloged procedure will contain default values for all of these overrides.

How much can the diversification in approaches to JCL construction be reduced? Basically, the problem reduces to a consideration of the number of distinct files necessary (or desirable) to run any PLATOFORM job. Any large modern MPS with a wide range of capabilities is going to require many "scratch" files with which to carry out its functions, for example, files on which to store working copies of the matrix, partial inverse (eta) records during optimization, scratch copies of the basis, and so on. These files are required only for the duration of a run and need not be of further concern. However, the size allocation for such files can benefit from indirect definition, which can be overridden at input time. Then there are library files required to store the MPS itself and any other procedures (FORTRAN, PL/I, etc.) required by the system.

Nevertheless, the files of interest in the present discussion are those more closely associated with the basic system structure design outlined earlier in this chapter. Figure 5.4, for instance, illustrates the use of four such files:

- A sequential input file.
- A file containing a precompiled PCL program.
- An ACTFILE containing precompiled DATAFORM programs.
- An ACTFILE containing a PLATOFORM data base.

Of these four, the input file will be a scratch file created during each run as output from the basic input translation module (PLATORD). The basic user input will come in through a system input file (see Figure 5.15).

The PCL program, which is generalized for all applications, can be distributed

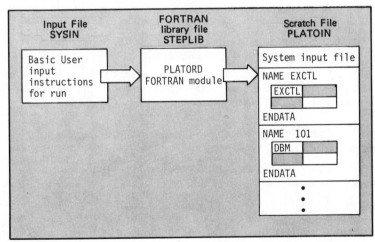

Figure 5.15   Creation of scratch input file for PLATOFORM run.

to all computer centers where it will be stored as a permanent data set (file SYSMLCP).

This leaves the two files most closely associated with the system structure—the DATAFORM program file and the data base. Both of these are DATAFORM ACTFILES and, as noted in Chapter 3, could be combined into a single physical file.

The first decision, then, is whether to combine the data base, substructured into ACTPROBs, with the DATAFORM program modules on the same physical ACTFILE. Although some very early PLATOFORM applications operated in this mode, it is clear that such an arrangement cannot be successful if the system is to support many applications. In Chapter 4 we have seen how each application (or group of related applications) requires its own data-base file. Not only would it be wasteful of disk storage space to store on each of these all the DATAFORM program modules, but central maintenance of these programs would be almost impossible. Clearly, programs and data have to be kept on separate files.

Although one data file is needed for each application, only one of these files will be required in any particular computer run. The standard JCL cataloged procedure needs to include only a single data definition of the data-base file, provided the data set name is substitutable as an override at input time. Later, we see circumstances where more than one data file may be necessary (e.g., to allow data transfer between files, or to create file copies as security protection during a run). These are discussed further in Chapters 7 and 8. The JCL implications are not relevant to the central discussion here.

When consideration is given to the file organization of the DATAFORM programs, issues that can affect the fundamental system design are encountered. As background, first recall that a system-unification strategy properly occurs fairly late in the life of a major, corporation-wide system. By that time, a deliberate prior policy of diversification has led to the accumulation of many discrete program modules (as many as 200 in PLATOFORM). Many of these have been locally developed and are locally maintained.

Section 5.3 introduced the classification of such modules into one of three types, depending on their degree of generality and their place of maintenance. This classification can be elaborated as:

Type-1. Fully general purpose modules, centrally maintained, distributed as precompiled modules.

Type-2. Open-ended modules, centrally maintained, distributed as source modules. Compiled locally with addition or substitution of local submodules.

Type-3. Application-specific modules, developed and maintained locally.

While the unification strategy will encourage a migration from Type-3 to Type-1 and some degree of consolidation within Type-2, we cannot expect to eliminate the need for a three-level classification scheme. Recall that compiled DATA-FORM programs are stored on a DATAFORM ACTFILE. Any number of programs can be stored on one physical file, provided their names are unique.

Should the standard JCL cataloged procedure, then, contain references to one program file or to three? This is not an easy, clear-cut decision, since it is not difficult to expand a centrally distributed program file with locally required additional modules of Types 2 and 3. The decision, perhaps, hinges on the number of applications that any one computer center can be expected to support, and the ease with which numerous data sets can be provided by the computer center. Influenced by the rapidly increasing number of applications and the growth of individual computer centers, Exxon has preferred the multifile approach. This approach has clear advantages for system maintenance.

A multifile approach implies that a standard cataloged procedure should contain references to three separate DATAFORM program files. Each center will install a single Type-1 file, perhaps two or three Type-2 files, and a Type-3 file for each application requiring local modules. Each application will then use JCL overrides to pick up its appropriate Type-2 and Type-3 files (see Figure 5.16).

But there is one more implication here. We have mentioned that a professional approach to system maintenance is necessary to support a major corporation-wide investment in system software. This implies an active, well-formulated maintenance and development policy under the control of a central group. This group must respond to requests for enhancements and "bug fixes" from a wide population of system users. The system must remain dynamic, expanding to

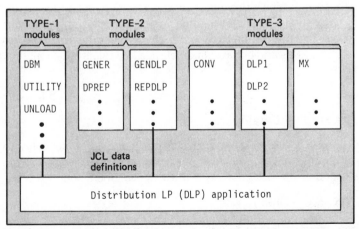

Figure 5.16   Allocation of DATAFORM program modules to separate program ACT-FILEs. Any particular application (DLP in example) selects the general-purpose Type-1 file plus one Type-2 file and one Type-3 file via three standard JCL data definition statements with overrides on data set names.

meet new requirements, but it must also remain dependable as more and more of the corporation's business comes to rely on it. New features must not be introduced too rapidly, lest they cause a well-established application to fail.

There are two ways in which the conflicting needs for both a dynamically expanding, yet secure, system can be reconciled. One is to ensure that each new enhancement honors the existing input syntax. This may sometimes lead to a slightly less than optimal overall syntax for a particular module, but it pays dividends in causing minimal disruption to existing applications when the new feature is introduced. This "fail-safe" policy is not always possible, but it should cover more than 95% of all minor enhancements.

The second way is to introduce each new version of a new module through a series of promotional steps. At first the new version will be introduced in "test" status. At this level, it is available only to the system development group that will subject it to a vigorous testing schedule. Later, it will be promoted to "user" status, where end users are invited to try it out, to check that it meets its design goals, and that it does not disturb existing applications. Only after this will the new version finally be promoted to "production" status and take its place as a recognized component of the total system.

This, of course, implies a need for further ACTFILES on which to store the "user" and "test-level" versions of each of the three program "types." The standard JCL-cataloged procedure must be expanded to contain reference to six program files: Types 1, 2, and 3; each at two levels (user and production). The test level must deliberately be omitted to restrict access to the system support group only.

However, a problem is now created that affects the basic system design structure. Suppose that a new version of DBM has just been promoted to user level and that user testing is desired. How can a user select DBM from the "user" Type-1 file while retaining access to other Type-1 modules at "production" level (see Figure 5.17)? The solution to this problem involves the use of a new catalog on the data base. This additional catalog plays a key role throughout the unification strategy.

5.4.3 The Module Catalog

To understand the role of a module catalog in providing flexibility to select discrete programs from different files, it is first necessary to see the way in which PCL itself handles this selection. When a PCL statement such as:

    EXDF ('NAME', 'DBM')

is executed, there is already an implicit need for the system to know on which file to seek the program whose name is DBM. In EMPS, this is controlled through another CR-cell (XDFPGFIL). The contents of this cell give the name of the ACTFILE to be used to locate this program.

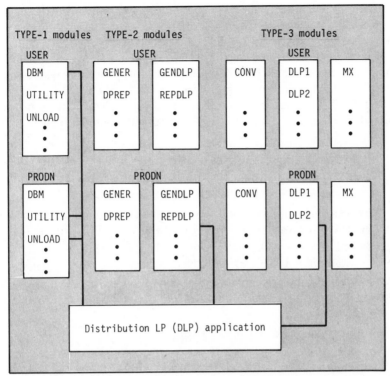

Figure 5.17   Allocation of DATAFORM program modules allowing two version levels (user and production). Six standard data definitions allow each application to select two files from each program type. The problem is how to select DBM from the USER file, and UTILITY, UNLOAD, and so on from the PRODN file.

In a multiprogram file environment, the contents of cell XDFPGFIL must be constantly changed as we progress through a sequential list of modules to be executed. One way to do this would be to give the program module PLEX an additional task: setting the contents of cell XDFPGFIL ready for the execution of the next module. But PLEX can do this only if it is provided with the required information.

One of the purposes of a module catalog is to provide this information. But a module catalog can also provide other information that will lead to a simplification of the "universal" PCL program. It has already been noted that the general PCL illustrated in Figure 5.12 was capable of further simplification by removing the need for IF statements to direct a branch to a label involving PCL procedures. The requisite branch label can be included in the catalog for each functional module.

An example of a module catalog that becomes an integral part of an application's data base is shown in Figure 5.18. This example is based on the program file environment illustrated in Figure 5.17; it represents part of the module catalog for the distribution LP application. This application needs the USER version of DBM but the PRODN version of other Type-1 modules. Thus in Figure 5.18 we can now see how the module catalog enables this fine level of discrimination to be achieved by allowing file names (column F1) to be associated with each individual module. PLEX will set (indirectly) the file name GLOUSER into CR-cell XDFPGFIL before passing control via PCL to the DBM.

Figure 5.18 illustrates the several types of functional modules previously discussed in this chapter, and it is useful to summarize them here:

- Modules that require only the execution of a single DATAFORM program. These can share a common branch label (DCEXDF) in the PCL program.
    Examples.   DBM, GENDLP, and so on.
- Modules that require a prior execution of a minor, parameter-setting DATAFORM program, followed by one or more PCL procedures. Each one requires a special branch label in the PCL program.
    Example.   UNLOAD (DCUNLD).
- Modules that require only the execution of one or more PCL procedures. No DATAFORM programs are involved, but each requires its own branch label.
    Example.   ABORT (DCSETAB).
- A fourth type, not previously discussed, that allows users to provide alias names for any functional module. In this way, a standard set of modules can be introduced, but users can continue to refer to them by names to which they have become accustomed. Figure 5.18 illustrates a definition of SAVE as an alias for UNLOAD.

These module catalogs are specific to each application (data ACTFILE) and, as suggested by the two-character head names (see Chapter 4), are eligible for maintenance by the user. Thus to access the USER-level version of DBM, the intersection of DBM in column F1 is changed from GLOFILE to GLOUSER.

MODULE.C	F1	L1	A1
ABORT		DCSETAB	
DBM	GLOUSER	DCEXDF	
DLP1	APPFILE	DCEXDF	
DLP2	APPFILE	DCEXDF	
GENDLP	PROFILE	DCEXDF	
REPDLP	PROFILE	DCEXDF	
SAVE			UNLOAD
UNLOAD	GLOFILE	DCUNLD	
UTILITY	GLOFILE	DCEXDF	

The column contents can be broken down as follows:

Column F1   —   File name for required DATAFORM program.

GLOFILE, PROFILE, APPFILE are used as file names for the production versions of TYPE-1, 2, and 3 modules, respectively.

GLOUSER, PROUSER, and APPUSER are the corresponding file names for the user-based versions.

Column L1   —   Label for branch in PCL program.

Column A1   —   Used for alias names.

Figure 5.18   Part of a module catalog.

It is possible, however, to design a PLATOFORM module, written in DATA-FORM, that will compose or update a module catalog automatically by scanning all the program files defined in the JCL and extracting the names of the programs contained on them. Such a program needs to honor certain hierarchical rules (i.e., PRODN versions have precedence over USER versions, etc.), and it must know all of the specific PCL branch labels. While it is not relevant here to discuss the full details of such a program, the fact that one can be developed is a striking illustration of the flexibility offered by the structure of EMPS/DATAFORM—aided, in this instance, by the basically similar storage structure (ACTFILEs) of both data and programs. Use of such a module relieves the user of setting up this catalog or modifying it whenever new programs are introduced, yet retains flexibility to select particular versions of a program through the standard update mechanism.

The introduction of a module catalog requires a somewhat different approach to the design of the PCL from that discussed previously. Since this is related to the specific nature of the PCL available in EMPS, it is not useful to discuss it further here.

5.5 DATE STAMPING

As we have followed the growth of the system structure from a simple beginning to the point at which a unification has been achieved in terms of a constant PCL program and a standard JCL, the growing importance of the control module PLEX became apparent in tying the system structure together. Consideration is now given to one more important role for this module.

In Chapter 4, reference was made to using a date stamp in the Contents Catalog to provide an audit trail for data table updates (Section 4.4). A similar date stamp was illustrated in the Problem Catalog (PROB. . .C) in Figure 4.20. Later chapters show the importance of such a facility as a control mechanism. It provides, for instance, the capability to print all tables updated since a given date (Chapter 7), and it also provides a means of developing an automatic generator that can regenerate those parts of a matrix affected by recent data changes by comparing the date of updates with the date the matrix was last generated (Chapter 9).

Of course, to be useful, this date stamp must contain more than just the calendar date on which an event occurred. Several machine runs can be, and usually are, made on a single calendar day. A convenient, eight-character representation of a date stamp could take the form:

YYMMDDRR

where:
  YY = last two digits of the year (82 = 1982);
  MM = a digital representation of the month (12 = December);

DD = day of the month;

RR = digital representation of run number on that day.

Thus the date stamp 82100603 refers to the third run on the 6th of October 1982. Note that by arranging the elements in this order, events can readily be sequenced and also compared to see if one occurs later or earlier than another. Such a system allows as many as 99 separate runs in any one day, but it can, and will, cause problems at the turn of the century.

Date stamps must, of course, be specific to a data ACTFILE (or application) and, hence, need to be recorded on the ACTFILE in a DATAFORM table. Figure 5.19 illustrates the form such a table can take.

The module PLEX is given the tasks of updating this table whenever PLEX is first entered or of creating the table for a new ACTFILE. To do this, PLEX makes use of DATAFORM's DATE function that returns the computer date (e.g., 821006). If this sorts higher than the date recorded in column DATE of Table RUNNO, the new date is substituted, the entry in column RUN set to 01, and the concatenation of these eight characters is entered in column DATERUN to form the reference date stamp. If the date returned by function DATE is the same as that in Table RUNNO, the RUN is incremented by one, and the new DATE-RUN formed as before. This provides a unique record on the ACTFILE of when the ACTFILE was last used. Subsequent modules (e.g., DBM, GENER) can extract this date stamp and transfer it to a column of the Contents Catalog or Problem Catalog or use it in decision-making.

Note that the Table RUNNO is a DATAFORM Z-table (the computer date is a character string) and that its contents are alphanumeric characters. The ease with which this mechanism can be incorporated into PLEX relies on two important DATAFORM functions: the DATE function and an INCR function that allows alphanumeric strings to be incremented. Later modules (e.g., DBM) rely on a DATAFORM capability to compare alphanumeric strings on a high-low basis in order to use the information effectively.

5.6 SUMMARY

In this chapter we have shown how a system, designed to support a wide spectrum of mathematical programming applications, can be both a single integrated system and yet retain modularity (and, indeed, submodularity) at a lower

RUNNO	DATE	RUN	DATERUN
A1	821006	03	82100603

Figure 5.19 A table recording date-stamp information. Such a table can be permanently stored on ACTPROB DATABASE and updated for each new run by module PLEX.

level. The single system is provided by a program written in the language of the MPS (PCL), and enables users to communicate with the system through a single input stream. Functionality is provided by an open-ended number of modules, or PCL procedures, that provide users with flexibility to carry out many different tasks in any sequence.

We have seen how a deliberately open-ended policy initially encourages system growth and diversification to ensure acceptance of the system throughout a large corporation. To control this growth, a careful maintenance policy needs to be adopted that leads to a need for a special open-ended design of some of the major modules. These modules can then be maintained for the most part centrally, with locally maintained extensions.

Having gained acceptance, the open-ended policy needs to be restrained as a more professional approach to system maintenance becomes mandatory. This leads to a unification strategy by which users are encouraged to consolidate diverse modules into more general, centrally maintained programs. The unification strategy must concentrate on providing a uniform environment under which all applications will operate. In this respect, it is possible to develop a single PCL program that essentially transfers all of the capabilities of the MPS out of PCL and into the "language" of the system. PCL can then be "buried." A similar approach to JCL leads to the development of a single cataloged procedure for use at each computer center. To provide sufficient flexibility in these cataloged procedures and to recognize the needs of a professional maintenance policy, a multiprogram file environment needs to be created. This leads to the need for a module catalog for each application, which can then also be used to simplify the PCL.

The emergence of a worldwide integrated system for PLATOFORM (WISP), the culmination of a unification strategy, occurred within Exxon some six or seven years after the first introduction of the basic system. It perhaps would not have succeeded had it been attempted any earlier.

So far, we have been looking inside a large, complex system designed to support many different types of applications. From this perspective the system may appear unduly complicated (but so might a simple pocket calculator if we examine, microscopically, a chip). The key to a successful system lies not in what is inside (although this, of course, is critically important), but what is outside. How a user views the system is the subject of the next chapter. Perhaps the primary aim of the complex interconnections within the system is to enable the simplest possible user interface to be developed. How do we cover up the chips and provide the user with a single button that will calculate a square root?

A. John Rowland

6. INPUT SYNTAX

6.1 INTRODUCTION

In Chapter 5, we saw that the modular nature of the PLATOFORM system, combined with the characteristics of the DATAFORM language, give rise to a structure for input that is hierarchically organized. An example of this system input (the file PLATOIN) was given, and it was noted that this format was not user oriented. In this chapter, we expand the example given in Figure 5.9 to illustrate more fully why this syntax is not suitable as a user interface. We then go on to discuss what can be done to provide a simple user interface.

Figure 5.9 showed the way in which the system is structured within its modular framework. In expanding the example at this stage, it is not our purpose to discuss the detailed syntax for controlling the system modules. Rather, Figure 6.1 highlights the thought processes that a user would need to exercise in order to produce this type of input, and thus illustrates the awkwardness of this syntax. In the example, five major modules are requested (DBM, GENER, OPT, RE-PORT, and UNLOAD), and each of these modules may require user-supplied control information.

We can now summarize the steps involved in setting up an input file of this type. The first data set read by the system control module (PLEX) contains a single table, EXCTL. This table lists the major modules in the execution order required by the user. The data set name given under the head DATASET is passed by PLEX to the major module, which is then able to pick up the correct control data set. Each of these control data sets is set up with the name (101, 105, 106, 107, 108) as given in table EXCTL. These data sets contain tables defining the major module controls.

For DBM, this control information consists of the definition of the ACTPROB (table PROBLEM) together with a list (table DBM) of submodules to be called. In the same way that table EXCTL defines names of control data sets for major modules, table DBM is set up with the names of submodule control data sets. The control data sets must then be named accordingly (102 for UPDATE, 104 for TPRINT). Finally, the actual data table updates are given in a data set whose name is obtained by adding 1 to the name of the UPDATE control data set (102—103).

```
NAME            EXCTL
EXCTL     ACTION     DATASET            ⎫  List of major modules to be called.
001       DBM        101                ⎪     For each module, the name of the
002       GENER      105                ⎬     data set containing the module's
003       OPT        106                ⎪     control information is given under
004       REPORT     107                ⎪     the head DATASET.
005       UNLOAD     108                ⎭
ENDATA

NAME            101                     ⎫  DBM control data set.
PROBLEM   PROBLEM    OLDPROB            ⎪
001       RSIDE      RBASE              ⎬     ACTPROB definition.
                                        ⎪
DBM       ACTION     DATASET            ⎪     List of submodules to be called.
001       UPDATE     102                ⎬     For each submodule, the name of
002       TPRINT     104                ⎪     the data set containing submodule
ENDATA                                  ⎪     control information is given under
                                        ⎭     DATASET.

NAME            105                     ⎫
GENCTL    ACTION     ARG1               ⎪
001       ACTPROB    RSIDE              ⎬  GENER control parameters.
002       FB         UF                 ⎪
ENDATA                                  ⎭

NAME            106                     ⎫
OPTCTL    ACTION     ARG1               ⎪
001       ACTPROB    RSIDE              ⎪
002       PERIOD     M1                 ⎬  OPT control parameters.
003       CASE       RSIDEM1            ⎪
004       SOLUTION                      ⎪
ENDATA                                  ⎭

NAME            107                     ⎫
REPCTL    ACTION     ARG1               ⎪
001       CASE       RSIDEM1            ⎬  REPORT control parameters.
002       BLEND                         ⎪
ENDATA                                  ⎭

NAME            108                     ⎫
UNLCTL    ACTION     ARG1       ARG2    ⎬  UNLOAD control parameters.
ENDATA                                  ⎭

NAME            102                     ⎫  UPDATE control data set.
UPCTL     TABLE      ACTION             ⎪     List of tables to be updated
001       SB.F.U.T     U               ⎬     and type of update.
ENDATA                                  ⎭

NAME            104                     ⎫  TPRINT control data set
TPCTL     STYLE                         ⎪     List of tables to be printed
SB***U*T  DDD                           ⎬     and print style.
ENDATA                                  ⎭

NAME            103                     ⎫
          UPDATE                        ⎪
SB.F.U.T    SUMX                        ⎬  Table update data.
  FLM1      2.5                         ⎪
ENDATA                                  ⎭
```

Figure 6.1   Expanded example of an input file (PLATOIN).

For other major modules, the control data set consists of a single table containing a simple statement of parameter requirements. If we now step back and review this format for the input file from the user's standpoint, some of the problems are obvious:

- There is a great deal of clutter in the input file.
- Less than half the input characters are actually concerned with system actions requested by the user.
- A lot of effort (and input) is made necessary by the cross-referencing of data sets.

In addition to these problems:

- The stub names used in control tables (to allow for duplicate action) have no meaning to the user.
- The specification of data set names (and their cross-referencing) and arbitrary numeric stub names is time-consuming and error-prone.

The hierarchical nature of the data set structure is illustrated in Figure 6.2. Consider the first action that is requested in this example. The first major module is DBM. The first DBM submodule is UPDATE. The first table to be updated is SB.F.U.T. The updates for this table are in data set 103. This is the last data set on the file PLATOIN (see Figure 6.1). What can be done to provide a more logical input format?

In a batch mode of operation where all input must be prepared in advance of job submission, we have the option to preprocess user instructions into the system syntax from any other format—providing that a one-to-one relationship can be defined. The design of such a translation process was set up at an early stage in the system's development. The primary aim was to provide a common syntax for all applications. To allow the execution of any module that might be added, this syntax should be complete, yet simple and easy to remember.

6.2 SYNTAX DESIGN

Input to a system such as PLATOFORM is not only the specification of data, but also a set of instructions to execute modules (and submodules) in a given order. The introduction showed us how the order of execution is given on PLATOIN in control tables. For the user, however, it is more natural to specify each action entirely before proceeding to the next. A sequential input logic can parallel the thought process involved in the job [e.g., change data, print data, generate linear programming (LP) model, solve, and report].

In PLATOFORM, sequentially organized user input instructions are automatically processed by a translation program in the first part of the computer job.

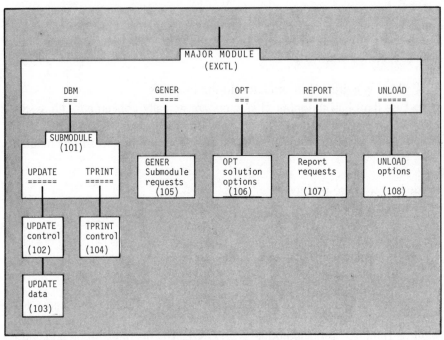

Figure 6.2   Hierarchical structure of PLATOIN example. The data set numbers are shown in parentheses (see Figure 6.1).

The design of this input syntax, the PLATOFORM "language," was therefore dependent on neither the system structure nor the underlying data-base structure. Given the freedom of design made possible by such a translation process, what factors influence the choice of a user input syntax?

One of the most successful types of format for input preparation is the preprinted form. In such a format, each piece of data is assigned a specific field on the form with text that explains what is expected by the program in the field. This structure can be used with LP-support programs for a specific model with well-defined boundaries. However, for a general system in which the contents of any data table are determined by the application, this is not possible. Furthermore, we need to recognize a fundamental difference between a data-base system and a traditional technical program that supports LP matrix generation. The input syntax we are discussing is not the input file to the matrix generator or any other program. (The input to the generator is the data in the ACTPROB itself.) The input syntax is designed to update data (primarily through use of the DBM) and to control execution of all other modules.

Return to the example input file (PLATOIN) shown in Figure 6.1 and consider only the input that is concerned with specifying the system actions. (We assume that the translation program, when creating the PLATOIN file, will handle all cross-referencing of data sets and control table stub naming.) Furthermore, we will reorganize the file so that the order of execution is defined as file order.

Figure 6.3 displays the results of this reorganization. The reorganized file is sequentially structured and has little redundant information. A mechanism can be defined by which this file can be processed to create the PLATOIN file required by the system. However, a translation program to do this would need to "know" which instructions refer to major modules and which to submodules. One way of doing this is to include such information in the program. This approach implies program modification whenever a new module or submodule is added and is therefore not acceptable. A more general approach is to include in the syntax itself a means by which the translation program can recognize major modules (and submodules). This method does not rely on any list of system components, and it has the added benefit of emphasizing the modular system framework to the user.

In PLATOFORM syntax, a "+" precedes the instruction name to indicate a major module (+DBM, +GENER, etc.). Within major modules, a "−" is used in the same way to call submodules (e.g., −UPDATE within +DBM) or to indicate control parameters.

Applying these refinements to the example in Figure 6.3 and, at the same time, eliminating any redundant information (the name of the table to be updated appears twice) results in the final PLATOFORM syntax illustrated in Figure 6.4. In the case of DBM, other syntax structuring has been used (e.g., to allow the

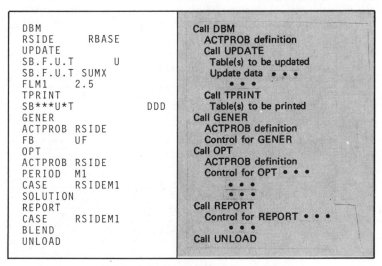

Figure 6.3   Reorganized input file.

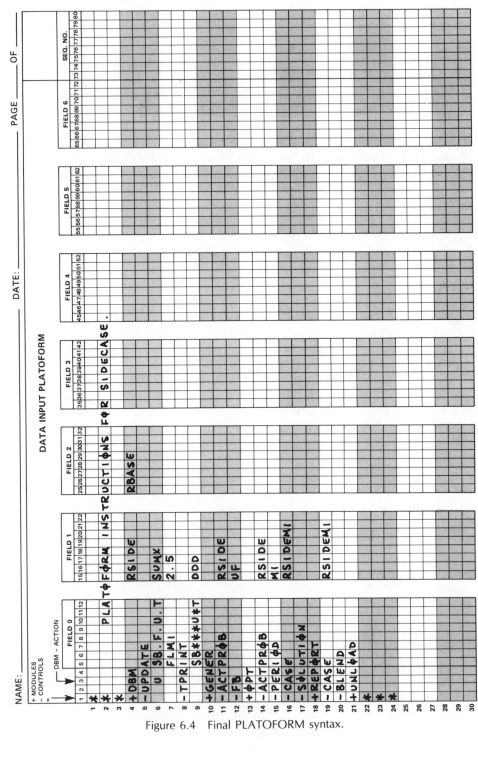

Figure 6.4 Final PLATOFORM syntax.

type of table UPDATE to be given in a fixed position on the input form—the U for table SB.F.U.T).

For all other modules, the amount of control information is small and can therefore be given as simple arguments on the submodule record. Any number of additional controls can be specified by key word (additional "−" records; e.g., −ACTPROB), leaving the major module to interpret this information.

A more recent extension (not illustrated in Figure 6.4) has been made to allow more extensive control input. Each "−" command is restricted to six parameters. To overcome this limitation, the syntax was extended to include an "=" command. With this syntax, subsequent dependent records can be submitted in table form, allowing an unlimited number of control parameters.

6.3 THE TRANSLATION PROGRAM (PLATORD)

The translation program, PLATORD (PLATOFORM read), is written in FORTRAN. In addition to translating sequential user instructions into hierarchical form, it checks syntax on the commands prepared by the user/analyst.

PLATORD is designed to operate in two modes. First, for the DBM instructions, it edits the input, and it outputs specific control tables. Second, for all other "+" instructions (all other major module requests), it builds a general control table containing the user-supplied parameters. Some detail of this process is given next to highlight the way in which this program can be made completely general, eliminating the need for modification when system components are added.

The syntax for the two DBM submodules UPDATE and TPRINT is handled in a special way, since both are followed by stylized data records and require a specific PLATOIN structure. For example, we have already seen that for table UPDATEing, the PLATOIN file includes a list of tables to be updated as well as the actual updates. This apparent redundancy allows the use of a powerful DATAFORM table-updating feature that is covered in Chapter 7.

All other DBM submodules are handled in a general way, as are all major modules except DBM. We will illustrate this generality by looking at the processing of the input records of any non-DBM major module. The control table produced has the following name and head structure:

xxxCTL ACTION ARG1 ARG2 ARG3 ARG4 ARG5 ARG6

The table name is derived by adding "CTL" to the first three characters of the module name. The body of the table is built up from the "−" records that follow the major module request. The stub is assigned arbitrary numeric names to allow for multiple "—" records to be input without causing duplicate stub names. (The "—" name goes under the head ACTION). Notice that the control table is output by PLATORD even if no "—" records are given. The module itself will pick up the control table and act accordingly. Parameter absence implies that a default action has been defined, as is the case for UNLOAD in the example. For all modules, PLATORD will include the module name in the execution control table in the order given in the user's input.

With this translation mechanism established, the system developer is able to design input syntax for new modules within the general "language" framework —knowing that PLATORD will handle it and produce output that can be READ-TABed and interpreted.

This general design has proved flexible enough to allow many modules and submodules to be incorporated into the system with no changes to the syntax-translation program. The only enhancements to PLATORD that have been made in recent years were to increase the amount of syntax checking performed and to allow comments to be included in the input stream ('*' records). The syntax checking performed is summarized in Figure 6.5. More DBM syntax checking is possible as a direct result of the special handling of this module by PLATORD.

6.4   AN OVERVIEW OF PLATOFORM CAPABILITIES

The example used in the discussion on user syntax is a realistic set of instructions for a side study, and provides an overview of the way in which PLATOFORM can be used to control the data and model-management aspects of mathematical programming.

Assume that the data tables for a refinery-planning LP model have been stored and a base-case LP matrix established on ACTPROB RBASE. We are now required to perform a side study to investigate a change in fuel oil specification.

---

1. +DBM record must include ACTPROB definition.

2. DBM submodule must be specified before any tabular input.

3. Each table UPDATE must have an entry in 'action' — the type of update to be performed on the table must be given.

4. Checks for blanks in table names, blank head and stub names.

5. Many checks for correct alignment of input data. (See Figure 6.4 which shows input 'fields'.)

Figure 6.5   PLATORD syntax checks.

Specifically, the model is to be rerun with a new maximum figure for fuel oil sulfur of 2.5 wt%.

+DBM	RSIDE	RBASE
−UPDATE		
U SB.F.U.T	SUMX	
FLM1	2.5	

User:	Set up new ACTPROB RSIDE from RBASE. Change sulfur maximum specification to 2.5 for new ACTPROB only.
System action:	The new ACTPROB is established by copying the CONT. . .C of RBASE to RSIDE. The LP matrix is also copied to RSIDE. Table SB.F.U.T is retrieved and updated. The updated version is filed on ACTPROB RSIDE only and the CONT. . .C of RSIDE updated. The version of this table used by RBASE is not affected.

−TPRINT		
SB***U*T	DDD	

User:	Print all product specifications for the United Kingdom. Use full dictionary translation.
System action:	All tables matching the mask SB***U*T are RECALLed via the CONT. . .C of RSIDE. These are then printed using decoding logic defined in GENTAB.C, DICT. . .C, and using text from the dictionaries.

+GENER		
−ACTPROB	RSIDE	
−FB	UF	

User:	Call the generator: modify model in line with data changes.
System action:	Retrieve the matrix on ACTPROB RSIDE and regenerate the fuel-blending structure (FB submodule) for location UF. File the modified matrix back on ACTPROB RSIDE.

+OPT		
−ACTPROB	RSIDE	

—PERIOD	M1
—CASE	RSIDEM1
—SOLUTION	

User:	Optimize side study.
System action:	Retrieve the matrix on ACTPROB RSIDE and prepare it for optimization of time period M1. Store the resulting solution as case RSIDEM1 and print this solution in standard form.

+REPORT	
—CASE	RSIDEM1
—BLEND	

User:	Print the BLEND report.
System action:	Retrieve solution case RSIDEM1. Run the BLEND submodule report.

+UNLOAD	
User:	Save the result of the previous actions.
System action:	The current ACTFILE is copied to the user's permanent file.

This brief overview highlighted some of the powerful features available in PLATOFORM. These include an effective side-study capability providing updating by exception, partial regeneration of LP matrices, solution-case storage, and data-driven reporting. However, the overview did not explain some other important functions that need to be included for file and data security. All of these items are described in subsequent chapters.

6.5 SUMMARY

This chapter has shown how a system, whose underlying software may be complicated, can be made easy to use by the provision of a translation program. The syntax used by the analyst is consistent and applicable to all PLATOFORM data bases. It is also complete in that all existing modules, as well as potential new ones, are included within this general language.

An overview illustrating the syntax of a typical side study was shown. From this example, we can also begin to see some of the interesting features of the major modules themselves. The example "run" used each of the four major system components that are at the center of mathematical programming (DBM, GENER, OPT, and REPORT). The first of these, the data management module, is the subject of the next chapter.

A. John Rowland

7. DATA MANAGEMENT

7.1 INTRODUCTION

It has already been shown that there are many advantages to be derived from a mathematical programming system (MPS) that allows the user to work problems at the data level. In PLATOFORM, this data is organized into a logically structured data base in the form of tables. Furthermore, these data tables are not just a collection of numeric values, but also include control information that determines the structure of the model to be solved. How, in practical terms, is this data to be managed? What factors affect the programming of a data-base manager (DBM)?

These questions are of prime importance, and they need to be considered early in system design. This chapter describes how PLATOFORM routines were developed that allow the user to maintain and display the data base.

The design structure of a PLATOFORM data base has already been described in Chapter 4. The data base is tied together by a system of catalogs and dictionaries. Some of these (e.g., PROB. . .C—the Problem Catalog, SOL. . . .C—the Solution Catalog) are totally maintained by the system, but others are application-specific and need to be set up before data input.

Some PLATOFORM utilities can be used to start up an application, and others enable parts of a master ACTFILE to be copied for the same purpose. At this stage, we assume that this one-time job has already been completed and that a basic PLATOFORM ACTFILE has been established. This ACTFILE will then have the structure illustrated in Figure 7.1. RBASE is an example of an already established user ACTPROB, and is used to illustrate PLATOFORM capabilities in the discussion that follows.

The functions required to manage the data base can usefully be segregated into two types, dependent on whether they address the whole file or just an ACTPROB. In PLATOFORM, these functions are grouped by type into two modules (DATAFORM programs). The UTILITY module can address the whole file, and the DBM module is designed to work with a single ACTPROB. They each contain a number of submodules concerned with different aspects of data-base manipulation. Some of these are for use by a data-coordination

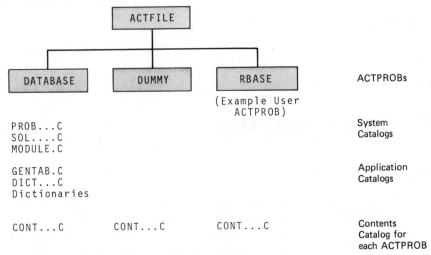

Figure 7.1   Basic PLATOFORM ACTFILE.

function in maintaining dictionaries and application catalogs. Others are directly concerned with updating and printing data tables.

7.2   TABLE UPDATING

Given that the basic ACTFILE structure for an application has been established, what table management is required? In any data-base system organized into tables, there are two basic types of table maintenance. Some actions are required on whole tables (add, delete, replace), and some are required on parts of existing tables (add dimensions, delete dimensions, change contents). All of these features must be included in a DBM.

We review the factors influencing the design of the table-updating features of PLATOFORM. There are two parts to this analysis. First, the data-base design seen in Chapter 4 must be maintained consistently, and, second, the DBM and UTILITY modules should fit the system structure described in Chapter 5.

7.2.1   Data-Base Structure Considerations

Data tables in PLATOFORM are grouped into table sets, and these sets are cataloged in GENTAB.C. A requirement of the data base is that all tables on the ACTFILE belong to one of the sets in this catalog. For any new table to be added to an ACTPROB, we must insist that the appropriate set be in GENTAB.C. Without this restriction, tables could be added that cannot be printed with text translation. Furthermore, the system is able to catch some input errors in table names by this check.

We could further insist that all stub and head codes be present in the correct dictionaries given in GENTAB.C. This type of code checking would be time-consuming from the system's viewpoint and, more importantly, would impose a severe constraint on the user. Table sets themselves are stable in the sense that, once established, extra sets are rarely needed in an application. However, this is not the case for table contents, where new codes may be added throughout planning studies. These should (and normally will) be added to the dictionaries, but a table update should not be rejected simply because a code is not in a dictionary; that would be too restricting to the user. All DBM facilities will still operate with these "nondictionary" codes, but the user will lose some advantages (e.g., in table printing and reporting), prompting correction of the omissions. The enforcement of such standards is a balance between user "friendliness" and ACTFILE consistency.

A high degree of integrity is needed when maintaining the system catalogs, as these provide the user (and the system) with control over the tables on the data base. The PROB. . .C catalog must be complete and consistent so that ACTPROBs have unique names. Without this, we cannot use the powerful cross-referencing of tables that provides PLATOFORM with its data-sharing power. This data sharing is made possible by the CONT. . .C catalog, which defines the tables associated with the model. The system must also ensure that the Contents Catalog on each ACTPROB is "clean" at all times.

The CONT. . .C may reference any number of ACTPROBs in pointing to the physical residency of tables. All RECALLs of tables by the system in all modules are made via the CONT. . .C. Thus it is of paramount importance that this catalog be consistent. (What we mean by this is that tables pointed to should actually be there!) This is not always as obvious as it sounds, because we need to consider carefully all the implications of table updating. What happens, for example, when tables are deleted? Some action must be taken on any number of CONT. . .Cs that may have pointed to the table. Another possibility is that the user may add a table that is already in the catalog (perhaps referencing another ACTPROB). What action should be taken to avoid the possibility of duplicate table names in CONT. . .C? (This is obviously not allowed, because it would be an ambiguous definition of the table to be used.)

We see how each of these issues is resolved later. First, however, we need to look at the DATAFORM language facilities that can be used in table updating and the overall system structure in which table updating is to be programmed.

7.2.2 System Structure Considerations

The DATAFORM language table-update verb (READTAB) can process an input file containing new tables and updates to existing tables already in core (i.e., already RECALLed from the ACTFILE). We can make use of this verb in the DBM by RECALLing all tables to be updated and executing a single READTAB verb. This is a fast procedure and eliminates the need for a program to perform the actual updates. The disadvantage is that no checking on data updates can be performed before these updates are processed into the existing tables. The READTAB verb does provide some error traps of its own (e.g., nonnumerics in T-tables, table stub/head deletions requested but not in table), and these can be used to set error conditions for use by the DBM. However, at this stage the update has already been made, and it is too late for corrective action.

Nevertheless, READTAB is used in the UPDATE submodule of DBM on the grounds of efficiency, and much of the syntax of —UPDATE is a direct result of this. (PLATORD performs the mapping from PLATOFORM syntax to READ-TAB syntax.) The error-checking implication is that in some instances updates to a table can fail because of a single error in one entry. The UPDATE submodule can only flag this error to the user and continue with the next table, bypassing all updates to the current table to avoid an erroneous update. The user must, in any case, correct the error, but the READTAB approach may imply the resubmission of all updates to this table.

7.2.3 DBM Actions in UPDATE/UPNAME

The table-updating features of the UPDATE submodule that are processed via READTAB are listed below. Each type of table update is assigned a single-character code that is used in the PLATOFORM syntax to request the relevant action:

• Add a new table	"A"-type action
• Replace an existing table	"R"-type action
• Modify an existing table	"U"-type action
• Delete heads/stubs from existing table	"D"-type action

In the case of U and D, the existing table is first RECALLed so that the table can be modified by the incoming READTAB data. For A and R, the new table is totally defined in the READTAB data.

We can now understand why PLATORD outputs two data sets for the UP-DATE submodule (see Chapter 6). The first data set lists the table names and the type of update. This data set is used to determine which tables to RECALL before the second data set of actual updates is processed. The deletion of a whole table is also included in UPDATE ("X"-type action). No update data as such is required, and the action is performed directly on the ACTFILE.

The process of renaming stubs/heads in existing tables is not available within the READTAB syntax, so a special procedure for this was developed in a separate submodule (UPNAME). As the change in name is done by a separate submodule (and not made by READTAB), we can perform additional checks before making the update. For example, the UPNAME submodule does not allow a stub/head name to be changed to the name of an existing stub/head.

Figure 7.2 summarizes the actions performed and the way in which the CONT. . .C catalog is used and maintained by the table-updating routines. Specific error checks are made before a new or modified table is filed on the current ACTPROB. These table-status checks are also summarized. From this summary, we can see that the CONT. . .C is the key not only to data-base structure, but also to DBM program design. The CONT. . .C definition of table residency is assumed to be correct at all times. All table-processing routines in PLATOFORM must recognize this and ensure correct maintenance in table-filing routines. The DBM itself ensures that the CONT. . .C is well defined by making the type of error checks discussed in Figure 7.2.

The way in which the table updating routines in PLATOFORM are programmed and the use of Contents Catalogs provide a large part of the system's data-sharing capability. We can now look at the implications of the UPDATE actions from the user's viewpoint to see how this comes about.

7.2.4 User View of UPDATE/UPNAME

Figure 7.3 provides an example of the PLATOFORM syntax used in table updating. Each call to the DBM must be for a particular ACTPROB referred to as

Update type	Normal DBM Action	Error Conditions	Error Action
A	File table on current ACTPROB. Add table name to CONT...C: - point to current ACTPROB - enter current date/run number	Table set not in GENTAB.C Table already in CONT...C	Table not added to ACTPROB.
R	File new table on current ACTPROB. Update CONT...C entry: - point to current ACTPROB - update date/run number	Table not in CONT...C	Default to 'A'-type action and checks
X	Remove table entry from CONT...C. If table is resident on current ACTPROB: - delete table - remove reference to this table from all CONT...Cs on other ACTPROBs	Table not in CONT...C	Table delete bypassed
D, AND NAME	Recall existing table from resident ACTPROB given in CONT...C. After update, file new version of table on current ACTPROB. Update CONT...C entry: - point to current ACTPROB - update date/run number	Table not in CONT...C	Table update bypassed

Figure 7.2   UPDATE/UPNAME actions and error checks.

```
+DBM              RBASE                              DBM on single ACTPROB
 -UPDATE
   A SB.F.U.T    V1MX      SUMX                      Add a new table
     FLM1        500       3                         called SB.F.U.T
     FLM2        500       3.5
   R SB.G.U.T    RVMN      D1MX      ROMN            Replace table SB.G.U.T
     GSM1        10        60        99.37
   X SB.D.UFT                                        Delete table SB.D.UFT
   U QB.D.UFT    SU        V1           ⎫
     VH          1         2.874        ⎪            In tables QB.D.UFT
     VD          1.7                     ⎪                    IB.F.UFZ
   U IB.F.UFZ    FL                      ⎪                  & BS...UFT
     VH          X                        ⎬           -Change body entries.
   U BS...UFT    M1FX      M1LO          ⎪            -Add stubs/heads.
     GS          0         35             ⎪
     FL˙         0         150           ⎭
   D FC...UFT    M2        M3                         Delete heads/stubs
     BA                                               (& entries) from table
                                                      FC...UFT
 -UPNAME
     BA.....T    HEAD      M2        M1              Change head/stub names
                 STUB      AS        AL              in table BA.....T
```

Figure 7.3   UPDATE and UPNAME submodules of DBM.

the "current" ACTPROB. The DBM checks that the ACTPROB is in PROB. . .C before proceeding. In addition, certain fundamental characteristics of catalogs are checked (e.g., the illegal definition of duplicate table sets—defined by the first two characters of table set names in GENTAB.C). Having established that the current ACTPROB is valid, the submodule retrieves the CONT. . .C of this ACTPROB. It is with respect to this definition of the contents of the ACTPROB that all updates are made.

All updates to a table already in the CONT. . .C of the current ACTPROB are processed in a similar manner. The original table is obtained from the ACTPROB indicated in CONT. . .C. Updates are made, and the modified table is filed on the current ACTPROB. CONT. . .C is updated to indicate that the new version is resident on the ACTPROB, and the DATECATG entry is updated to the Date/Run number of the computer job. The original version of the table is modified only if it was resident on the current ACTPROB. In this case, it is, of course, changed by implication for all ACTPROBs that may be referencing it.

For an update to a table that is not resident on the current ACTPROB, the original version is not modified. This is the essential feature that provides the user with a way to make any number of side studies without affecting the base case. The base case remains unchanged and can be used by other analysts for their own studies.

We have seen some of the ways in which the DBM ensures adherence to the structural rules of PLATOFORM. This mnemonic discipline allows tables to be automatically translated when printed and powerful routines to be developed to replace the mnemonic coding for certain entities. We now demonstrate this

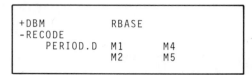

```
+DBM              RBASE
-RECODE
     PERIOD.D   M1        M4
                M2        M5
```

Figure 7.4   RECODE entity codes.

recoding capability in some detail as an example of what is possible in a well-structured data base.

7.3   RECODING A DATA BASE

An ACTPROB (or the entire ACTFILE) can need recoding for several reasons. For example, during data-coordination activities, several codes may need to be changed to bring them in line with a circuit model that includes a representation of the same planning area. Also, as planning cycles move forward to a new set of time periods, a recode of planning time periods can save a lot of work in updating numerous tables.

The cataloging of data tables can be used to great advantage in meeting these needs. The RECODE submodule of DBM utilizes the GENTAB.C to find all occurrences of a given code in table stubs and heads. Where found, the code is changed to a specified new code. RECODE will change only those codes in tables that are physically resident on the current ACTPROB. This design is deliberate; to make a copy of a table on another ACTPROB and to RECODE only the copied version is self-defeating in terms of mnemonic consistency.

7.3.1   User View of RECODE

Figure 7.4 is an example of the PLATOFORM syntax for RECODEing an ACTPROB. This syntax is the user input that, when processed by PLATORD, will create the control table shown in Step 1 below.

7.3.2   System Design of RECODE

We now show in overview the steps performed during execution of the RECODE submodule. This demonstrates the power of the combination of the DATAFORM language and the structural data-base rules. The steps are illustrated using example catalogs as derived in Chapter 4.

0. DBM manager has already recalled the catalogs GENTAB.C, DICT. . .C, parts of which are shown below:

GENTAB.C	S1	S2	H1	H2	M1	F1	D1 - D3
SB.PCLOT	PRODUCTD	PERIOD.D	QUALITYD	MAXMIN.D	XXXAABBX	D3	BLEND SPECIFICATIONS
. . .							

DICT...C	D1 - D3	C1
PRODCLSD	PRODUCT CLASSES	PC
PRODUCTD	PRODUCTS	PR
QUALITYD	QUALITIES	QU
MAXMIN.D	MAX-MIN	MM
PERIOD.D	TIME PERIODS	N2
LOCATN.D	LOCATIONS	LO

1. Read control table for RECODE from PLATOIN file:

RECCTL	ACTION	ARG1	ARG2
001	PERIOD.D	M1	M4
002		M2	M5

This is a request to RECODE time period code M1 to M4 and time period code M2 to M5.

2. For next code to be changed, pick up dictionary name, old code, new code, and proceed to following steps:

Dictionary name = PERIOD.D, old code = M1, new code = M4

3. Find all occurrences of the dictionary under heads S1, S2, H1, H2 in GENTAB.C. Make a list of the table sets for which the dictionary is one of the stub/head entities. In our simplified example, this list will just contain the single set SB.PCLOT (which has PERIOD.D under head S2 in GENTAB.C). Time periods are the second segment of stub names in SB tables.

4. For each table set in the list, determine the position(s) within stub/head where the entity occurs. This is done by determining the total structure of the stub/head from GENTAB.C together with a knowledge of code lengths as defined by the generic codes in DICT. . .C.

For SB tables, a stub name consists of two segments. The first segment is a code from PRODUCTD (S1 entry in GENTAB.C); the second is a code from PERIOD.D (S2 entry in GENTAB.C). In DICT. . .C (under head C1) is a code that provides the generic reference to the dictionaries.

> PRODUCTD   Generic code PR (two-character code)
> PER I OD.D   Generic code N2 (two-character code)

Hence, in generic form, a stub in an SB table has the structure PRN2. Time period codes occur in positions 3, 4.

5. Examine CONT. . .C; list all resident tables belonging to one of the sets involved. RECALL these tables. One of these tables would look like this:

SB.F.U.T	SUMX	V1MX
FLM1	3	500
FLM2	3.5	500

6. For the next table, look for old code using position information derived in step 4. Ignore any nonstandard (wrong-length) stubs/heads. In the example table, M1 occurs in the stub FLM1.

7. Change old code to new. In tables with complex (two-entity) stub or head, this may change any number of stubs/heads. In the example table, FLM1 becomes FLM4.

8. Repeat for next table. Go to step 6.

9. Repeat for next code change. In the example table, FLM2 becomes FLM5. Go to step 2.

10. For each table, check that code changes have not resulted in duplicate stubs/heads. The final status of example table is:

SB.F.U.T.	SUMX	V1MX
FLM4	3	500
FLM5	3.5	500

11. Enfile tables on current ACTPROB. Update CONT. . .C.

RECODE will not examine the table name itself to look for the entity code. This feature may be added (all the information required is in GENTAB.C and DICT. . .C), but it would necessitate a search through the CONT. . .Cs of all ACTPROBs to find pointers to the renamed table.

7.4   KEEPING TRACK OF DATA CHANGES

PLATOFORM has all the flexibility needed to maintain data and logic tables describing planning problems. Another important question is to what extent it is possible to provide an audit trail. Several features of the system, in two classes, fall into this category. First, the system, through use of the system catalogs, can automatically track such key facts as date of update and data residency. Second, the system provides for comments on any data table to be added by the user.

Chapter 4 showed us the table design by which the annotating feature of data tables is implemented. The comments are stored in an alphanumeric table with the same name as the host table (the table in which the actual data is stored), except for the last character of the table name (which becomes X). While it is not necessary for the user to be concerned with this storage mechanism, the implications for the system's developer are significant. It is no longer possible to consider the data table alone when making updates or performing CONT. . .C maintenance. The possibility of a comments table needs to be included in all table-filing logic, and this adds considerably to the complexity of the procedures. An example from UPDATE illustrates this.

Figure 7.5 describes the CONT. . .Cs of two ACTPROBs as they would appear before and after the update shown. When filing the new version of table BS. . .UFT on ACTPROB RSIDE, the original version must be checked for the existence of an associated comments table. If present, it is also copied over to RSIDE so that the pairing of host and comment is complete. For design considerations, the two tables are considered one piece of data. Their actual separation gives rise to additional programming to avoid a logical separation. Leaving the BS. . .UFX pointer that points from RSIDE to RBASE would imply that any change to the comment on RBASE is equally applicable to RSIDE, which has a different version of the host table. However, these complications do not concern the user, who is able to treat data and comments in the same way, using a simple syntax (to modify comments) that is made possible by special handling in PLATORD.

7.4.1   Input Syntax for Data-Table Comments

Modification or updating of comments is another function provided by the UPDATE submodule of the DBM. "C"-type updates are used to add comments to any table. Individual comments are identified by a "line number" having the format L01-L99. The comment itself is written in plain text across the record. In modifying existing comments, use is made of the line numbers to define

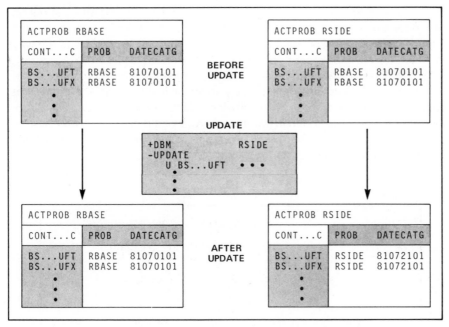

Figure 7.5   CONT. . .C maintenance for comments tables.

editing requirements. In the example of table BS. . .UFT (Figure 7.6), L01 is
replaced and a new line placed between lines L01 and L02. (L015 is interpolated
between L01 and L02.) Existing lines are deleted by giving only the line number.

7.4.2   System Design of Comments Processing

The C-type updates are processed into UPDATE as part of the READTAB input
stream along with data-table updates. A special format on the input file PLA-

```
+DBM            RBASE
-UPDATE
*    ADD A NEW COMMENT
  C SB.F.U.T
     L01         DATA TAKEN FROM CABLE 10 DEC 1980 FROM FAWLEY
     L02               SULFUR RELAXATION OBTAINED FOR FEBRUARY
*    MODIFY EXISTING COMMENT
  C BS...UFT
     L01         DEMANDS CONFIRMED IN CABLE.
     L015        CONSISTENT WITH MARKET OUTLOOK AND WITH TELCON
```

Figure 7.6   UPDATE syntax for comments.

TOIN is generated by PLATORD to take advantage of a READTAB input option that does not remove imbedded blanks from text. Thus the user can write comments in plain English on the input form without any knowledge of the way in which they will be stored (in blocks of eight characters).

UPDATE processes a set of comments by sorting them into line-count order, removing blank lines, and renumbering them. It is important, therefore, that comment updates are made with a knowledge of current status. As an aid to this, the date on which the comments were last processed is provided by the system as a first comment line called DATE. Comment tables are printed automatically whenever the host table to which they refer is printed.

```
TPRINT                          ACTPROB RBASE      DATERUN 81072101

SB.F.U.T                             UPDATED 81072001    RESIDENT ON RBASE
========                    BLENDING SPECIFICATIONS  FUELS     U.K.

                                     SULPHUR    V.CS100
                                     MAX        MAX

                                     SUMX       V1MX
FUEL      JAN         FLM1           3.000      500.000
FUEL      FEB         FLM2           3.500      500.000

DATE                                 81072101
L01      DATA TAKEN FROM CABLE 10 DEC 1980 FROM FAWLEY
L02            SULFUR RELAXATION OBTAINED FOR FEBRUARY


QB.D.UFT                             UPDATED 81070102    RESIDENT ON RBASE
========                    BLENDING COMPONENT QUALITIES     DIST    FAWLEY

                                     SULPHUR    VISC
                                     WT.PERC    CS
                                                AT 100F
                                     SU         V1
VIRGIN HEATING OIL      VH           1.000      2.874
VACUUM DISTILLATE       VD           1.700      7.420
CAT DISTILLATE          CD           1.500      5.630


IB.F.UFZ                             UPDATED 81070501    RESIDENT ON RBASE
========                    COMPONENTS TO BLEND               FUELS   FAWLEY

                                     FUEL OIL

                                     FL
VIRGIN HEATING OIL      VH           X
VACUUM DISTILLATE       VD           X
CAT DISTILLATE          CD           X
REFINERY FUEL OIL       RS           X
```

Figure 7.7   Table printing with text translation.

7.5 TABLE PRINTING

The data-base tables on the ACTFILE are stored and updated using codes for stub and head names. Chapter 4 indicates that, although this is efficient for this purpose, it is not the way in which we would like data printed. A method of printing tables using dictionary translation was mentioned as the way by which the PLATOFORM system overcomes this. The example given in Figure 7.7 shows the output from a TPRINT (table print) of three data-base tables.

We now review the mechanism by which the system "decodes" tables to produce the illustrated output. In this discussion, we use the example catalogs developed in Chapter 4 and already mentioned in this chapter in Section 7.3.2. Using these catalogs, we can follow the steps performed by TPRINT with SB.F.U.T as an example table to be printed.

7.5.1 System Design of TPRINT

1. DBM manager has already recalled GENTAB.C, DICT. . .C.
2. Read control table for TPRINT from PLATOIN file.
   Recall CONT. . .C from current ACTPROB.
3. For next table to be printed:
   Recall table (and comments table, if any) from ACTPROB indicated in CONT. . .C.

4. Decode table name:	*Example*
Store table name.	SB.F.U.T
Find generic name by matching first two characters with GENTAB.C entries.	SB.PCLOT
Look up mask in GENTAB.C.	XXXAABBX
Use mask to break up table name	F.
	U.
and generic table name.	PC
	LO
Look up generic codes in DICT. . .C.	PC = PRODCLSD;
	LO = LOCATN.D
Table name is decoded as:	
D1 through D3 entries from GENTAB.C.	BLENDING COMPONENTS
M1 entry from relevant dictionary for each part of table name.	FUELS
	U.K.

5. Find stub/head dictionaries from
   GENTAB.C:
   Look up S1 and S2 entries in GENTAB.C   PRODUCTD PERIOD.D
   as stub definition.
   Look up H1 and H2 entries in GENTAB.C   QUALITYD MAXMIN.D
   as head definition.
6. Look up DATECATG in CONT. . .C for
   date last updated.
7. Print table with dictionary translation.
   Look up F1 entry in GENTAB.C for
   printing format.
   Print comments.
8. Repeat for next table.
   Go to step 3.

The actual printing of tables (step 7) makes use of the powerful DATAFORM verb TABULATE (see Chapter 3). This single program statement controls all aspects of table printing, including overflow of large tables into multiple strips, rounding of values to a given number of decimal places, and format/position on the output page. In addition, it provides the translation mechanism for stub and head by automatic reference to a given "dictionary." Here "dictionary" is used to mean any table with codes as stub and text under heads. The TPRINT submodule prepares these working tables (in step 5) from the PLATOFORM dictionaries (which owe some of their characteristics to their use in this context and hence to TABULATE design). The TABULATE verb also uses the "dictionary" to control which table stubs/heads are printed and the order of printing.

The use of TABULATE for table printing is highly efficient both in execution time and ease of programming. Its only minor drawback is the limitation that all columns must be printed with the same format. Its advantages, however, are so great that almost all table printing in DBM and REPORTs is handled this way.

7.5.2 Use of Dictionaries in TPRINT

An important result of the design of TPRINT is that a standard order for codes can be established in a dictionary and used whenever a table with that entity class is printed. This relieves the user of any concern over stub/head positioning during UPDATE. (It also means that the capability to order codes within dictionaries is a required aspect of the system.)

When using text from dictionaries to translate stub or head codes, the dictionary order of codes is used in the printed table. Where no translation of stubs or heads is requested, codes are presented in table order. A table printed with no text translation is thus displayed exactly as stored on the ACTFILE.

In Figure 7.7, note that all types of table prints include the code in addition to any translation. This enables the analyst to use the codes in table updates. Stripping off the translation from any TPRINT shows the table as stored, except for reordering caused by dictionary translation.

The question arises as to what action should be taken for codes that are present in a table to be printed, but are not in the appropriate dictionary. The dictionaries in PLATOFORM are regarded as providing text translation and standard order, but not as providing a complete list of valid codes. The omission of a code from a dictionary is regarded as an oversight that should not result in an incomplete table being printed. To this end, any code not given in the dictionary is "flagged" (i.e., the code is filled on the right with asterisks), and printed at the end of the stub or head list. This good fallback procedure allows analysts flexibility in defining codes for use, for example, in rush studies. However, complete dictionaries are maintained whenever possible in order to take full advantage of TPRINT and other reporting routines.

7.5.3 Syntax for TPRINT

The TPRINT submodule allows any table or group of tables in the CONT. . .C of the current ACTPROB to be tabulated in a variety of printing styles. Figure 7.8 illustrates this capability.

Selection of multiple tables to be printed can be made by mask and further controlled by date of last update. Using this latter feature, an analyst can determine which tables have been modified from any prior time, for example, from the date of the last base case, or even any changes made while the analyst was on vacation.

All tables referenced in the CONT. . .C can be printed by use of the keyword ALL. This can produce a large volume of output and requires that the tables be printed in a well-defined order so that individual tables can be found. TPRINT allows a SORT option that sorts by table name before printing.

Each page of output is carefully laid out to provide clarity and maximum information per page. Consideration is given to printing of "wide" tables (in terms of number of heads per strip to be printed), alignment of heads with printed contents, and so on. These details are important, since the TPRINT output is the main working document for users of PLATOFORM-supported models.

7.5.4 "Ordered" Print of Data Tables

A TPRINT of all the tables in the CONT. . .C may cover data for many locations and for many aspects of the planning model. A print by sort order of table name is useful in finding a specific table, but generally will not group tables by location or planning responsibility.

```
+DBM            RBASE
-TPRINT
    SB.F.U.T    DD              ⎤  Individual tables
    QB.D.UFT    DD              ⎬  (see TPRINT output
    IB.F.UFZ    DD              ⎦    Fig. 7.7)
-TPRINT
    SB*****T    MMD             ⎤  Multiple tables specified
    QB.D.**T    DD              ⎦    by mask
-TPRINT
    SB*****T    DD     810101   ⎤  Controlled by date of last
    QB.D.**T    DD     810101   ⎦    update
-TPRINT
    ALL         CC              ⎤  All tables in CONT...C
    SORT                        ⎦
```

```
The print style is controlled by a code input by the user following the
table name. This code consists of up to 3 characters:
    1st  Character  -  refers to printing of stubs
    2nd  Character  -  refers to printing of heads
    3rd  Character  -  refers to printing of table name text

Three levels of text translation are available in dictionaries and each
of these is assigned a symbol for use in the print code:
    D  -  translate using 24-character descriptor from D1-D3
    M  -  translate using descriptor from M1 only
    C  -  use code without dictionary translation
```

Figure 7.8   Table printing.

An enhancement to TPRINT whereby a user can prespecify a group of tables and print them with a single request is available via the OPRINT submodule of DBM. The group of tables is specified by listing them in order as stub entries in a "driving" table that is itself stored in the data base. Furthermore, the tables can be specified generically, in that substitution in positions 6, 7 of the table name can be made when the OPRINT submodule is run. This assumes that table names are structured with a major entity occurring consistently in this position. In Exxon applications, it is common for location codes to be included in positions 6, 7 of table names. An example of the use of this enhancement is given in Figure 7.9.

The OPRINT request is for two groups, RO and FD, perhaps representing refining options and financial data, respectively. For the first of these, the group of tables for each location associated with the ACTPROB is to be printed using print style DD. The second group is to be printed for one location only (UF) using the default style specified in the driver table.

Each page of the output is assigned an index number that is used in an OPRINT index printed following the tables. The index is sorted alphabetically by table and includes the text translation of the table name as well as the

Example 'driver' table for OPRINT – stored on ACTFILE

ZZ...ROZ	STYLE
SB.G.LOT	MM
IQ.G.LOZ	
IB.G.LOZ	
SB.D.LOT	
IQ.D.LOZ	
IB.D.LOZ	
SB.F.LOT	
IQ.F.LOZ	
IB.F.LOZ	
QB***LOT	

Table ZZ...ROZ defines the 'RO' OPRINT

The entry under STYLE is the default print style for all tables in the OPRINT.

Example PLATOFORM syntax for OPRINT request

```
+DBM          RBASE
-OPRINT
      RO      ALL         DD
      FD      UF
```

Figure 7.9   OPRINT for user-specified order of tables.

cross-reference to the index number of the page where it can be found. Thus an OPRINT has the advantage of a prespecified order as well as an index by table name.

7.6   PRINTING COMBINED TABLES

All of the table-printing capabilities discussed so far are alternative ways of viewing data tables defining planning problems. In all cases, the data is viewed in terms of the "dimensions" of the set to which it belongs. Often, data that is essentially three-dimensional has to be arbitrarily spread among two-dimensional tables. For example, in a multirefinery model, data describing the quality of distillate streams is a function of the specific quality (sulfur, gravity, etc.), the distillate stream, and the location. In setting out this data in tables, the organization chosen might be stream versus quality in a set of tables by location (as in the QB-table shown in the TPRINT, Figure 7.7). A view of this data from another angle would be useful, for example, in comparing all the sulfur qualities for each location "side by side." This is a capability included in the system via the XPRINT submodule of DBM as shown in Figure 7.10.

The printed table is the result of extracting all the SU data from each member of a group of tables and combining them into a single display in tabular form. This is a powerful technique for data control and coordination. In the example, a stream not in the product dictionary (CF) is flagged with asterisks. This type

Example data base tables

QB.D.UFT	SU	V1	GS	• • •
VH	1	2.874		
VD	1.7	7.42		
CD	1.5	5.63		

QB.D.UMT	SU	V1	GS	• • •
VD	1.72	7		
VH	1.1	2.8		
CF	1.55	5.9		

Output from XPRINT

```
                                 ACTPROB RBASE      DATERUN 81072201
QB.D.**T
========              BLENDING COMPONENT QUALITIES     DIST     SULPHUR

                                 FAWLEY      MILFORD
                                             HAVEN

                                  UF          UM
VIRGIN HEATING OIL      VH        1.000       1.100
VACUUM DISTILLATE       VD        1.700       1.720
CAT DISTILLATE          CD        1.500         .
CF******                                      1.550
```

Figure 7.10   Combining tables with XPRINT.

of output could indicate a data error or a difference in modeling between the two locations that may need review.

7.6.1 System Design of XPRINT

XPRINT is the first example of what could be called a data-base report as opposed to a display of the actual data tables. It was the first type of data report available in PLATOFORM. Since that time, a generalized report-writing tool has also been developed that includes many other data-table reporting functions. This ADHOC report writer is covered in Chapter 11.

The XPRINT submodule takes advantage of both the cataloging of table sets in GENTAB.C and the CONT. . .C definition of the tables associated with an ACTPROB. The mode of operation is to prepare a working table that can then be passed to TPRINT for printing, thus avoiding any repetition of table-decoding logic. The following steps show how this is done by the program.

1. DBM manager has already
   recalled GENTAB.C, DICT...C.
2. Read control table for XPRINT
   from PLATOIN file.
   Recall CONT...C from current
   ACTPROB.
3. For next request: *Example*
      Store user mask of table set to QB.D.**T
      be combined.
      Check that set is in   QB.PCLOT (PC = PRODCLSD,
      GENTAB.C.   LO = LOCATN.D)
                                       XXXAABBX
4. Remember/store GENTAB.C   S1 entry = PRODUCTD (generic
   entry for this set.   code PR)
      (To be changed and stored   H1 entry = QUALITYD (generic
      again later.)   code QU)
5. Make temporary change to
   GENTAB.C entry for this set—
   reflecting new view of data.
      Change generic name.   QB.PCQUT
      Change stub (head) definition   H1 entry = LOCATN.D
      S1, S2 (H1, H2).
      This is essentially an exchange
      of entity between table name
      and stub/head.
6. Search CONT...C for all tables
   matching mask.
      Create working table by   SU head extracted
      extracting specified stub/head
      from each table.
      Store in working table by   Stored by location
      entity extracted from table
      name.
7. Use TPRINT to print working
   table.
8. Restore GENTAB.C entry to
   original form.
9. Repeat for next request. Go back
   to step 3.

7.6.2   User View of XPRINT

Figure 7.11 gives an example of the PLATOFORM syntax for XPRINT as seen by the user. The group of tables to be included is given in a mask. This is followed by the definition of the dimension to be extracted. (SU is in the head of this table set.) DD is the print style for the resulting table output.

7.7   MAINTAINING DICTIONARIES AND CATALOGS

All data-table printing and system-report writing use the dictionaries to provide text translation of codes. Maintaining these dictionaries (and the application catalogs) is therefore an essential feature of the system. Dictionaries are tables with some special characteristics. They always have a predetermined head structure. Each head is two characters long, and the number and names of heads are known for any particular catalog and set of dictionaries. This type of head structure has influenced the design of a syntax for "updating" dictionaries. Other design factors are the need for positioning of the stubs and input of the text translation of codes. This syntax is available in the UPDICT submodule of DBM and is markedly different from that of UPDATE. UPDICT cannot alter head structure, but is concerned only with stub "entries" and the associated "row" of dictionary text.

7.7.1   System Design Considerations

The need for special routines for handling dictionary maintenance is a function of their structure and limitations in the DATAFORM verbs READTAB and TABU-LATE. We can understand the design of UPDICT by looking at a particular dictionary update. In the XPRINT example (Figure 7.10), we observed what happens in table printing when a code is not in the relevant dictionary; the code CF is flagged with asterisks.

What is involved in updating the PRODUCTD dictionary to include CF? First, since the order of table stubs and heads (in printing) is controlled by dictionary order, we will want to specify the position within PRODUCTD for the new entry. This can be achieved by reference to an existing entry after which the new entry is to be placed. Positioning before an existing entry is also possible.

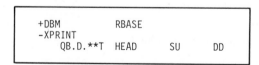

```
+DBM             RBASE
-XPRINT
      QB.D.**T  HEAD      SU       DD
```

Figure 7.11   Input syntax of XPRINT.

Second, we will want to input the text translations of CF. The full twenty-four-character translation is stored under heads D1, D2, D3 (and the shorthand form under head M1). In table printing, the translations appear as continuous text, but in UPDICT we need to recognize that they are composed of blocks eight characters long. Each block (D1, D2, D3) can be modified, so we must specify the head under which each block is to be stored. Figure 7.12 shows how this can be done.

Other dictionary and catalog maintenance functions can be carried out via "key words" in the syntax. Functions included are adding new dictionaries, "scrubbing" dictionaries, renaming stub entries, and deleting entries. Catalogs CONT. . .C, PROB. . .C, SOL. . . .C cannot be changed in any way by UPDICT, as these are totally maintained by the system.

7.7.2 Dictionary Printing

A UTILITY routine that recognizes the special structure of dictionaries is provided to print them (see Figure 7.13). The contents of entries under D1, D2, and D3 are concatenated and the dictionary heads translated.

7.8 DATA SHARING

The data-base management discussed thus far has been general in that the functions of the DBM and UTILITY modules apply to any application. We now briefly discuss another aspect of data management related to data sharing. As we see more fully in the chapter on matrix generation (Chapter 9), data-table

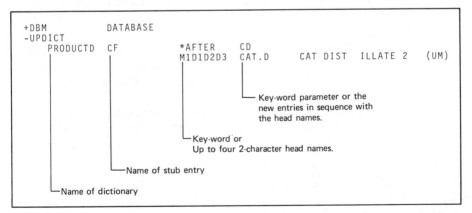

Figure 7.12   Modifying dictionaries with UPDICT. CF is placed in PRODUCTD after the existing entry CD. The mnemonic and descriptor for this new entry are added.

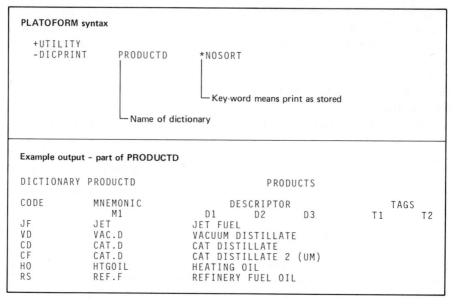

Figure 7.13   Dictionary display obtained by DICPRINT.

design is dependent on the generator that will process it. Much data in planning models is time dependent, and this can present difficulties in the direct sharing of data tables across models for different time frames.

Let us suppose that a matrix generator requires that product specifications be given in SB tables as illustrated in Figure 7.14. The table set groups specifications by product class (F.—fuels) and location (U.—UK). Heads are qualities concatenated with specification type (e.g., SUMX—sulfur maximum), and stubs are product grade concatenated with model time period (e.g., FLM1—fuel oil, month one).

Many models will need specification data for products. Local refinery models, head office planning models, and regional circuit models all follow quality of product, but they typically have different planning cycles covering different time spans.

For SB tables, it is not possible to share the data tables directly, as each model has a different set of time periods. The periods "M1" and "M2" (used in defining

SB.F.U.T	SUMX	V1MX
FLM1	3	500
FLM2	3.5	500

Figure 7.14   A product specification table.

specification data in the example SB table, Figure 7.14) are defined only in terms of a single planning model. Time period M1 may be quite different or may not be defined in any other model. A more general format is required, as shown in Figure 7.15, one that can be converted to model-specific format as needed.

Each SS table is the definition of the specifications for a single product over time. The time span over which a value applies is specified absolutely by date, and not with respect to any model. This time definition is given in terms of starting and ending dates: starting month (MS), starting year (YS), ending month (ME), and ending year (YE). The value is given under head V1. Changes in specification are indicated by the inclusion of "continuation stubs" (SUMX.1, SUMX.2). Other table syntax is available to input seasonal specifications and to allow changes to occur on any day of the month.

We can now share the SS tables directly across models both on the same ACTFILE and on different ACTFILEs. However, we need to convert the general data to model-specific format. To provide this conversion, submodules can be written and included in a separate major module (perhaps called DPREP—Data Preparation). The SPEC submodule of DPREP will extract data from several SS tables (i.e., for each product belonging to the product class), and combine these data into a single SB table. Specifications for each model time period are calculated, based on the overlap between model period and specification definition in SS tables.

7.9 SIDE-CASE CAPABILITY—ACTPROB ORGANIZATION

As we have seen in Chapter 4, ACTPROBs are the first "branch" in a hierarchical structure that allows tables with the same name to be resident on two or more ACTPROBs. PLATOFORM design utilizes this fact and uses the Contents Catalog (CONT. . .C) to associate data table "residency" with any given model. Thus the tables used in building a model can reside on several ACTPROBs throughout the ACTFILE. This has benefits in data organization in that data common to more than one model can reside on a single ACTPROB with updating responsibility

SS.FLU.Z	V1	MS	YS	ME	YE
V1MX	500				
SUMX	3			JAN	1982
SUMX.1	3.5	FEB	1982	FEB	1982
SUMX.2	3.1	MAR	1982		

Figure 7.15 "Generalized" product specifications (SS tables). SS tables contain numeric data (under V1) stored in alphanumeric form. This data is converted to floating-point form when needed in an arithmetic expression.

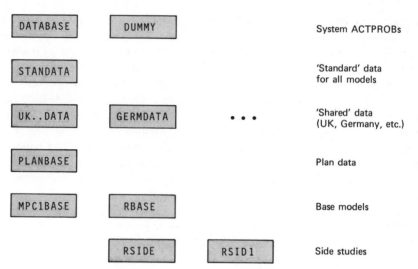

Figure 7.16   An example ACTPROB structure.

assigned to a specific individual or planning group. Figure 7.16 shows an example ACTPROB structure that allows this type of data organization. Each "level" shown in the figure represents the different degree of generality of the data resident on the ACTPROBs.

The DATABASE and DUMMY ACTPROBs will be present on any PLATO-FORM ACTFILE. (The purpose of ACTPROB DUMMY is to start up a new ACTPROB, as described later.) The other ACTPROBs show a possible organization for an analyst group responsible for refinery planning. STANDATA contains data tables on standard blending correlations and qualities of pure components. This data is maintained by a central headquarters group with responsibility for consistent formulations of linear programming (LP) models throughout the company. UK..DATA and GERMDATA store data on product specification and plant capacity provided directly by the location involved. PLANBASE defines the yields and refinery operating conditions that are available for all planning cycles. RBASE is the base-case model for the current planning cycle, while RSIDE and RSID1 are side studies on the current plan. MPC1BASE is the final model from the previous cycle and is available for comparison reporting.

This is just one potential organization for the planning group, and it serves to demonstrate the flexibility of the PLATOFORM approach to data-base structure. The flexibility derives from the hierarchical ACTFILE structure combined with the use of CONT. . .C to "define the model data." The CONT. . .C of RBASE in Figure 7.17 will have pointers to STANDATA, UK..DATA, GERMDATA, and PLANBASE.

CONT...C	PROB	DATECATG
KQ.F...Z	STANDATA	76122101
QC.....T	STANDATA	76122101
SB.F.U.T	RBASE	81070101
SS.FLGHZ	GERMDATA	81010102
SS.FLU.Z	UK..DATA	81010101
YP.C.UFT	PLANBASE	81020301

Figure 7.17   Part of RBASE CONT...C.

7.9.1 Creating New ACTPROBs

The DBM is used to create ACTPROBs by copying the CONT. . .C from an existing ACTPROB. ACTPROB DUMMY is provided for starting up a new ACT-PROB with an empty CONT. . .C. Figure 7.18 shows how a side study is created and the status of the CONT. . .C of ACTPROB RSIDE following UPDATE.

   This illustration is the same as that used in Chapter 6 to show PLATOFORM syntax. It is necessary to give only the data that defines the side study, all other tables being located as in RBASE. The system actions are:

1. Copy CONT. . .C from RBASE to RSIDE.
2. Copy the matrix from RBASE to RSIDE.
3. Update PROB. . .C to include new ACTPROB:
     Record date of creation.
     Record presence of a matrix.

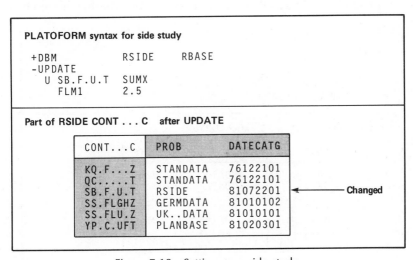

Figure 7.18   Setting up a side study.

4. Proceed to —UPDATE. UPDATE will modify one table (SB.F.U.T), filing the new table on RSIDE. RBASE is not affected in any way.

The CONT. . .C of RSIDE is different from that of RBASE only in the entry for SB.F.U.T. Since all modules access data tables via CONT. . .C, this change can be transmitted to the model when the matrix generator is called.

7.9.2   Adding Pointers

This ability to cross-reference data leads to the need for other functions, for example, to reference data added to a "higher" ACTPROB after the creation of a dependent ACTPROB. Suppose a new table is added to PLANBASE that defines an additional process now available. The table might be YP.B.UFT. All future plans should incorporate this data in the planning model. In particular, RBASE for the next planning cycle should include this table in CONT. . .C. Reference to new data tables on other ACTPROBs is achieved using the COPY submodule of DBM (Figure 7.19).

7.9.3   Removing ACTPROBs

The organization of an ACTFILE depends not only on the type of model but also on the data responsibility of the planning groups using it. Many functions are required to manage the total file and ensure its consistency. One of these is the removal of an old ACTPROB that is no longer required. This is achieved by use

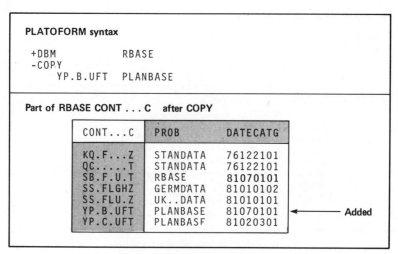

Figure 7.19   Adding a pointer.

of the UTILITY submodule SCRUB. From the preceding discussions on data sharing, it should be evident that care needs to be exercised in deleting ACT-PROBs. An ACTPROB is not an independent "subfile," and many other ACT-PROBs could be referencing data contained in it. On the other hand, a side case at the end of a "branch" may be totally independent and thus could be removed without affecting any other part of the file (e.g., RSID1). Removal of RBASE is not so straightforward, since RSIDE depends upon it.

From the system's viewpoint, we can simply remove an ACTPROB in a single statement by "pruning" the hierarchical ACTFILE at the ACTPROB level. However, in PLATOFORM we must also perform all the catalog maintenance associated with ACTPROB removal. This includes removing the ACTPROB entry from PROB. . .C and removing solution cases on the deleted ACTPROB from the Solution Catalog SOL. . . .C (this catalog is covered in more detail in Chapter 10).

But how can we handle the fact that other ACTPROBs may be pointing to the removed ACTPROB for data tables? We must provide a way to rescue these tables before the ACTPROB is deleted. The steps performed by the system to achieve this are as follows.

For each ACTPROB in PROB. . .C (except the one to be deleted):

1. Recall CONT. . .C from ACTPROB (call this "current" ACTPROB).
2. Search CONT. . .C for all pointers to ACTPROB to be deleted; make list of tables so pointed.
3. Physically copy each of these tables to current ACTPROB.
4. Update CONT. . .C to show tables now resident on current ACTPROB.
5. Go back to step 1. Repeat for next ACTPROB.

Suppose that at the end of the planning cycle, RSIDE is the final plan and RSID1 is no longer required. Furthermore, RSIDE is to be the starting point for the next plan, and RBASE is no longer relevant. ACTPROB RSID1 is first deleted (see Figure 7.20). The standard RESCUE action (for RBASE) moves all tables used by other ACTPROBs to these other ACTPROBs and changes the CONT. . .C pointers accordingly.

In the example given in Figure 7.20, RSIDE will receive all the tables that were pointed to on RBASE. Only when all required tables have been "rescued" is the

```
+UTILITY
 -SCRUB     ACTPROB    RSID1     PURGE
 -SCRUB     ACTPROB    RBASE     PURGE      RESCUE
```

Figure 7.20   Deleting ACTPROBs.

named ACTPROB scrubbed. In a file with many ACTPROBs dependent on a scrubbed ACTPROB, RESCUE can create copies of the same table resident on many ACTPROBs. It is possible to avoid this with "rescue" to a named ACT-PROB. In this case tables are moved only to a named receiving ACTPROB and other ACTPROBs pointed to it.

Incorrect use of SCRUB can produce inconsistencies in the ACTFILE, and SCRUB should only be used with a knowledge of the ACTPROB dependencies. Further, we have seen that all DBM and UTILITY actions discussed in this chapter assume that the system catalogs are complete and consistent. We need to be sure that an ACTFILE is "clean" and to remove inconsistencies that may be introduced by errors in the programs or by the actions of inexperienced users. Utilities can be written that will show the interrelationship between ACTPROBs, will check total file consistency, and will reorganize as necessary. Some of these utilities are discussed in Chapter 8.

7.10 SUMMARY

In this chapter, we have seen the way in which the data for LP models is updated and displayed in the PLATOFORM system. Included are all the table-maintenance functions of modern data-base systems. The factors influencing their provision in this system have been covered.

The data-base rules, established in Chapter 4, are imposed by the UPDATE facilities within the system. The advantages of this can then be realized in the provision of powerful table printing (TPRINT, XPRINT) and total file RECODEing. The data-sharing features of the system design are ensured by the automatic system maintenance of PROB. . .C and the Contents Catalog, CONT. . .C, of each ACTPROB. By insisting that all modules work with this CONT. . .C definition of ACTPROB data, we can provide a unique side-study facility not seen in many other LP support systems. Another result of this use of CONT. . .C to define the data is that ACTPROB organization can be varied in any number of ways to suit the planning organization it serves. We have seen one example of this and touched on some of the resulting file-management needs (e.g., SCRUB for ACTPROBs). A full treatment of these functions is the subject of Chapter 8. Following Chapter 8, we go on to look at the way in which matrix generators are written to map data into LP matrices.

Jeremy D. Sammes

8. FILE MANAGEMENT

8.1 INTRODUCTION

Earlier chapters described some of the fundamental design considerations that need to be addressed during the formulation of a large data-base system. We have seen how the design of the PLATOFORM data base offers solutions to some of the major organizational problems associated with storing data in an accessible and maintainable form. This design is characterized by a set of dictionaries and catalogs that define the ground rules for the use of the data base. Although at first these rules may appear to impose some restrictions on access to the system, we have seen how, in practice, they determine how easy it is to use and how powerful and flexible its capabilities are.

In this chapter, we extend the discussion to some of the major factors influencing the design of the broader file-management capabilities. Some of these issues are common concerns to the designer of any integrated software, and well-established solutions are available. For instance, we expect to provide some automatic backup/security function for our ACTFILE data bases, and perhaps provision also for a read-only password capability. A modern data-processing environment routinely offers the basic building blocks for such facilities. These may include a procedure for automatic generation of a cataloged tape cycle to maintain the most recent versions of the ACTFILE and a fireproof safe to store selected tapes. Passwords may be incorporated at the ACTFILE level and submitted at run time to provide overall protection.

Besides these issues, however, a number of more specialized concerns derive from the nature of the mathematical programming system (MPS) and from the infrastructure of the data base. In these areas, we are breaking new ground and need to establish robust structures to service the needs of the user community.

One of the most significant issues lies within the relationship between the user/analysts and the data. The data may be needed by one user or by several, either separately or in a shared fashion. Also, users may need to access data that resides on more than one ACTFILE. The provision of tools to address this situation turns out to be nontrivial. In keeping with the design philosophy of the system, writing the control statements to perform a logically simple task should also be simple. As we shall see, in practice, operations between ACTFILES can become involved, and in some situations apparently reasonable requests translate into ambiguous or conflicting actions.

The second part of this chapter describes some of the more important utility capabilities that evolved in response to the file-management needs of the users. These include modules that perform housekeeping, ensure overall file integrity, and report potential inconsistencies to the user. These inconsistencies arise precisely because the rules embodied in PLATOFORM are not sacrosanct and may on occasion be broken. As we have seen in Chapter 4, this flexibility is invaluable, but it demands a mechanism for tracking and, if necessary, correcting any subsequent problems.

Finally, we turn our attention to a particular concern to the PLATOFORM user —the provision of sound techniques to guarantee ACTFILE survival in the face of an abnormal job termination. TILLER is the underlying file-handling program for PLATOFORM and is responsible for all updates to the ACTFILE. For efficiency, however, TILLER operates on in-core sections of the ACTFILE, and it completes updating of the physical device only when it closes it (at the end of a PLATOFORM module such as the +DBM module). If the run terminates before this close, the ACTFILE is rendered unusable. Rather surprisingly, this is a relatively straightforward problem, and its remedy is not costly compared to the efficiency offered by this approach.

8.2   THE ENVIRONMENT OF THE APPLICATION DATA BASE

Chapter 5 has explored the idea of a "single integrated system" and shown how such a system allows us to tie all the separate functions needed to a single input stream and a single data base. We have seen how this consolidated data base is central to the development of a powerful data-management capability and, in fact, that it underpins the entire system structure. Let us look, now, a little more closely at the ways in which the user may wish to relate to this data base.

One of the immediate advantages offered by a data-management system operating on a central repository of information is the number of users who can access it. This facility for users to share data has been discussed in Chapter 4, and it has been a hallmark of the success of the system. In addition, communicating between separate data bases can also be addressed, which opens a wide range of opportunities for data extracting and merging. For convenience, these different operating environments can be classified into three broad categories:

- One analyst   Personal data base
- Several analysts   Shared data base
- Several teams   Multiple data bases

All three situations exist, today, within individual applications, and the relationship may evolve further with time. The career of a new application may begin by involving data for a long-range planning study, and subsequently it may be extended to a shorter time-frame study as well. Alternatively, two user teams working on related but separate ACTFILEs may identify a common set of information that they should share. Both of these situations have organizational implications, in addition to the system design challenges that they pose. Where a data base is shared, we have to provide a way for users to share data amicably. In the multiple data-base case simple, understandable, extract-and-merge routines are necessary to facilitate communication between ACTFILEs.

The single-analyst/single-data-base application is of course the simplest situation and is often the most appropriate for the user organization. The analyst is entirely responsible for the data base, and the layout of the individual ACTPROBs is largely a matter of convenience. It may be quite acceptable to place all of the basic information onto a single ACTPROB and to generate subsequent ACTPROBs only when necessary, to begin a new plan or to investigate a side study. The analyst has no immediate need to partition the data in any other way, although it may be prudent to consider whether this situation may change significantly over time. This option, then, is straightforward, with no further data-layout analysis needed.

8.3   THE SHARED DATA-BASE FILE

What about the more complex situation involving more than one user? Two key concepts were introduced in Chapter 4. First, we can physically partition the information on an ACTFILE into ACTPROBs. This partitioning may reflect different states of the data (base case versus side study) or organizational needs (long-range planning versus short-range). The second key idea centered around the use of the CONT. . .C catalog, which provides the cross-referencing and allows us to construct other kinds of ACTFILE organization besides a strictly hierarchical one. These two powerful concepts give us the tools we need to arrange our data base in the best way to suit the users who will be accessing it.

Let us return to the example illustrated in Chapter 4 (Figure 4.13) and pursue the implications of data sharing a little further. This diagram (reproduced here as Figure 8.1) illustrates how the data for two separate users can be logically partitioned into common data and data that is of specific interest to each user. The specific data can then be further segregated into base-case information and side-study information. In the subsequent discussion within Chapter 4, we saw how this logical arrangement could be transformed into an effective hierarchy of ACTPROBs that minimized duplication of data yet allowed maximum control over each user's view of the data (Figure 4.17).

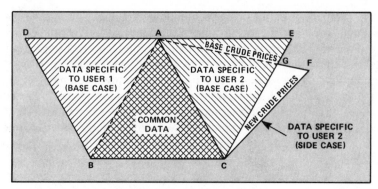

Figure 8.1   The concept of data sharing.

The layout we adopted was, of course, one of several equally acceptable alternatives. We chose to represent the arrangement via three ACTPROBs which split the diagram as follows:

ACTPROB	Graphical Area Represented
BASE1	ADB+ABC
BASE2	AEC with pointers to ABC
SIDE2	CGF with pointers to (AEC−AEG)+ABC

One implication of this approach is that the BASE1 user is responsible not only for his or her own base case (area ADB) but also for the shared data (area ABC). This is seen most clearly from the contents of BASE1 (Figure 4.14). We could, instead, have included the shared data ABC with the BASE2 ACTPROB, in which case the two tables YP. . . . .T and QB. . . . .T would have resided on BASE2, and all the CONT. . .C references would point to this ACTPROB. In strict terms, BASE2 is now at the top of the hierarchical tree, with BASE1 and SIDE2 dependent on it. In practice, the shared data may need coordination as a separate function, and this could be reflected by splitting area ABC entirely into a separate ACTPROB (e.g., BASEDATA), giving the modified structure:

ACTPROB	Graphical Area Represented
BASEDATA	ABC
BASE1	ADB with pointers to ABC
BASE2	AEC with pointers to ABC
SIDE2	CGF with pointers to (AEC−AEG)+ABC

We still have a hierarchy, of course, with BASEDATA at the top of the tree, BASE1 and BASE2 as dependents, and SIDE2 dependent on BASE2. Although this layout is more complex, it more closely mirrors data responsibilities within the organization. The custodian of BASEDATA can oversee the activities of the BASE1 and BASE2 users and can coordinate the details and timing of changes made to the shared set of data ABC.

Let us see what happens if we extend this structure one stage further, by subdividing ABC. Consider a simple hypothetical business environment involving two related linear programming (LP) models that are being used within an affiliate organization. The models describe two planning periods (BASE1 and BASE2). Each LP model is used by a separate group, and the BASE2 group is composed of analysts who are working on a side study as well as the base-case model. Let us also suppose that some of the information about these models comes from a regional organization that coordinates with the affiliate.

Figure 8.2 illustrates an ACTFILE organization that could be used to serve the needs of these two planning groups. The structure in this case is not a simple hierarchy (although it could be), but it reflects the levels of data stored. The regional data has been collected onto an ACTPROB we have called REGN-DATA, and it is maintained uncontaminated by any affiliate or local data. The remaining information about the models that is not specific to either one has been concentrated onto the affiliate-shared ACTPROB called AFFDATA. This data is the primary domain of the affiliate's Data Coordinator. The remaining tables that define the specific parameters of the base-case models reside physically on BASE1 and BASE2, respectively.

If we apply this organization to the ACTPROB structure we have been developing, we see that it condenses to a separation of the common data ABC into two subdivisions. The split arises because responsibility for the "common" data

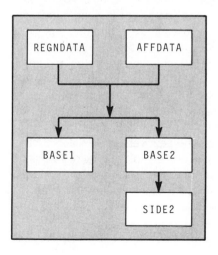

Figure 8.2   An ACTFILE organization serving two planning groups.

resides not with one person or group, but in this case with two—the region and the affiliate. Our original simple example consisted of four tables, two of which were shared.

Let us suppose that the region is ultimately responsible for coordinating processing yields (YP. . . . .T), and the affiliate has charge of product qualities (QB. . . . .T). We arrive at the allocation of tables displayed in Figure 8.3. This will lead us, ultimately, to an ACTFILE arrangement (Figure 8.4) that involves five ACTPROBs whose logical relationship closely reflects the pictorial representation of Figure 8.2.

In summary then, we can develop ACTFILE structures with considerable flexibility, in response to the changing needs of the organization. These structures can continue to evolve as requirements and responsibilities shift. The ACTFILE layout is summarized by the CONT. . .C tables, and it can be modified by suitable data-base management functions. We can move data tables up or down the hierarchy, as well as between siblings, and we can also tailor the CONT. . .C cross-references as necessary.

8.4 MODIFYING THE SHARED DATA

Now that we have a set of ACTPROBs that more faithfully model the needs of our organization, can we exploit this arrangement further? We most certainly can, and in a way that is not easy to duplicate with simpler data-base structures. We shall see that the apparent increase in ACTFILE complexity arising from the data segregation needs, in turn, simplifies the mechanism for modifying data shared by several users.

Consider the information on REGNDATA. This was provided by the region presumably on a tape (and introduces the concept of data sharing *between* files that is discussed later). Let us assume that the region has just issued a replacement version of the data on REGNDATA to be used in the future (i.e., the yield data table YP. . . . .T). The BASE1 analysts are about to begin a new study for which they need this data, but the BASE2 users, inevitably, are in the middle of their planning cycle and must finish it first.

How can we address this conflict? At least the regional information is all in one place; the situation would be much more complicated if this shared data were merged with BASE1 itself (Figure 4.14). What is needed, here, is a mechanism for allowing the BASE1 users to address the new version of REGNDATA, while still allowing the BASE2 (and SIDE2) users access to the previous version. There is no way to avoid having both versions of REGNDATA on the data base, but how should we incorporate the addition in a way that will allow us to unravel the effect later? When the BASE2/SIDE2 users have completed their study, they will also want to change over to the new REGNDATA.

We can attack the problem in stages. As a first step, let us copy REGNDATA,

Data Table	Description	Residence
YP.....T	Process Yields	Common data (regional)
QB.....T	Product Qualities	Common data (affiliate)
BS.....T	Demand Data	Specific to BASE1 and BASE2
FC.....T	Crude Prices	Specific to BASE1, BASE2 and SIDE2

Figure 8.3   Allocating responsibility for the tables.

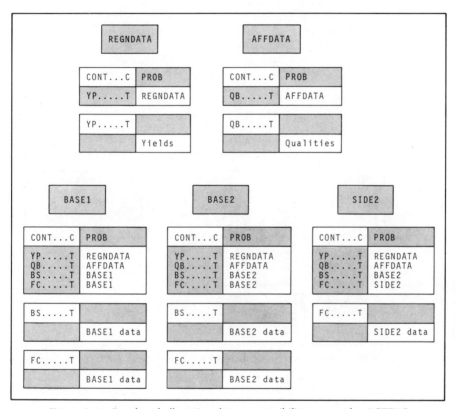

Figure 8.4   Results of allocating data responsibility across the ACTFILE.

including *all* the tables on it. We could call the copy OLDREGN. Now we have two ACTPROBs that are identical in their physical contents, with CONT. . .C tables that "point" to themselves (see Figure 8.5).

OLDREGN, of course, is not referenced by any other ACTPROB at this stage, so although it duplicates REGNDATA, it is isolated from the rest of the data base. The next step (Figure 8.6) is to "point" BASE2 and SIDE2 to this ACTPROB instead of to REGNDATA, leaving BASE1 as it is.

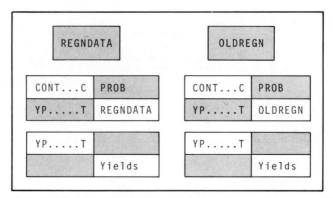

Figure 8.5   Duplicating REGNDATA to isolate the BASE2/SIDE2 analysts.

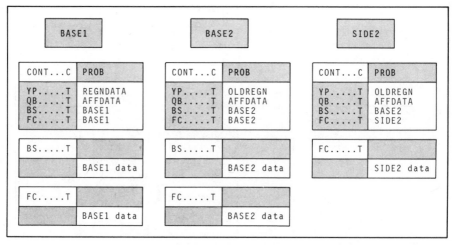

Figure 8.6   "Repointing" the BASE2/SIDE2 analysts to OLDREGN.

Notice that this second step involves only a direct update to those entries in table CONT. . .C that previously referenced REGNDATA. No other table modifications are required; the disturbance to these two ACTPROBs is minimized. The BASE1 ACTPROB is not modified in any way. The third and final step is to replace the REGNDATA ACTPROB with the new version. We look at this final process in more detail in the next section, but in principle we can see that we have achieved our goal: to decouple one set of users from a common ACTPROB with minimal disturbance to the rest of the data base.

These three steps in the decoupling process are accomplished in PLATO-

FORM by a DROP module that identifies the ACTPROB to be "dropped" and the dependent ACTPROBs that are to be rerouted to point to the copy, as demonstrated in Figure 8.7.

We also need a mechanism to unravel the effect of the DROP at the end of the study. In practice, this usually reduces to the simple process of archiving and then deleting the current versions of the ACTPROBs concerned. In our example, SIDE2 will be deleted at the end of the study anyway. If BASE2 will be needed in the future, the simplest strategy is to make a temporary copy of it before the DROP. This copy (say BASE2D) will contain only the CONT. . .C table, and it can be safely rerouted to point to OLDREGN. At the end of the study, BASE2D and OLDREGN can also be deleted, and the data base is once again "clean."

8.5 THE DATA-COORDINATION ROLE

We have made oblique references in this chapter to a data-coordination role. The need for such a role is clearly very much dependent on the complexity and diversity of the data that makes up the data base. Where the data involved crosses company or departmental boundaries, there is often a need to clarify both the roles and the working relationships between the groups concerned. In such situations, it can be advantageous to appoint a person or group who can arbitrate and manage the interfaces between the affected parties.

This data-coordination function, whether it resides with a separate individual or is merely an implicit part of a user/analyst's job, comprises a number of separate functions. These can be summarized as:

- Clarify responsibilities for the individual data.
- Design and implement the ACTPROB structure of the file.
- Review the data residence as organization evolves.
- Monitor the daily use of the data base.
- Coordinate data sharing between data-base files.

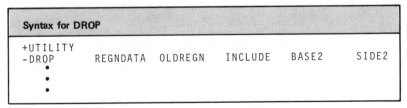

Figure 8.7   PLATOFORM syntax for DROP.

The design of the file, as we have seen, flows directly from perceived responsibilities for the underlying data, and the human role here as referee and adjudicator is clear. This is not, however, a once-only activity. In a real business environment, priorities and directions constantly change, and this is reflected in shifting organizational and functional needs. At the planning data-base level, these forces may be felt in new and changing data as well as in changing responsibilities. Information that was once model-specific may become fixed by common consent. Historically constant data may become more transient. The recognition and formalization of this movement are important coordination functions.

Perhaps one of the most significant functions of this role, especially in a large, multiuser planning environment, is monitoring of daily activities on the data base. In practice, some of this monitoring may be cure, not prevention; there is a need for all users to adhere, wherever possible, to the rules implied in the original design of the data base. The fact that the coordinator is human allows rules to be bent or even broken, on occasion, and this is an important element if the tool is to be used effectively. A human arbitrator may be less efficient but is better able to weigh a temporary data degradation against the need for a quick result.

We shall see how the horizons of data coordination are widened by the data-management problems associated with transferring information to and from other PLATOFORM files.

8.6 SHARING DATA BETWEEN ACTFILES

A number of situations demand an ability to be able to move information between data bases. In this discussion, we are not concerned with interfacing PLATOFORM data bases to other types of data bases via bridge programs. Instead, we focus on those situations that require data movement between PLATOFORM ACTFILEs. These can arise in a number of ways. Within Europe, for instance, a common set of information is needed for all the refinery LP models run within the region. Such information might include processing-unit yield patterns that have been agreed upon by common consent. This shared data could be held on a central ACTFILE, perhaps segregated into a series of reference data ACTPROBs. We clearly need a mechanism for communicating with the user ACTFILEs in the field, possibly in both directions.

Inter-ACTFILE sharing can also arise when the number of individuals who need to access the same data base becomes sufficiently large that a backlog of batch jobs develops. This translates into an increased turnaround time that may prove unacceptable to the user pool. One alternative is to segregate the ACTFILE into several smaller data bases, with the shared or common ACTPROBs carefully duplicated and monitored. Updates to the common copies then have to be controlled and executed in a coordinated manner. Once again, a reference ACTFILE may help to clarify responsibilities, and once again, we need to address the actual transfer of the physical data.

If we turn again to our simple example, we can trace this process as it affects the REGNDATA ACTPROB that is to be replaced by a new version on another ACTFILE. As will become clear when we look at the details of this transfer, the segregation of the portion of data to be replaced into a separate ACTPROB greatly simplifies and mentally clarifies this process. The idea, then, is to consider the ACTPROB a single item for cross-ACTFILE transfer. This means that we will replace one entire ACTPROB on one data base with a copy of the same-named ACTPROB from a second data base. So, one immediate prerequisite for the copy is that the ACTPROB must exist on both ACTFILEs.

As we shall see, several additional structural demands need to be satisfied in order to simplify the communication as far as possible and to avoid confusion to the users. Consider the affiliate ACTFILE and a regional reference ACTFILE, both containing copies of REGNDATA as illustrated in Figure 8.8. Using this structure, the region can periodically send a tape of its ACTFILE to all affiliates, and the REGNDATA ACTPROB can be copied onto the local data bases on an agreed date. We have seen how DROP allows this replacement without affecting current studies within the participating files.

Let us now look more closely at the actual transfer mechanism itself. The process is not as simple as it appears. One complicating factor, the isolation of the transferable ACTPROB within the ACTFILE, has already been stressed. This means, in effect, that the logical view of this ACTPROB (the tables ''associated'' with it) is the same as its physical contents. Put another way, the CONT. . .C table contains no references to any other ACTPROBs; all the pointers are internal. For convenience, we refer to this kind of isolated ACTPROB as a ''master'' ACTPROB. There may be any number of master ACTPROBs in a data base. Together, they constitute the highest level of data, and they could in principle be merged to a single, shared-data ACTPROB.

A little reflection shows that an attempt to formulate the rules for transferring

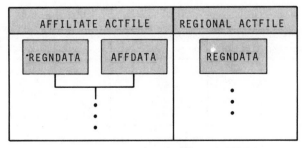

Figure 8.8   Regional and affiliate ACTFILEs.

an ACTPROB that does have external pointers is dangerous. A prime concern is to minimize disturbance to the rest of the data base, and this turns out to be impossible if "associated" tables are also involved in the transfer process.

Given an isolated master ACTPROB, we also need to consider its contents on a table-by-table basis. Three situations can arise during the transfer: the same table may exist on both data bases, on the donor only, or on the receiver (host) only, as demonstrated in Figure 8.9.

Few problems arise if the table resides on both ACTFILEs (Option 1). The new table simply replaces the old one completely. If the incoming table is new (Option 2), it is physically added. The new table will not yet be referenced by any dependent ACTPROBs, of course, because these will not "know" that it exists. Users must consciously decide to include this new table (by adding a cross-reference to table CONT. . .C), which then changes their modeling view of the data base.

The remaining situations concern the table that exists on the receiving, or host, ACTFILE but not on the donor. We need to decide what to do about any other ACTPROBs that currently point to this table. One solution, of course, is simply to delete the table together with any such CONT. . .C pointers (Option 3). This certainly cleans up the ACTFILE, but it also changes the view these ACTPROBs have of the data base without allowing users any flexibility. Nevertheless, this action is the simplest and often the most practical action, and it is the default in PLATOFORM.

Two alternatives exist. We could suppress the deletion altogether (Option 4, keyword 'SAVE'). No disturbance of the host data base is now necessary, but, of course, the two ACTFILEs are now no longer synchronized—which is one of the prime goals of the entire operation. The host ACTFILE contains tables that have no counterparts on the donor. Instead, we can move the offending tables to another part of the data base (Option 5, keyword "RESCUE"). This entails naming an ACTPROB that is to receive them and rerouting the CONT. . .C pointers of all the affected ACTPROBs. Although more complex, this achieves our two major requirements: to synchronize the ACTFILEs and to avoid wholesale changes to the user view of the ACTPROB. Users can then decide at leisure

	Table Resident on Donor ACTFILE	Table Resident on Host ACTFILE	Control Key Word	Action on Host ACTFILE
1	YES	YES	—	REPLACE
2	YES	NO	—	ADD
3	NO	YES	—	DELETE
4	NO	YES	SAVE	RETAIN
5	NO	YES	RESCUE	TRANSFER

Figure 8.9   Table of possible actions that can be taken during the table transfer between ACTFILEs.

whether any of these tables are necessary and can delete them at their own ACTPROB level.

This discussion has centered around the need for moving data across data bases in a complex, multiuser environment. To achieve this, several restrictions on the structure of the ACTPROBs are essential, and a good understanding of the process is needed by the data-base coordinator. Other situations, however, involve cross-ACTFILE communication and are much simpler; for these, correspondingly simpler methods are needed.

For example, an analyst may wish to retrieve a block of tabular data that was used in a previous side study but has now been archived. Is there a simple tool that can retrieve this information without keying it in a second time? The tables are quite likely to be distributed across several ACTPROBs on the archived data base that are not master ACTPROBs at all. This retrieval can be achieved, of course, but only by relaxing some design constraints. The ACTPROB can no longer be the logical unit of transfer; the table must adopt this role. Nor can we avoid changing the host ACTPROB's view of the data base. Both of these assumptions are often quite acceptable, and a simple CONT. . .C-driven table extract-and-merge routine can be designed that fulfills this need. Even so, there are possible complications. For instance, the incoming table may be referenced in the host CONT. . .C, but the host table may be physically resident elsewhere. These and other potential inconsistencies warrant constructing a set of utility modules to help in the auditing and tracking of changes made to the data base.

8.7 FILE-MANAGEMENT UTILITIES

The PLATOFORM utility routines evolved alongside the data-base management capabilities in response to file-management needs. Several of them developed as data-coordination responsibilities became clearer and as it became necessary to provide tools for individuals or groups to track the activity on their data bases.

8.7.1 The New ACTFILE

One early need was to be able to bootstrap the PLATOFORM ACTFILE into existence. Having allocated the space on a direct access device, we need to make the new data base a recognizable ACTFILE and to set up the system tables that the Data-Base Manager (+DBM) will need. DBM expects to find an ACTPROB called DATABASE on which the dictionaries and system catalogs will reside, and we also need another dummy ACTPROB to use as a starting point for building our first data ACTPROB. The output from the STARTUP module, then, is an ACTFILE that comprises ACTPROBs DATABASE and DUMMY. Skeletal system catalogs are built on DATABASE (PROB. . .C, SOL. . . .C, DICT. . .C, etc.), and an empty CONT. . .C table is constructed on both ACT-

PROBs. This module is called once and only once for each grass-roots application. (An attempt to run STARTUP on a current ACTFILE is trapped by the system, and the entire run is promptly terminated without damage to the data base.)

Following this utility module, a call to +DBM on ACTPROB DATABASE allows us to build up the catalogs (GENTAB.C, DICT. . .C, etc.) via −UPDICT (Figure 8.10). At this point, a second call to +DBM allows us to create our first user ACTPROB (say BASE1) from DUMMY, and the bulk of the data can be entered via −UPDATE (see Chapter 7).

8.7.2 ACTFILE Inconsistencies

We have acknowledged that the PLATOFORM design philosophy allows for some of its own structural rules to be broken, in order to preserve operational flexibility. Several of these rules are embodied in tables that are entirely in the hands of the user/analyst (e.g., GENTAB.C and DICT. . .C), rather than being system-maintained. It is quite possible, therefore, that some internal inconsistencies can creep into the data base. These may take a variety of forms.

For example, a dictionary added to the data base may not be entered correctly into the dictionary catalog (DICT. . .C). Alternatively, DICT. . .C may contain code entries (under the C1 head) that duplicate other codes. This could lead to unexpected output from the table-printing (TPRINT) module of DBM. These and similar examples can be trapped by a carefully designed CLEANUP module whose task is to report the inconsistencies concisely. This is a necessary first step, since the correction of the problem may depend on its context, and a decision must be made by the user.

Consider a new dictionary that was not added to DICT. . .C. Do we delete the dictionary, or do we add a dummy entry in DICT. . .C? The former action could be drastic, and the latter requires a description of the dictionary that only the user can provide. In practice, the user probably wants to keep this table and

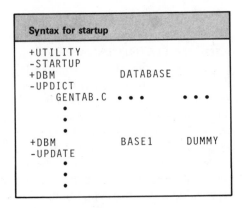

Figure 8.10 PLATOFORM syntax to create a new ACTFILE.

will make the addition to DICT. . .C if reminded about it. Thus a reading of the CLEANUP report will lead to a series of updates by the user and to a rerun to check for further inconsistencies. As a final step, CLEANUP can be run with a PURGE key word, which will cause it to resolve any remaining ACTFILE errors by making deletions according to a well-documented set of rules.

8.7.3 Changing the System Catalogs

Once the STARTUP module has been run, the function and context of the system catalogs is fixed for the life of the application. A new catalog, for instance, cannot easily be added to the data base. As a large system such as PLATOFORM evolves over time, however, additional system requirements may require new catalogs or additions to existing ones. The problem, then, is to incorporate the changes. Modifying STARTUP clearly solves the problem for new applications, but what about all the existing data bases?

A CATCHUP module is needed here, one that can be run on occasions as directed by the systems-support group. It can act as a focus for a variety of system enhancements that are once-only changes needed to bring the ACTFILE in line with the latest version of the PLATOFORM code. CATCHUP is an important component of the system, and it reflects the dynamic nature of the business.

8.7.4 ACTPROB Interrelationships

In a large, multiuser data-base environment, we have seen that a number of specialized data-coordination functions are necessary. The powerful data-management facilities allow users to move data around with great flexibility, and ACTPROB dependencies may easily become unclear. The coordinator needs a summary of these dependencies in order to understand which ACTPROBs may have become redundant or which have CONT. . .C pointers that may be incorrect.

A report showing the relationships between ACTPROBs is easy to construct from a scan of the constituent CONT. . .C tables on the ACTFILE. The output (see Figure 8.11) is a pictorial representation of the connections currently active within the ACTFILE. The table illustrated has several interesting properties. All the master ACTPROBs (with no external pointers) are immediately recognizable. Their rows have only one entry *on* the diagonal (REGNDATA, AFFDATA, and DATABASE). Entries *below* the diagonal characterize ACTPROBs hierarchically dependent on the diagonal entry above them (BASE1 is dependent on REGNDATA and AFFDATA). Entries *above* the diagonal (there are none in this example) correspond to relational dependencies. These can arise, for example, when a table is copied from a child to its parent. This report, then, provides a simple

```
ACTPROB DEPENDENCY TABLE
                  D  D  R  A  B  B  S
                  A  U  E  F  A  A  I
                  T  M  G  F  S  S  D
                  A  M  N  D  E  E  E
                  B  Y  D  A  1  2  2
                  A     A  T
                  S     T  A
                  E     A

DATABASE       X
DUMMY
REGNDATA          X
AFFDATA              X
BASE1             X  X  X
BASE2             X  X     X
SIDE2             X  X     X  X
```

Figure 8.11 Pictorial representation of ACTPROB interdependencies.

but highly effective visual summary of the ACTPROB relationships. Anomalous cross-references can be recognized readily.

8.7.5 Monitoring ACTFILE Size

It falls to the Data Coordinator to accept responsibility for issues affecting the data base as a whole. Among these is the physical size of the ACTFILE (i.e., how much disk space it can consume). Because the underlying TILLER software allows the analyst to add elements to all parts of the file, it is continuously growing. The deletion of unwanted ACTPROBs, tables, and solution cases does not automatically release this file space. Therefore, some kind of compression procedure is necessary, and one is provided as part of the Enhanced Mathematical Programming System (EMPS). This procedure (COMPRESS) is packaged into a PLATOFORM module, which then is run regularly to keep the overall data-base dimensions within bounds. The frequency of running COMPRESS is a function of data-base size, activity, and the fraction of reserve space that the users are prepared to pay for. The syntax is illustrated in Figure 8.12.

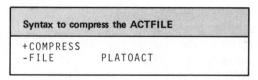

```
Syntax to compress the ACTFILE

+COMPRESS
-FILE          PLATOACT
```

Figure 8.12   PLATOFORM syntax used to compress the ACTFILE.

8.8 PROVIDING AN AUDIT TRAIL

The file-management tools that we have been describing are all concerned with housekeeping activities on the data base. Although these are indispensable, they are not sufficient by themselves. What is missing, so far, is any kind of audit or tracking capability embracing all changes made to the data base from all sources. We have seen how individual programs, in particular the Data-Base Manager, provide their own automatic auditing facilities. These include date stamping of updates and flexible comments tables. However, many other modules update the data base and therefore might be responsible for introducing errors and inconsistencies.

The simplest solution is to store the control record images of every run on a separate file. Updates to this LOG file occur in the sequence of the jobs on the computer, and it is simple to backtrack through this log to the point of interest. In addition to the PLATOFORM control records, we also need to store identification for the run. This might include the analyst's name or the name of the job in addition to date and time of day. Figure 8.13 provides an example of one possible run log.

This simple but effective device is not only a useful memory aid to the recently executed runs, but it can serve in a disaster to rebuild the ACTFILE directly from a historical backup. The analyst needs to identify the log file sections that

```
*
*                *****************************************************
*           **                                                   **
*           **                                                   **
*           **                    PLATOFORM  RUN  LOG            **
*           **                                                   **
*           **        JOBNAME         =   RSST0027               **
*           **        DATE            =   79.270 (27/09/79)      **
*           **        TIME            =    9 HR 24.12 MIN        **
*           **                                                   **
*                *****************************************************
*
+DBM          RSIDE      RBASE
-UPDATE
  U SB.F.U.T   SUMX
     FLM1      2.5
+GENER
-ACTPROB      RSIDE
-FB           UF

     •
     •
     •

+STOP
```

Figure 8.13   Printout of a PLATOFORM run log.

correspond to updates made since the last usable backup data base. By deleting the "STOP" records plus any passive commands (such as reports) that do not WRITE to the data base, this log can be converted to an input control data set and a single run submitted.

The PLATOFORM log can thus provide useful additional security for analysts who neglect to back up their ACTFILE at the end of each run, and who may not keep good account of their data-base activity. It is necessary to remember, however, that this log file is not a TILLER file and that it needs to be edited regularly, since it grows with every run.

For a multifunctional application data base, the need for a structured audit becomes almost mandatory. The log file is a useful tool to the Data Coordinator, since it can track all the changes to the data base. These changes may be made by different analysts working on different ACTPROB "views" of the data base and geographically separated. The Coordinator can unravel the effect of a recently discovered error by one analyst and identify who else may be affected.

8.9 FILE-PROTECTION PROCEDURES

A discussion of the data-coordination role would not be complete without a review of basic security and accessibility requirements. Within the framework of PLATOFORM, we have access to three main categories of data protection:

- Data Center default device protection.
- READ-only protection afforded by TILLER.
- PASSWORD module available within PLATOFORM.

The PLATOFORM analyst can take advantage of the device READ and WRITE protection capabilities that form part of the normal service of the Data Center. By requiring a password with batch jobs, they provide security against unauthorized access or damage to the data base. A related kind of elementary protection is offered by PLATOFORM's underlying file-handling software TIL-LER, which forbids a WRITE to an ACTFILE that has a Job Control Language (JCL) disposition of SHARE. This can occasionally be an advantage (e.g., to provide a simple interactive table-browsing capability for PLATOFORM analysts).

Both these facilities generally protect the file against accidental or malicious damage. However, a more selective password protection may be useful, one that reflects the different responsibilities of the working analysts. This needs to be provided at the ACTPROB rather than the ACTFILE level and may be used in conjunction with a job password. For this reason, the ACTPROB-level password can be restricted to WRITE protection. Perhaps the major advantage to be gained from this facility is in protecting the inexperienced users from themselves.

Consider Figure 8.4. An inexperienced analyst working on SIDE2 needs to

example, the Data Center billing algorithm for job submission can have a major effect. Comparative storage costs for disk versus mass storage versus tape affect the medium of choice and therefore raise different risks. Also, the number of job steps in the run can be influenced by the charge levied for consuming a tape drive, for instance. (This may be significant in a run requiring five or six hours of elapsed time.) Other issues may be highly specific to a particular Data Center's operating environment. The job setup for PLATOFORM evolved in response to all of these forces, and it is subject to periodic modification as the computing environment changes. The important design issue, here, is to adopt a job framework that is reasonably adaptable and universally usable and to be prepared to change it as external factors place an increasing strain on the current layout.

8.11 SUMMARY

Having now reviewed in some detail the major influences in the design and operation of the PLATOFORM ACTFILE, we must turn again to the main task. We have seen how an analyst can plan and organize data using powerful and flexible structures. Having that data now in place, how do we create from it the LP matrix and solve the business problem at hand? That is the task of the next chapter.

A. John Rowland

9. MATRIX GENERATION

9.1 INTRODUCTION

The matrix-generation stage in a mathematical programming system (MPS) is the essential feature of a linear programming (LP) application, linking the fundamental data and the matrix representation of the real-world problem.

The matrix generators previously used in Exxon had been developed or were set up as required by specific projects and typically were tailored to the needs of individual models. Since there was no common data-base approach, data for any one model had its own format and input design. Furthermore, some model-specific data were sometimes included in the programming of the generator. While this was efficient and convenient for the user/analyst, it presented considerable maintenance problems to the support staff. It was common for program updates to be required not only when model-formulation changes were needed, but also when model data changed.

The PLATOFORM data organization provides a common approach to the storage and maintenance of basic planning data, and all generators written within the system can be totally data driven. This type of generator can then be used operationally without constant system-staff support. Maintenance requirements can be further reduced by providing a generator with not only the data for a specific matrix, but also with the logical structure of the model. In this way, a model is built by reference to tables describing the incidence of standard structures. Data for these structures is extracted from data libraries. Using this philosophy, an analyst can expand a model not only to include, for example, an additional crude oil, but also to cover many sections of the company's activities.

The range of application of mathematical programming techniques in Exxon is extensive. It covers both traditional refinery-planning models and large supply-and-distribution models as well as specialized ones, for example, in energy conservation and plant scheduling. Many types of generators have been developed and used within the PLATOFORM system, which is able to support any number of them within its modular framework. The system is convenient for any application, since the Data-Base Manager is general and can be used immediately in setting up a new application. The choice of a matrix generator for a new model depends on the type of planning-data organization that the analyst will accept, as well as on the functional requirements of the model itself.

A large subset of Exxon LP applications is in supply planning, which covers refinery operations, crude and product distribution, and marketing. In this general area, the range of model types includes:

- Single blending plant, single refinery, multiple refineries.
- Single planning time period, multiple periods (not linked).
- Linked time periods with inventory structures.
- Mixed-integer investment models.
- Distribution models.

Many generators could have been produced by individual system groups to model these areas for different planning groups. However, Exxon systems staff has developed a standard supply generator that now supports 90% of the analysts using LP in supply-planning groups. The benefits are:

- A sound basis for data compatibility across applications dealing with different aspects of a company's total supply operation.
- A single style of formulation for matrices.
- Minimal program maintenance.

The development of this program has been touched on (in Chapter 5) and is discussed further in this chapter.

Many applications of the standard generator require no additional programming, and application-development time is a function of only data collection and validation. The model structure is defined in the logic tables, supporting data is input in the data base, and user-specific reports are also defined via tables on the ACTFILE, as described in Chapter 11.

Other applications use some of the standard submodules, but require some specific submodules of their own. These can be written by the applications group and included with standard submodules in a generator. This is made possible by the open-ended approach to module design described in Chapter 5 (Figure 5.11).

Other generators have been developed to meet needs not covered in the standard generator. Most significantly, PLATOFORM has made it possible to greatly reduce the time required to put together applications requiring special solution techniques and formulations. It is possible to develop a specific generator in a matter of days, and with PLATOFORM all other aspects of an LP project are available "off the shelf."

9.2 MATRIX STRUCTURE

Any matrix generator needs not only a consistent structure of matrix row and column names, but also standard formulation techniques. To learn what formulations are possible and what factors influence the choice of appropriate standards, we use an example of product blending.

Figure 9.1 illustrates a planning task for blending two streams to produce a finished grade of product that must meet a single quality restriction on sulfur content. In addition, there is a restriction on the availability of one of the streams and a maximum demand for the finished grade. Finally, the costs of the streams and the realization on the product are given. Notice that we have already assigned a two-character code to each stream. The rationale for such codes was discussed in Chapter 4 with particular reference to table structure. We see shortly how they are used in naming the matrix rows and columns.

One possible formulation of this example is given in Figure 9.2, where three variables are used to represent the three streams RS, CD, and FL. The formulation is an accurate representation of the blending problem, but is it suitable as a standard approach? What is an appropriate organization for the basic data in the real-world problem? Finally, how can the formulation be expanded to include other ways of meeting demand (from imports or inventory)? These are important considerations in developing a generalized LP matrix generator.

One factor that has a large influence on the answers to these questions is the data organization (in PLATOFORM this is equivalent to table structure) suited

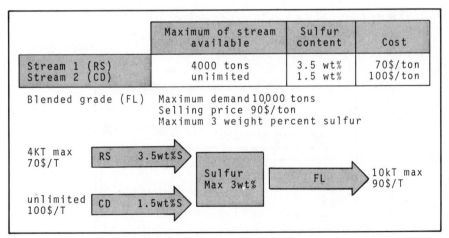

	Maximum of stream available	Sulfur content	Cost
Stream 1 (RS)	4000 tons	3.5 wt%	70$/ton
Stream 2 (CD)	unlimited	1.5 wt%	100$/ton

Blended grade (FL)   Maximum demand 10,000 tons
Selling price 90$/ton
Maximum 3 weight percent sulfur

4KT max
70$/T   RS   3.5wt%S

Sulfur
Max 3wt%   FL   10kT max
90$/T

unlimited
100$/T   CD   1.5wt%S

Figure 9.1   A product blending example.

	RS	CD	FL	Right-Hand Side
COST	70	100	-90	Minimize
AVAIL	1			≤ 4
DEMAND			1	≤ 10
SULFUR	3.5	1.5	-3	≤ 0
BALANCE	1	1	-1	= 0

Figure 9.2   Formulation 1 of the blending example.

to the real-world problem. One view of the blending-problem data is that it consists of three parts: stream availability (quantity and cost), product demand (quantity and realization), and blending data (stream qualities and finished-grade specification). This organization is naturally suited to analysts. Additionally, it has some system advantages in that it organizes the data into logical blocks from which a matrix-structure mapping might be achieved.

To review the interdependence between table structure and matrix structure, we start with some sets of tables that have been introduced in previous chapters. We can then discuss the mapping of this data into an LP matrix.

The structure of the SB table in Figure 9.3 was derived from Chapter 4. We showed that six prime attributes are required to fully qualify a product specification. In our blending problem, the attributes describing the number "3" are:

1. Quality class—SB table set (specifications).
2. Quality—SU (sulfur).
3. Max or min—MX (max).

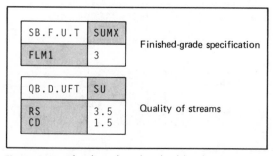

Figure 9.3   Blending data for the blending example.

4. Product—FL (fuel oil).

5. Location—U. (U.K.).

6. Time period—M1 (January).

In the formulation of the blending problem (Figure 9.2), this number becomes the coefficient of variable FL on row SU (SULFUR). In this formulation, the number "3" has only two attributes. Thus we have "lost" four attributes in this mapping of specification data into the matrix. In the same way, we can see that the stream qualities have four attributes in the data table but only two in the matrix. In a generalized matrix generator, we must have a mapping between data and matrix that does not lose information, that is, that includes all attributes of the data in the row and column names. How is such a mnemonic structure for matrix identifiers established?

9.2.1 Matrix Mnemonic Structure

In EMPS, as in many other LP systems, row and column identifiers have a maximum length of eight characters. This is not a serious limitation in working directly with a single matrix, but for the developer of a general matrix generator it limits the design of a mnemonic structure. In this context, a mnemonic structure is a well-defined naming convention by which LP variables (columns) and restrictions (rows) can be recognized. This is not only useful to the LP analyst, but is essential for the generator in defining an organized mapping of data (and logic) tables into the matrix structure.

Returning to the example of product blending, how should we utilize the sixteen characters available (eight in the column name and eight in the row name) to describe any matrix coefficient? We have already noted that the specification on sulfur has six attributes—all of which must be included in the coefficient "description." We can review these attributes to see which belong as part of a column name (describing an activity) and which belong to a row name (describing a restriction on activities). At the same time, recognizing that row and column names are at maximum eight characters, we will plan to use short data-base codes for each attribute within the identifier names.

Consider first the column name. This activity is a function of the location, the product, and the time period. Furthermore, it is a particular type of activity: that of making product to meet a specification (sometimes called a "spec vector"). The row is a restriction on the maximum sulfur permitted in the blend. This restriction is a function of the product, the location, the quality, and the MAX or MIN. The table below summarizes this assignment of attributes to row and column. Note that we will use a specific location code (UF—U.K., Fawley) rather than the general countrywide code (U.—U.K.), as the latter allows for blending at many locations within the country.

Attribute	Code	Column	Row
Quality class	SB	X	
Quality	SU		X
Max or min	MX		X
Product	FL	X	X
Location	UF	X	X
Time period	M1	X	

We now have some flexibility in mapping these attributes into the row and column names. The important consideration is that this mapping be general so that each type of activity (and restriction) has a well-defined mnemonic structure. A generator module can then be written to generate the structure from the logic and data tables. A good mnemonic structure is the key element and has several other advantages, particularly in report writing, which is discussed in Chapter 11.

The mnemonic structure adopted in many PLATOFORM models defines classes of columns (and rows) by their first character. Within each class, the structure of the name is defined by fixed positions for each attribute. Figure 9.4 shows a mnemonic structure of this type for the blending problem. The maximum sulfur specification is the coefficient of the column QFL1WTUF on the row XSU.FLUF. Writing these column and row names generically, we can see how the data attributes are represented.

Qprtwtlo	Contents
Q	Variable class (specification vectors)
pr	Product
t	Time period
wt	Extension of variable class (see below)
lo	Location

Xqu.prlo	Contents
X	Row class (maximum specification on quality)
qu	Quality
.	Invariant
pr	Product
lo	Location

	BRS.FLUF	BCD.FLUF	QFL1WTUF	Right-Hand Side	
FAT1..J.	70	100	-90	Minimize	
ARS...UF	1			≤	4
DFL...UF			1	≤	10
XSU.FLUF	3.5	1.5	-3	≤	0
BFL.WTUF	1	1	-1	=	0

Figure 9.4   Formulation 2 of the blending example.

The mentioned extension of variable class refers to the fact that more than one type of specification vector may be required to allow for the different blending methods of product qualities (blending by volume, weight, etc.). The stream-blending variables also have a well-defined structure:

Bpr.prlo	Contents
B	Variable class (blend vectors)
pr	Product (stream into blend)
.	Invariant
pr	Product (finished grade)
lo	Location

From this example, we can see that many attributes need to be represented in column names. As a result, it is sometimes necessary to map a two-character data-base code into a single-character form for use in the matrix mnemonics. This is done in many PLATOFORM models for time-period codes.

The objective function (FAT1. . J .) is also time period dependent and is formulated for minimization by the optimizer. The other rows in the LP problem have a mnemonic structure as follows:

Apr. . .lo	Availability of product
Dpr. . .lo	Demand for product
B pr.WTlo	Weight balance for product

Some similarity should be noted, at this point, between the mnemonic structure for matrix identifiers and the mnemonic discipline established for data tables in Chapter 4. Just as we were able to catalog table sets (GENTAB.C), it is possible to catalog row and column names in such a way that they can be fully translated in solution prints. We return to this in Chapter 10, Optimization.

9.2.2 Modular Matrix Design

We have seen that a good mnemonic structure is needed to define the mapping of data tables into the matrix structure. The DATAFORM language matrix-generation verbs can then be used to generate the matrix. These verbs are very flexible, allowing the matrix to be built by column, row, or coefficient. Any part of the matrix can also be revised in the same way.

One way, therefore, of developing a generator is to recall all of the data tables associated with an ACTPROB and to deal with each table in turn, placing the data into the appropriate structure. This approach will work for a specific model but will not serve a general system in which many applications will require different parts of the total matrix design. A modular design is required with a mapping between sections of the data and blocks of the matrix. How can these blocks be organized to allow users flexibility in building any type of model within the scope of the generator? A review of the data sections highlighted for the blending problem shows that each section is concerned with specific activities: making components available, meeting product demand, and blending. If matrix blocks with these activities can be isolated, a logical mapping can be set up. One convenient way of doing this is to "open up" the formulation, grouping all data items for a given activity within a well-defined set of columns.

Figure 9.5 illustrates a common design technique in PLATOFORM matrices. A block of the matrix is defined that includes all data for a particular activity/lo-

	Fuel Blending at UF			Demand at UF		Component Availability at UF	
	BRS.FLUF	BCD.FLUF	QFL1WTUF	DFL...UF	ERS...UF	ECD...UF	R.H.S
FAT1..J.				-90	70	100	Minimize
BRS...UF	-1				1		=0
BCD...UF		-1				1	=0
BFL...UF			1	-1			=0
XSU.FLUF	3.5	1.5	-3				≤0
BFL.WTUF	1	1	-1				=0
M1BOUND UP				10	4		
LO				0	0		

Figure 9.5   Formulation 3 of the blending example (PLATOFORM-style formulation).

cation combination. Balance rows (Bpr . . . lo, where "pr" is product, "lo" is location) provide the connecting logic between blocks. The right-hand side is all zeros, and the data representing demands and availabilities is transferred to bounds on the appropriate variables.

The advantage of this formulation is that the generator can now be written in a modular way. Each submodule can deal with a specific set of columns (defined by mask as a result of the established mnemonic structure for variable names). The data tables required for this submodule will be a logical subset and can be recalled as needed by each submodule. A user now has building blocks with which to define the allowed activities in the model. For example, the blending problem can now easily be extended to allow components to be made available from inventory or to include another grade of fuel oil with different specifications.

9.2.3 Model Time Periods

Some other MPS features underlying PLATOFORM have affected the type of mapping between data and matrix in matrix generators. EMPS allows multiple objective functions and bound sets to be embodied in a single matrix, with a particular pair selected at optimization time. Thus the matrix stored on an ACTPROB can represent a real-world problem with several sets of constraints on the variables (bounds) and several sets of costs (objectives). With suitable formulation standards imposed, these different data for the same problem can represent different planning periods. Most data that varies over planning time periods (product demand/availability, plant capacity, etc., and associated costs) can be formulated as bounds and costs on variables. There are some exceptions, wherein a data item that varies over time is part of a model restriction relating to more than one variable. We have already seen the example of product specifications (see Figure 9.5).

Another feature of LP systems (including EMPS) allows time-dependent variables to be included in the matrix on the ACTPROB. This feature is sometimes called "flagging." Using this technique, a group of columns matching a mask can be ignored for a given optimization of the matrix on an ACTPROB. In the example of product specifications given in Figure 9.6, the generator can set up

	QFL1WTUF	QFL2WTUF	QFL3WTUF
BFL...UF	1	1	1
XSU.FLUF	-3	-3	-2.5
BFL.WTUF	-1	-1	-1

Figure 9.6 Time-specific variables for three planning periods.

time-period-specific columns to carry the specifications in the different periods. The matrix on the ACTPROB will have all these columns. The appropriate one is selected when that planning period is optimized (by flagging out the periods not relevant).

Many PLATOFORM generators are designed to produce a matrix containing valid representations of several planning periods. An optimizable model for any one period can thus be extracted from the matrix on the ACTPROB. The user has only to provide the time-period name. The objective function and bound set can then be selected by the system, and the matrix variables for other periods flagged out. Notice that this flagging procedure requires a consistent matrix mnemonic structure. PLATOFORM generators usually reserve position 4 of the column name for the time-period code (with "." meaning that the column is valid for all time periods).

9.3 MATRIX GENERATORS

Some of the challenges in developing a generalized data-driven generator have already been discussed in this chapter. A standard generator is modular, consisting of a manager (GNM, the Generator Manager) and a number of submodules, each of which can generate a block of column activities by location. The essential feature of this approach is that the formulation style gives a one-to-one mapping between sections of data and the matrix block generated by each submodule. The submodules can, therefore, be viewed as independent subroutines that, together with the GNM, form a complete PLATOFORM module. There are some fifteen submodules maintained centrally as part of the standard supply generator.

This particular matrix generator covers all aspects of supply/refinery planning and is used more widely in Exxon than any other. The aim in its development was twofold. First, the desire for a common data base that can be shared across applications implies a standard data-table structure from which to build the models. Of course, the same tables could be used by more than one generator, but the additional advantages of a standard formulation in solution analysis are so large as to justify the commitment to a standard approach. Second, a single generator program offers formulation consistency with a minimum of program maintenance, and it can be supported by a central systems group.

However, it was recognized that some groups may require additional structure (within the general formulation framework) that is not provided in the standard submodules. In this situation, the group can develop a locally maintained submodule and incorporate it into the generator for local use. In this view, a generator is a control program (GNM) and a collection of submodules that can be put together in any number of ways to provide flexibility in extending the range of applicability. It should be noted that the number of locally developed submodules is now decreasing, reflecting the move toward more centralized support. An additional factor contributing to this decrease is that the individual submodules are data driven, and they include many options within the formulation block they serve.

The design of an "open-ended" generator presents unique challenges of coordination among supporting groups. What functions should the GNM perform? How is a generalized submodule written? What can be done to ensure that locally written submodules do not conflict with the central code? The data-base organization and programming features of the DATAFORM language provided the answers to these questions.

9.3.1 Generator Design

Generation of the matrix in large LP problems (several thousand columns) can take significant time. However, with a modular matrix structure it is not necessary to regenerate the entire matrix for each side study. Suppose, in a large problem, that a base-case plan has already been established and that the required side study is to change fuel oil specifications. The only data change is to one table (perhaps SB.F.U.T), and this table is used only in the fuel-blending submodule of the generator. This submodule generates a specific block of the matrix. We must consider what action to take for each block requested by the user.

When a block already exists in the matrix and is to be changed, it is not possible to know beforehand which columns of the block will be modified, which no longer will be required, and what new columns may be inserted. A solution to this is simply to delete the entire block and regenerate from basic data. This is supported by the fact that each block is reasonably small. In addition, the established mnemonic structure for column names gives an efficient way of defining by mask all the columns in any block.

The DATAFORM language has a deletion procedure for columns. This ERASE facility is efficient because of the way in which the matrix is stored on the ACTPROB (hierarchically by column name). Thus all the variables in the block are first deleted and then rebuilt according to the logic and data tables for the submodule. This partial regeneration capability is included in PLATOFORM generators and is extensively used in side-study work.

Each submodule (in a partial regeneration mode) could, therefore, as its first action, delete the columns of the block. This is a method used in some generators, but it repeats similar program statements in each submodule. Furthermore, because of the "open-ended" nature of the generator, it is essential that each block (each submodule), whether centrally maintained or locally written,

```
B***F*        Blend-component variables
QF****        Finished-product variables
K***FB        Control variables
```

Figure 9.7   Definition of a matrix block. Each mask has the location added in positions 7, 8 to form a complete description of all variables in the block.

generate a unique set of columns. A single program location for the entire set of block definitions has the advantage of better control over submodule development. For these two reasons, the definition of submodule blocks and the matrix delete logic are included in GNM. Matrix blocks are simply defined by a set of masks for columns in the block (see Figure 9.7). The definition of the block for a locally written submodule can be included in the GNM logic by a "hook" to an application subroutine of predefined name, as discussed in Chapter 5.

9.3.2   The Generator Manager (GNM)

In the development of a control program for the ordered execution of a set of submodules, there are a number of obvious requirements. These are to read in and check user instructions and to call for the execution of each submodule in turn. However, in an open-ended generator, submodules are not fixed in name or number, and the validity of a submodule request is application-dependent. In addition, some requirements are a result of the formulation style adopted—the definition of the column set for each block and the control of the partial regeneration facility. A less obvious need results because some submodules may produce data required by other submodules (for instance, the quality data of streams from a crude distillation calculation will be input to blending submodules). This implies that a specific order of execution must be established.

The definition of valid submodules and the order of execution could be defined within the system, but this would require a program change in each installation of the generator, and would soon lead to divergent versions of the program. A single GNM meeting all of the needs mentioned above is essential in maintaining control over a generalized system's evolution. This can be achieved by placing such GNM control data on the data base. The list of submodules available in the generator used by the ACTFILE might, for instance, be contained in a dictionary or catalog. In many PLATOFORM generators, this is achieved in the GENRMODD dictionary with stub entries consisting of submodule names in the order dictated by their interdependence. This list is complete for all uses of the generator on the ACTFILE.

GENRMODD	M1	D1-D3
CA	CRAVAIL	CRUDE AVAILABILITIES
BO	EXTERN	EXTERNAL IMPORT/EXPORT
CI	CAPINV	CAPACITY&INVESTMENT
PE	PSTLL	PIPESTILL
PR	PROCESS	SPECIAL PROCESSES
GB	GBLEND	MOGAS BLENDING
DB	DBLEND	DISTILLATE BLENDING
FB	FBLEND	FUEL BLENDING
RE	RECIPE	MISC.BLENDS (RECIPES)
DM	DEMAND	DEMANDS
TR	TRAN	PRODUCT TRANSPORTATION

Individual ACTPROBs have tables describing the scope of the model such as the valid locations, the planning time periods, and the valid submodules for each location.

An example definition of model time periods:

P2.....Z	T1
M1	1
M2	2

The entry under T1 is the one-character equivalent of the two-character time-period code and is used solely in position 4 of time-dependent row and column names.

An example definition of locations and valid matrix blocks (submodule/location combinations):

IG.....Z	UF	UM
CI	X	
PE	X	
PR	X	
GB	X	
FB	X	
RE	X	
DM	X	
BO		X

Together, these control tables define the overall scope of the matrix; the number of blocks is equal to the total of "X"s in the IG table, and the time-period structure is fixed in the P2 table. We can now summarize the functions of GNM in Figure 9.8.

9.3.3 Generator Submodules

The standard generator has a number of submodules that have been developed for general use in Exxon mathematical programming work. The original set was developed for the Esso Europe supply application. However, it was written generally and has found use in several other applications. Since that time, more modules have been added and some have been extended, so that today most LP applications require little if any programming by system staff.

The set of submodules, with a brief description of the activity covered, is shown in Figure 9.9, together with an overview of the mnemonic structure established for each submodule.

Each call to a submodule must generate a specific column block for the location passed as an argument from the GNM. In doing this, a number of tables will be recalled defining the data for the block. The formulation style for the

```
1. RECALL key catalog and dictionary. PROB...C and
   GENRMODD.

2. Read and check user instructions.
   Check that ACTPROB is in PROB...C.
   RECALL matrix (if MODEL entry in PROB...C = YES)
   Set partial mode for existing matrix.

3. RECALL CONT...C for ACTPROB.
   RECALL, via CONT...C, P2.....Z and IG-table.

4. Make list of blocks to be regenerated.
   Check validity of user's request against GENRMODD
   and IG-table.

5. Delete existing structure for these blocks.

6. Regenerate (or generate) each location in turn.
   Call submodules requested (in GENRMODD order) passing
   name of location as subroutine argument.

7. File matrix back on ACTPROB.
   Update and file PROB...C.

8. File CONT...C back on ACTPROB. Print CONT...C.
   (May have been updated by submodules filing data tables.)
```

Figure 9.8 Generator manager system design.

	Purpose	Matrix Block	Delete Masks
PE	Atmospheric and vacuum distillation	Pcr.unlo Where un=D*,V*	P***D*lo P***V*lo
RF	Powerforming	Pfd.unlo Where un=R*,P*,Q*	P***R*lo P***P*lo P***Q*lo
GB	Mogas blending	Bcp.prlo Qprtwtlo Where pr=G*	B***G*lo QG****lo
DB	Distillate blending	Bcp.prlo Qprtwtlo Where pr=D*	B***D*lo QD****lo
FB	Fuel blending	Bcp.prlo Qprtwtlo Where pr=F*	B***F*lo QF****lo
PR	Other processes	Pfd.unlo Where un is not D*,V*,R*,P*,Q*	P*****lo (excluding those in PE, RF)
RE	Miscellaneous blends (recipes)	Mfd.relo	M*****lo
CI	Capacity and investment	Ctf.unlo Itf.unlo	C*****lo I*****lo
TR	Product transportation by importer	Tpr.lolo	T*****lo
CT	Crude transportation	Ocr.lolo	O*****lo
CA	Crude availability at load port	Acr...lo	A*****lo
BO	External product import/export	Epr.lolo	E*****lo
DM	Product demand and purchase/sale	Dpr.dmlo	D*****lo

Key to mnemonics		Example Dictionary for entity
lo:	location	LOCATN.D
cr:	crude	CRUDES.D
pr:	product	
cp:	blend stream	} PRODUCTD
fd:	feed	
un:	unit	UNITS..D
t:	time period	As given in P2.....Z
wt:	weight/volume	MISC...D
re:	recipe	RECIPE.D
tf:	capacity type	CAPTYP.D
dm:	demand type	DEMTYP.D

Figure 9.9 Standard generator submodules.

block is fixed by overall matrix design. We can show the relationship between data and matrix with the simplified example from the FB (fuel blending) submodule used in the discussion on matrix structure.

The two tables in Figure 9.10 are sufficient to define the matrix structure shown on the right. The submodule can process each table in the following way. First, the SB table defines a single sulfur specification (maximum) for the finished grade FL in time period M1. This information is mapped to the single specification vector QFL1WTUF and the single quality row XSU.FLUF. Second, the QB table defines the sulfur quality for the two streams RS and CD. This information is mapped into the two blending vectors BRS.FLUF and BCD.FLUF. Thus the tables define both the data and the required matrix structure. Is this a good approach to the formulation logic for a generalized blend submodule? The answer is no—if we consider the data-sharing goal of PLATOFORM. Since the tables provide the logic for the matrix structure, we cannot add additional data (perhaps required by some other model on the ACTFILE) without affecting the generated matrix on the current ACTPROB. The data tables are bound to the required matrix structure and are thus so specific that data sharing is impossible.

An answer to these problems may be to segregate the data for the model from the logic of its structure. In many PLATOFORM generators, this is done by utilizing control tables (Z-tables) to define the logic for a given model, thus freeing data tables from this role. Data tables are then treated as libraries that can be shared across models and applications. This approach provides much more flexibility to individual users in that data for all uses of a model can be permanently stored in data tables. The data's use in any model is specified in the control tables. This flexibility in turning on (and off) standard structures is the essential feature of the standard generator that has led to its widespread applica-

DATA		MATRIX		
			Fuel Blending at UF	
QB.D.UFT	SU	BRS.FLUF	BCD.FLUF	QFL1WTUF
RS	3.5			
CD	1.5			
		BRS...UF	-1	
SB.F.U.T	SUMX	BCD...UF		-1
FLM1	3	BFL...UF			1
		XSU.FLUF	3.5	1.5	-3
		BFL.WTUF	1	1	-1

Figure 9.10   Data/matrix mapping for fuel blending example.

bility in Exxon. The control tables are the logical link between a shared data base and a generalized matrix structure. This linkage makes it possible to provide a standard generator.

In Figure 9.11, the IQ.F.UFZ table is the primary driving table for fuel blending at UF. The single X requests a single grade FL following one specification SUMX (sulfur maximum). The second driver table (IB.F.UFZ) defines the streams for this pool as RS and CD. Notice that other grades are defined here but are not present in this model. For example, the grade FH is not requested in the IQ table. With a full IB table, grades can be switched in and out of the model by simple updates to the IQ table only. The SB and QB tables are now providing only data and have no model logic associated with them. The full power of PLATOFORM data sharing can now be applied to these sets of data.

In the actual blending submodule of the standard generator, a number of other control tables define, for example, which qualities blend by weight, which blend by volume, and so on. In addition, conversion of units from basic refinery qualities to blending numbers is specified in a completely gen-

DATA				MATRIX		
Logic - Control Tables					Fuel Blending at UF	
IQ.F.UFZ	FL					
SUMX	X			BRS.FLUF	BCD.FLUF	QFL1WTUF
IB.F.UFZ	FL	FH		BRS...UF	-1	
RS	X	X				
CD	X	X		BCD...UF		-1
Data Library				BFL...UF		1
SB.F.U.T.	SUMX	V1MX				
				XSU.FLUF	3.5	1.5 -3
FLM1	3	500				
FHM1	3	600		BFL.WTUF	1	1 -1
FCM1	2.5	500				
FLM2	3.5	500				
QB.D.UFT	SU	V1	W1			
VH	1	2.874				
VD	1.7	7.420				
CD	1.5	5.630				
RS	3.5		44			

Figure 9.11   Data/matrix mapping for the fuel blending example using the PLATOFORM standard generator.

eral way. This submodule is in fact one of the largest (3,000 DATAFORM statements) and is completely general-purpose within the formulation style used. The same control-table approach is used in all standard submodules. As a result, we can summarize the design philosophy of generator submodules in Figure 9.12.

9.3.4 Exception Structure and General Bounds

When formulating an LP model, a specific control may be required that the submodules were not designed to handle. One solution is to provide a direct matrix REVISE. However, this would mean that not all matrix structure is represented in the data base. Another alternative solution is to allow the user to specify such matrix controls in data tables for each submodule.

The exception structure referred to in Figure 9.12 is a routine called by each submodule to map these additional tables to the matrix structure of each block, as illustrated in Figure 9.13.

The set of EB tables covers all generator submodules and locations with the generic structure EB.gmloT, where gm = generator submodule and lo = location. Head names are variables in the matrix block, and stubs are matrix row names. Thus an EB table is an exact picture of the additional structure required in the matrix.

Extra control variables can be used but must conform to a mnemonic structure identifying the block to which they belong. K***gmlo is the convention used, wherein gm = submodule and lo = location. In this way, the delete logic of the GNM can include these variables (see Figure 9.7).

The set of BG tables (BG.gmloT) is available for additional bounds, wherein

1. Definition section: tables required for this submodule. Pick up location argument from GNM.

2. Recall tables for this location.

3. Perform specific calculations (e.g., in Blend – conversion of units, scaling, blending tolerances, etc.).

4. Using control tables as drivers, prepare matrix structure in terms of fixed-formulation style.

5. Place data into structure outline.

6. Place structure on matrix.

7. Call 'exception structure'.

8. Return to GNM.

Figure 9.12   Generator submodule design.

Exception structure table

EB.FBUFT	BCD.FLUF	QFL1WTUF
GCD.FLUF	1	-0.10

A row (GCD.FLUF) is added to the matrix and specifies that 'CD' must be more than 10% of the finished grade (FL).

General bounds table

BG.FBUFT	M1UP	M1LO
BRS.FLUF	2.5	1

The stream RS in grade FL is bounded between 1 and 2.5 in time-period M1.

Figure 9.13   An example of exception structure and additional bounds.

the stub contains variable names and the head is the concatenation of time period and bound type.

In the early days of the PLATOFORM standard generator, large EB tables were commonly used to overcome some shortcoming in the applicability of the module. This, in turn, prompted the system staff to enhance submodules. This was done by incorporating extra structures into the submodules and adding extra table sets to carry the required logic and data. Today, most Exxon LP models can be generated with few EB/BG tables.

9.4 INPUT SYNTAX

The input syntax for PLATOFORM generators is usually simple, involving only the specification of the ACTPROB name and the modules to be executed. Some examples are given in Figure 9.14. By using the IG table (valid location/sub-

```
+GENER
-ACTPROB    RBASE    Partial regeneration
-FB         UF       Fuel Blending at 'UF' only

+GENER
-ACTPROB    RBASE    Partial regeneration
-PE         U*       Pipestill at all UK locations
-PR         UF       Process at 'UF'

+GENER
-ACTPROB    RBASE    Full generation
-ALL        **       All locations
```

Figure 9.14   Generator input syntax. A block of the matrix is regenerated by requesting the relevant submodule/location combination. (ALL = all submodules for a location; the location can be input by mask.)

module combinations) in the GNM, we are able to provide a shorthand input for a request for several submodules.

9.5 DATA ECONOMY

We have seen in earlier chapters how the contents catalog (CONT . . . C) is used to reduce data redundancy. For example, product-quality data can be stored in a QB table on one ACTPROB and referenced by other ACTPROBs. An example QB table for gasoline qualities at UF might have the following dimensions:

QB.G.UFT	RO	R3	R6	GS
C4	94	98	99.5	.
CA	88	94	98	0.625
XH	89	90	91	0.82
P4	88	90.6	92	0.792
P7	97	100.9	103	.

This data can then be accessed by the matrix generator's blending submodule. Since all PLATOFORM programs recall data tables via the CONT . . . C, the same data can be used by head office models of all U.K. refineries or by local models of the individual refinery.

In a model of two refineries (UF, UM) two tables are required—one for UF and the other for UM. The tables QB.G.UFT and QB.G.UMT may contain the same numbers for certain entries (this is certainly the case for pure chemical streams used in blending). It is possible to avoid this duplication by providing a U. table (QB.G.U.T) that contains data common to all locations whose code has U as the first character.

An override table can still be used containing location-specific additions/exceptions to the more general data. The matrix generator must, of course, recognize that the U. table is subject to this data "overlay" in order to take the appropriate value.

Data table overlay is easily written in DATAFORM, and the use of the procedure is widespread in the standard-generator submodules. We can illustrate the full extent of this capability in Figure 9.15. The generator can recognize three levels of data by "location." These levels are default data (location . .), country-specific data, and location-specific data.

The overlaying table may change existing data, add new data, and add stub and head entries. Overlaying data is used only when it is nonzero (T-tables) or nonblank (Z-tables). With this technique, the same data can be shared by many

QB.G...T	R0	R3	R6		Default data (location . .).
C4	94	98	99.5		
CA	88	94	98		
XH	88	90.6	92		

QB.G.U.T	R0	R3	R6	GS	Country data (location U.).
XH	89	90	91	0.82	

QB.G.UFT	R0	R3	R6	GS	Refinery data (location UF).
CA	.	.	.	0.625	
P4	88	90.6	92	0.792	
P7	97	100.9	103	.	

	R0	R3	R6	GS	Data used by submodule for location UF (after overlay).
C4	94	98	99.5	.	
CA	88	94	98	0.625	
XH	89	90	91	0.82	
P4	88	90.6	92	0.792	
P7	97	100.9	103	.	

Figure 9.15   Data overlay example.

locations and exceptions specified with no duplication of common data. The overlay routine available to generator submodules is general and can be applied to any table set as required by the logic of the data. While the table remains the unit of data, this allows individual items of data to be shared without involving the Data-Base Manager.

9.6   AUTOMATIC REGENERATION

We have seen that PLATOFORM generators allow the user to define LP problems at the data level by modifying data and logic tables. A typical side-study run was shown in Chapter 6 (Section 6.2) with the following DBM and GENER request:

```
+DBM                    RSIDE              RBASE
 −UPDATE
     U  SB . F.U.T      SUMX
         FLM1           2.5
 +GENER
 −ACTPROB  .            RSIDE
 −FB                    UF
```

The user has made the data changes and then called the correct standard-generator submodule to ensure that the new data is placed in the fuel-blending block of the matrix. Historically, this has been one of the most error-prone areas of the system. PLATOFORM has made it easy for analysts to define LP problems at a distance by dealing with planning data in tables, but the user is required to know the relationship between data and matrix in order to make the correct GENER call. The user must also consider other factors in matrix regeneration:

- Has data been changed in previous runs?
- Has data on an ACTPROB referenced by the current model changed?
- Do generator submodules create data used by other submodules?

A simple way around all of these problems is to regenerate the entire matrix (−ALL   **) every time, but this becomes very expensive in large models. Can anything be done to assist the user?

It would appear, at first glance, that all the necessary system components are available in PLATOFORM to determine which tables have been modified since the last matrix generation. If this is so, then since in any generator the mapping of tables to matrix structure is well defined, we have the potential to use the generator to run the correct submodules automatically. What, then, is involved in tracking all data modified since the last +GENER?

A PROB . . . C entry can contain the date/run number of the last generation of the model on an ACTPROB, and a similar date stamp is maintained by the DBM in CONT . . . C for each table associated with the ACTPROB. Simply by comparing dates, every new and modified table can be found. But there is no reference to tables deleted since the last generation. This information is required by an automatic generator, as removing structure is also a valid reason for submodule regeneration.

An alternative way of capturing all table changes can be established. The technique adopted is to have the generator file out the CONT . . . C status at the end of generation as a logic table OLDCONTZ. The next time the generator is called, it will compare the current CONT . . . C with OLDCONTZ and thus trap all table changes—new, modified, and deleted. It is necessary to use a Z-table rather than an additional system catalog to ensure that a new side-study ACTPROB can find OLDCONTZ via the CONT . . . C.

The first automatic generator of this kind has been working successfully for a number of years. With the future inclusion of similar logic in the standard generator, it is conceivable that users will be able to create new models using only the DBM.

9.7 MATRIX DISPLAY

Having generated an LP model, the user needs to see the resulting matrix. We must, therefore, provide an option for matrix display via the system syntax. The modular design of PLATOFORM allows a link between user instructions and an LP procedure with a simple DATAFORM program acting as the interface. For example, a display of a generated matrix can be requested, as in Figure 9.16.

The usual masking features of most LP procedures can delimit the amount of the matrix displayed. In PLATOFORM, this is particularly useful in displaying columns by activity block as a direct result of the mnemonic conventions that are imposed.

The output resulting from a matrix display can be very large, as can the solution print from an optimization. It is essential to consider the order of rows and columns within these basic LP procedure printouts. The order within the matrix on the ACTPROB will be used by all LP procedures within PLATOFORM.

The matrix generator operates by deleting column blocks and replacing them at the request of the user. DATAFORM causes new or replaced blocks to be positioned at the end of the matrix. Furthermore, this is an unpredictable order that depends on which submodules are executed. Within matrix blocks, it is possible to generate columns in a specific order. However, this involves serious loss of flexibility, especially in extending or modifying the programs. The result of these considerations was the invention of a procedure that will allow a standard order of row and column names to be established after generation of the matrix.

This Exxon-developed LP procedure (ORDER) will reorder the internal index of row and column names without moving the physical structure of the matrix on the ACTPROB. Subsequent LP procedures use this index to recall the matrix and thus will use the desired order. ORDER is automatically called as part of +MATRIX and within the +OPT module, providing a standard order for the basic LP printouts. The sort key can be given by the user, but a default is also

```
+MATRIX                      Request for display of
-ACTPROB    RBASE            entire matrix on
-TRANCOL                     ACTPROB RBASE


+MATRIX                      Request for partial display
-ACTPROB    RSIDE            of matrix on ACTPROB RSIDE
-TRANCOL    B***F*UF         (only fuel blending variables
                             at location 'UF')
```

Figure 9.16   A matrix display. The term TRANCOL is usually associated with the updated matrix following optimization. In +MATRIX (before optimization), a TRANCOL will produce a display of the original matrix.

provided by the system. One result of this order is that all variables for a location are together, which reflects the way in which users like to see them, but not the way they are generated.

9.8 SUMMARY

In this chapter, we have discussed the way in which PLATOFORM generators are designed. We noted that data table structure for a particular model is determined by the generator that will use this structure as input. By organizing this data into functional groups and as a result of some of the features of EMPS, we can derive a mapping of data tables to model by matrix "block."

These blocks can be defined in terms of matrix column mnemonics (built from codes used in the data tables) in such a way that partial regeneration of matrices is possible. Each submodule of a generator can then work with the subset of data mapped to a single block.

In the Standard Supply Generator, we have been able to provide many general-purpose submodules (under the control of a single GNM) that are used in most Exxon LP models dealing with refining, supply, and transportation. This has been made possible by the data-sharing design of the data base (extended by data overlay in the generator itself) and by the separation of data from model logic. This generator has effectively bridged the software gap between our ability to handle LP data and solve LP problems.

Finally, we looked at the matrix-display features that must be included in any LP system. Now, Chapter 10 turns our attention to the way in which the user can access the optimization algorithms within EMPS to convert a representation of a real-world problem into a solution.

Jeremy D. Sammes

10. OPTIMIZATION

10.1 INTRODUCTION

Traditional mathematical programming system (MPS) packages used within Exxon viewed the optimization phase as the hub of the system. Indeed, in the earliest programs, the optimizer *was* the system. The problems associated with persuading the machine to "massage" the linear program (LP) and produce numbers far outweighed the time spent preparing the data and building the LP matrix. In the early 1960s, computing power was the major limiting factor. Even with adequate computing power, highly skilled LP and systems analysts were required to deliver usable results within a reasonable timeframe.

Since that time, the perception and role of optimization have changed considerably. It has become part of a broader framework that includes data management and the presentation of easily assimilated reports. Today, we wish to process a wide range of different kinds of mathematical programming models. This variety encompasses not only differences in the types of business problems and hence matrix formulation, but also differences in the mathematical strategies needed to solve them.

In this chapter, we review the major factors that directed the evolution of the optimization phase within PLATOFORM to its current form. We begin with a discussion of the place of optimization in the overall problem framework and how this influenced the expected functions. The nature of the MPS imposes some design constraints on the provision of these functions. The Program Control Language (PCL) and DATAFORM components that together make up optimization have a relationship with each other that is quite complicated and that is very different from the relationship we have seen in other PLATOFORM modules. In fact, it becomes clear during the ensuing discussion that more than one DATAFORM program will be needed in this situation.

Subsequently, we describe how to build a framework that allows algorithmic extension in response to changing business opportunities. This framework needs to encompass a variety of techniques including:

- Group upper-bounded (GUB) formulations.
- Mixed-integer optimizations.

- Nonlinear (recursive) methods.
- Iterative sequences of linear optimizations.
- Parametric algorithms.
- Networks.

The ability to add new techniques or to modify existing ones without major changes is a challenge for any system. The secret, of course, is to modularize or decouple these methods as much as possible. We cannot, however, simply build a set of DATAFORM programs or subroutines to achieve this modularity. The complicating factor, here, is that the PCL contains some basic capabilities that we need, not the least of which is the optimizing algorithm itself. How can we develop a modular framework when DATAFORM must return to the PCL for crucial parts of the optimization phase? As we shall see, full modularity is not possible without enhancing the underlying MPS, and, in consequence, there are strong incentives for making some basic system changes.

Finally, we delve a little deeper into the internal workings of the software to see how a number of key issues can be addressed. One issue concerns the management of solution cases that are the major products of optimization. A catalog of solutions is introduced that can support this function and that provides the link between optimization and reports. This catalog can also fulfill other requirements, including the management of advanced starting bases for the optimizer. The control of data elements within a PLATOFORM catalog can impose some mnemonic constraints on the names of the elements. Is there a need for a solution case-naming structure, then; if so, should it be mandatory?

In Chapter 9, we described how the matrix row and column mnemonic structure can be developed from the data-table codes. Is it possible to provide a dictionary translation of these row and column names that is similar to the TPRINT table printing we saw in Chapter 7? Once again, a little reflection indicates the need for a way of generically cataloging the matrix identifiers.

10.2 OPTIMIZATION ENVIRONMENT

Previous chapters have described the precursor phases of data management and matrix generation. Figure 10.1 demonstrates the position of optimization relative to other stages of model development. Output from the GENER module is a mathematical programming model stored on the analyst's data base. The primary task of the optimization phase is to find an "optimum" solution to that model by minimizing or maximizing the value of a selected objective function row. This process may involve a single pass by a suitable optimizing algorithm for a simple LP. A mixed integer or nonlinear formulation, however, may require several passes with intermediate modifications to the matrix (bound set or intersection changes). Some sort of convergence/termination test is also necessary

in this more complex situation. Once the optimization criteria have been satisfied, we need to decide how best to present the basic results. The report
programs will provide comprehensive reporting capabilities, but we may expect
the optimizer to tabulate the raw values and also to provide basic printout for
analysis of matrix formulation and data errors.

So, optimization occupies an intermediate ground between matrix generation
and report writing. Input to this phase is a mathematical model and a set of
instructions that will direct the course of the optimization. Output consists of one
or several solution cases representing passes through the optimizer or parametric
algorithm. We may wish to print these cases immediately in raw form, in
addition to storing them for access by any subsequent reports (Figure 10.2).

Let us look at the input side of the process first. What additional information
do we need to provide that is not contained within the matrix itself? This data
falls into two broad categories:

- Selective "tailoring" of the matrix.
- Optimization strategy data.

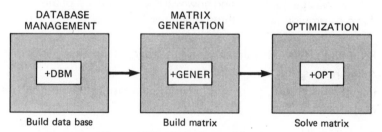

Figure 10.1   Steps in model usage.

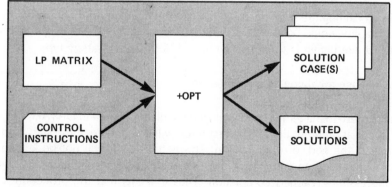

Figure 10.2   Summary of information flow through the +OPT module.

A simple example best illustrates what we mean by "matrix tailoring." In a majority of PLATOFORM models, there is a need to distinguish a single-time-period study from one that crosses several time periods (with inventory structure providing the links). This is normally done by encoding the time period into the matrix identifiers. In a single-time-period model, we are interested only in rows and columns for this period. Given the time-period code, it is possible to eliminate any matrix identifiers that are specifically oriented to other time periods. This "tailoring" of the matrix can also be extended further to eliminate selected portions of the model that are not to be included in the current optimization.

An example of strategy information is an instruction to change a matrix coefficient parametrically. Both kinds of data can be expected to change with each optimization, so the input-control stream seems the obvious vehicle for defining this information. In practice, stored tabular data may also be needed (i.e., to direct a mixed-integer search) that can be triggered by the user's control stream.

We can see intuitively that this control information will be needed at the very start of the optimization process, to prepare the matrix and to select the appropriate algorithm. This implies, in turn, a manager segment to read and process all the control information, a theme that is now a familiar one to the reader. But what will this manager look like, and how will it fit into the rest of the framework? It will have to provide at least three functions:

1. Read and check the validity of the user's input stream.
2. Apply necessary matrix tailoring.
3. Identify the strategy to be used and pass control to the appropriate program segment.

We might suspect that there will be further tasks required of this segment, and indeed there are. Let us turn briefly to the right-hand part of Figure 10.2—the output side of the process. We have hinted at the need for a case-management facility to catalog the solution cases that are the major product of optimization. Such a facility would be expected to form a logical part of the manager segment, since the alternative is to repeat it for each individual strategy type:

4. Manage a catalog of solution-case names.

All of these manager functions will necessarily be written in DATAFORM, but they occur at different times. The first three will be handled at the start, and the last one at the end, of optimization. This is because it will not be known for certain what solution cases have been produced until the optimizer algorithm has successfully reached completion. Thus it seems that the procedure must enter DATAFORM at least twice to solve a given model.

What we have achieved, so far, is a broad understanding of the input needs and output expectations of optimization. Already, we can see that a single DATAFORM program embedded in PCL will not be sufficient to address these requirements. Let us try to build a practical design that will work and that can also evolve as the environment changes.

10.3 BUILDING A FLEXIBLE FRAMEWORK

Chapter 5 introduced the PCL component, and we saw how it provides the links between the PLATOFORM control program (PLEX) and the individual modules that the user wishes to run. These modules may simply involve DATAFORM programs (e.g., +DBM). Others require a direct call to an Enhanced Mathematical Programming System (EMPS) procedure, with no DATAFORM element. A number of modules require calls to one or more EMPS procedures as well as to a DATAFORM program, and for these the structure of the resulting PCL is very much a function of what facilities are provided by the high-level EMPS procedures used. They are linked together in the necessary sequence, and the task of the DATAFORM part becomes heavily oriented toward providing the necessary PCL switches and procedure arguments.

10.3.1 Basic Requirements

Let us begin with a fragment of PCL extracted from the embryo program that was illustrated in Figure 5.12. The optimization segment of this (Figure 10.3), following the call to DATAFORM (EXDF), calls upon three procedures that, respectively, copy the matrix onto a work file (READY), solve it (WHIZARD), and tabulate the "optimal" values (SOLUTION).

This is a start, but it is clearly a long way from a generalized matrix-optimization capability. If we begin by understanding what information is needed by the

```
ZPLEX       EXDF        ('NAME','PLEX')
             •
             •
             •
ZOPT        EXDF        ('NAME',XCHAR10)
            READY       ( - - - - )
            WHIZARD
            SOLUTION
            GOTO        (ZPLEX)
             •
             •
             •
```

Figure 10.3 Optimization segment of a PCL program.

PCL procedures, we can clarify the task of the DATAFORM front end and the input control records needed to drive it. READY needs to know the name of the ACTPROB that contains the model to be optimized. WHIZARD needs the name of the objective function row (there may be several candidates), the column bound-set name, and the right-hand-side vector name. We could, of course, obtain all this information directly from the user.

The example in Figure 10.4 illustrates an input stream that requires the user to explicitly identify all these variables. Can we do better, since this already appears burdensome? In Chapter 9 (Section 9.2.3), the encoding of the time period into the objective function, bound set, and certain matrix column names was introduced. This permits some simplification. The only variable part of the objective function name "FAT1. . .J." is the fourth character code for time-period 1. This corresponds to the first two characters M1 in the bound-set name. Hence, knowing the time period, we can deduce these two names within DATAFORM. PLATOFORM also operates with an all-zero right-hand-side vector (Section 9.2.2) that can therefore also be given a unique default name.

To complete the mapping of time-period code to matrix identifiers, a data table (P2. . . . .Z) is needed to provide the link between all one- and two-character codes. This table (see Figure 10.5) can also be used to drive the "tailoring" of the matrix for this time period. Chapter 9 established the fourth-character position in the matrix row/column name as the convention for storing a time-period code. A period (".") in this position indicates a matrix identifier that is valid over all planning periods. The task of the program in the example, therefore, is to retain only those identifiers with a "." or "1" in this position. In practice, this is best achieved by eliminating the remainder (i.e., those containing a "2" or "3" in our example).

How is the tailoring to be accomplished? It is pointless to generate a matrix with several objective functions merely to delete all but one for the first optimization. Rather, we wish to flag (temporarily ignore) these parts of the matrix without modifying the version on the ACTFILE. EMPS achieves this with READY in conjunction with another procedure, EDITW, that applies the flags passed as arguments. The working copy of the matrix (to be used by the optimizer) is constructed by these procedures, omitting unneeded portions. In practice, DATAFORM has to fill a set of PCL cells with masks corresponding to the matrix

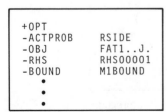

```
+OPT
-ACTPROB    RSIDE
-OBJ        FAT1..J.
-RHS        RHS00001
-BOUND      M1BOUND
   •
   •
   •
```

Figure 10.4   Preliminary syntax for +OPT.

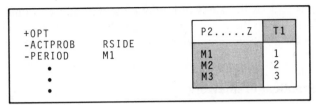

Figure 10.5   Modified +OPT syntax, using the time period table.

identifiers to be flagged, and these are then passed to EDITW. This gives us the first illustration of a recurring problem: the interdependence of the PCL and DATAFORM that results from the need to control an EMPS procedure from a DATAFORM program. This mutual dependence (or coupling) degrades the flexibility of both components and complicates their support.

READY alone is insufficient for other reasons. One of the simplifying design criteria for the generator program was to ignore the *sequence* in which identifiers are constructed. (They are stored—and printed—in this sequence.) A fast resequencing procedure is necessary, here, to sort the column and row names for readability. This ORDER task naturally follows after +GENER, and before READY copies the ACTFILE version onto the work file, and it can be accomplished via a procedure called from the PCL. Finally, in most situations an advanced starting basis will be needed to reduce cost and turnaround time. How this basis is generated and stored is considered later, but for now a method for inserting it is required. Several methods may be available, depending on the basis, but all will necessarily be PCL procedures, since this activity must fall after READY and before WHIZARD. This, therefore, further complicates the PCL.

We can thus see an evolution from a simple READY statement toward a more complex PCL segment (or macro) whose task it is to PREPARE (i.e., set up the matrix in preparation) for the optimizer. We do not explore this macro in depth, because the details are highly dependent on the specific nature of EMPS. The major issue here is that a practical matrix-preparation phase for PLATOFORM depends on a set of PCL procedures that cannot be directly accessed from DATAFORM. Rather, they are driven by a pool of common variable cells set up by a DATAFORM "front-end." (These communication cells were introduced in Chapter 3 and are available to all programs within the EMPS environment.)

What about the optimizer portion? Surprisingly, this phase is responsible for the most involved sections of the PCL. The complicating factor, here, is the handling of off-normal demands (points during optimization when the algorithm is triggered to pass control back to PCL). Reasons for doing this include the handling of infeasible or unbounded terminations (specialized printouts), or of digitally unstable matrices that may require special treatment. Once again, the details are very much dependent on the algorithmic and language capabilities of the MPS, as well as the kinds of model most typically encountered. In any case, we can envisage the development of an OPTIMIZE macro to serve these needs, as the PREPARE macro served the model-setup requirements.

10.3.2   A Practical Framework

Referring to Figure 10.3, there are still some unresolved issues to pursue for even our simple example. It is unlikely that a printed SOLUTION will be required for every run. In some cases summary reports may be sufficient, or the optimization may merely be the starting point for a parametric study. Besides SOLUTION, what other output is possible? The user may prefer to see other kinds of postoptimal reports (e.g., the cost of changing a row or column activity and the range for which this cost is valid). Such reports, available at the PCL level, may be selected via the setting of yet more PCL cells and lead ultimately to a DISPLAY segment or macro.

Another problem with this example is that no provision has been made to store the results of WHIZARD computations. We saw that this was one of the basic functions of +OPT (Figure 10.2) and that each stored solution case requires a name. A PCL statement is needed to accomplish this RECORD of the case, and a case name must be supplied by the user. This yields a control syntax as displayed in Figure 10.6.

This syntax seems to provide an acceptable level of control, without overly burdening the user with unnecessary detail. We see, shortly, that we can simplify Figure 10.6 further by adopting a case-naming convention. There may, of course, be a considerable number of additional user-control parameters that are optional or model-specific. For instance, it may be necessary to optimize a model that does not fully conform to our naming conventions. A −RHS record will allow the user to define a right-hand side other than the default (RHS00001), for example. We can choose which postoptimal reports, if any (−SOLUTION in the example), we may require. Many optional variations are possible.

Before we summarize the results of this discussion in a diagram (see Figure 10.7), we must consider one last function of the PCL: the catalog management of the stored solution. Since this can be achieved only in DATAFORM, an additional EXDF is necessary before control is passed back to PLEX.

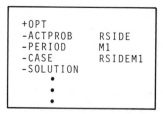

Figure 10.6   Practical syntax for +OPT.

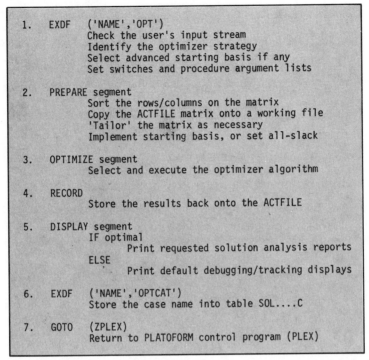

```
1.   EXDF   ('NAME','OPT')
            Check the user's input stream
            Identify the optimizer strategy
            Select advanced starting basis if any
            Set switches and procedure argument lists

2.   PREPARE segment
            Sort the rows/columns on the matrix
            Copy the ACTFILE matrix onto a working file
            'Tailor' the matrix as necessary
            Implement starting basis, or set all-slack

3.   OPTIMIZE segment
            Select and execute the optimizer algorithm

4.   RECORD
            Store the results back onto the ACTFILE

5.   DISPLAY segment
            IF optimal
                    Print requested solution analysis reports
            ELSE
                    Print default debugging/tracking displays

6.   EXDF   ('NAME','OPTCAT')
            Store the case name into table SOL....C

7.   GOTO   (ZPLEX)
            Return to PLATOFORM control program (PLEX)
```

Figure 10.7 Summary of a practical PCL framework linking DATAFORM with the necessary optimization procedures.

It is apparent, already, that even with the relatively simple outline presented by Figure 10.7, fairly intensive communication is needed between PCL and DATAFORM. As more procedures are introduced, this coupling increases substantially. PLATOFORM utilizes over 200 cells to provide the necessary interface. The Communication Region (CR) supplies approximately forty of these, the remainder being defined as local PCL variables. This intensive communication causes considerable systems problems. The software becomes obscure and thus less easy to support. The PCL and DATAFORM components are more dependent on each other, and source program modifications require a coordinated approach. Used in this context, PCL is a rather low-level programming language that is not easily adapted to provide the range of flexibility needed. As a consequence, almost 700 statements (out of a total of 1,500) are consumed within the OPT phase of PLATOFORM.

This brings us to a turning point in the discussion. So far, we have looked at only a relatively simple OPT framework, one that can support a limited set of standard LP models. Already, the PCL is large and closely coupled to the (two)

DATAFORM components. As the role of OPT is extended, this situation can only worsen. The structure we have adopted was largely dictated by the needs of the several EMPS procedures we are obliged to use, which can only be called from PCL. The next step, then, is to determine if these procedures might be called from DATAFORM directly. As we shall see, this objective permits considerable simplification, but it still does not allow us to work with only a single DATA-FORM module.

The sequence of operations required to support a simple LP application is summarized in Figure 10.8.

10.3.3 Other Mathematical Programming Strategies

How can we build upon this framework to encompass other kinds of models? As we have observed earlier, a variety of more complex mathematical programming problems can be processed by a fairly straightforward iterative approach:

1. Check for convergence/termination criteria.
2. Make necessary matrix adjustments.
3. Optimize.
4. Save key results.
5. Return to step 1.

Mathematical programming strategies that can be applied within this general framework include:

- Mixed-integer (branch-and-bound) techniques (MIP).
- Nonlinear (recursive) optimizations (NLP).
- Iterative sequences of linear optimizations (ILP).

```
ZPLEX       EXDF          ('NAME','PLEX')
              •
              •
              •
ZOPT        EXDF          ('NAME',XCHAR10)
            PREPARE
            OPTIMIZE
            RECORD
            DISPLAY
ZEND        EXDF          ('NAME','OPTCAT')
            GOTO          (ZPLEX)
              •
              •
              •
```

Figure 10.8 A PCL sequence that supports a simple LP application.

Depending on the strategy, of course, there will be significant differences in the way the steps are implemented. In the case of a mixed-integer model, we may expect the first iteration in the loop to be a continuous optimization (all integer variable bounds relaxed). However, initial updates to the matrix may be necessary for the other categories (representing a first approximation or guess). The criteria for terminating the loop and the information saved to set these criteria will also be highly dependent on the specific technique.

Looking at the steps outlined above, we can see that these strategy differences occur in steps 1, 2, and 4. If we can accommodate these within DATAFORM, we shall be able to maintain the PCL transparent to the particular technique we are using. The test for termination and the matrix update steps can only be written in DATAFORM. How should we do this?

For a start, we might expect a logically separate segment for each gross strategy. Should these be DATAFORM subroutines or independent subprograms? For practical reasons, individual subprograms are easier to support, since they tend to evolve independently. Most of the rest of the DATAFORM may be relatively stable, and a large, single DATAFORM program is not the best solution here. Is it still possible to maintain the PCL transparent to the technique, given separate DATAFORM subprograms? We can see that by extending the PLEX control philosophy (Chapter 5) to the OPT DATAFORM manager program, we can avoid explicit references within the PCL (see Figure 10.9).

In Figure 10.9, we have included the strategy definition in the +OPT control stream. The =RECURSIV record causes the first DATAFORM program OPT to set the CR cell XCHAR10 to RECURSIV. The PCL then passes control to the RECURSIV DATAFORM program via the second EXDF call. RECURSIV applies the necessary convergence test and, if satisfied, sets XCHAR10 to RETURN to terminate the iterative loop. Otherwise, this program updates the LP matrix, as necessary, and the next iteration is entered (label ZOPT10). With this design, we can easily add new subprograms without modifying the PCL or recompiling the OPT manager program, since all DATAFORM subprograms are defined via XCHAR10.

The remaining problem is where to close the iterative loop beginning at ZOPT10. We can obviously update the Solution Catalog once, right at the very end, for any or all solution cases we have stored on the ACTFILE. For simplicity, we can also print the solution displays only for the final pass through the loop. This assumes that the user will not routinely want to see a printout of the intermediate results. If these are needed later, they can be obtained from the stored cases. So the loop can be closed following the RECORD operation that saves the results of the current suboptimization (see Figure 10.10).

Once the main loop is entered, the second DATAFORM program need not modify XCHAR10 until the final pass is completed. However, there is a powerful capability hidden here. By careful manipulation of this CR cell, we can simulate nested levels of the main loop without adding to the PCL.

Consider a model that contains integer variables and also some nonlinear

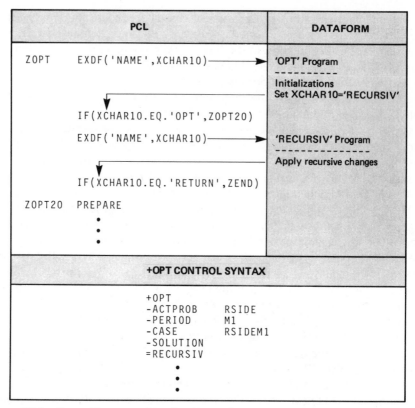

Figure 10.9   Controlling execution of an OPT subprogram (RECURSIV) without modifying the PCL or OPT itself.

characteristics. Let us assume that this is to be solved via two subprograms that are geared, respectively, to the mixed-integer aspects and the linear-approximation aspects of the matrix. Within each node in the branch-and-bound search, we can enter a pseudo-loop to complete the recursion to convergence for this node. This can be managed by the two subprograms (for mixed integer and recursion), resetting the contents of XCHAR10 to "point" to one another as appropriate. The current status of the search can be stored in reference data tables available to both subprograms.

10.4   EXTENDING THE FRAMEWORK

We now have a rather flexible design framework. However, a number of significant problems remain. One is the close coupling of the DATAFORM and PCL

```
ZPLEX       EXDF        ('NAME','PLEX')
              •
              •
              •
ZOPT        EXDF        ('NAME',XCHAR10)
            IF          (XCHAR10.EQ.'OPT',ZOPT20)
ZOPT10      EXDF        ('NAME',XCHAR10)
            IF          (XCHAR10.EQ.'RETURN',ZEND)
ZOPT20      PREPARE                                         MAIN
            OPTIMIZE                                        LOOP
            RECORD      ( - - - - )
            IF          (XCHAR10.NE.'OPT',ZOPT10)
ZEND        DISPLAY
            EXDF        ('NAME','OPTCAT')
            GOTO        (ZPLEX)
              •
              •
              •
```

Figure 10.10   A portion of PCL showing how to close the iterative loop that begins at label ZOPT10.

that we discussed earlier, which in practice greatly obscures the PCL flow. Another arises within the RECORD segment that stores the solution values back onto the ACTFILE. We might expect to store *all* the values at the completion of the whole process, but this turns out to be unnecessary and inefficient for the intermediate steps. Here, it is better to store only the data necessary to check for termination and to update the matrix for the next iteration. However, the nature of this data is highly dependent on both the technique we use and the business problem itself.

A menu of RECORD statements within the PCL offers a compromise, but it adds to the complexity and binding of the PCL. A third drawback to this design is due to the need, following DATAFORM, to PREPARE the matrix from scratch again. This is because DATAFORM frees available core for its internal buffer needs, so that the old matrix (with its associated basis) is lost.

If we pause for a reflection here, it becomes apparent that we have become constrained by certain key physical limitations in the EMPS software. The switching from PCL to DATAFORM is rather artificial. It is necessary only because the large algorithms cannot be driven directly by DATAFORM. The PCL is a rather inflexible control language, and it would be convenient to be able to replace it with DATAFORM itself. How should we explore this possibility?

10.4.1   The Need for SLEUTH

The central issue is whether DATAFORM can be persuaded to communicate directly with the working copy of the LP matrix (as well as the ACTFILE version).

If this issue can be solved, PREPARE can be moved out of the PCL loop and be executed only once at the start of the process. If both READ and WRITE communication with the working matrix can be provided, the RECORD solution-storage piece of the PCL can also be transferred into DATAFORM.

To achieve this in practice, it is necessary to configure the core region such that DATAFORM, the optimizer (WHIZARD), and the working matrix all coexist. Another way of looking at this from a systems-design viewpoint is to imagine that WHIZARD and RECORD are now callable from DATAFORM. In EMPS, the procedure SLEUTH provides this confluence of procedures. SLEUTH is essentially DATAFORM with some additional verbs that allow access to and from the working matrix. SLEUTH can also call WHIZARD directly via an OPTIMIZE verb. These basic capabilities provide us with the tools to construct a greatly simplified optimizing framework that now focuses on a high-level DATAFORM environment rather than on a low-level PCL macro.

Figure 10.11 represents a simplified plan that is made possible by SLEUTH. No main loop is now visible; it is contained within the SLEUTH program itself. A linear optimization is executed in the same manner as before. For more complex models, however, the DATAFORM subprogram (e.g., RECURSIV) is executed via the second EXDF to set the scene for the PREPARE. RECURSIV is then reentered under SLEUTH control and is now responsible for all subsequent matrix updates, optimization, retrieval of results, and termination tests. SLEUTH will thus remain in control until the outer loop is satisfied. Before completing, this program must also ensure that the matrix on the ACTFILE is updated to reflect the working copy that will be lost. This matrix is then re-solved from a basis passed by SLEUTH, and the final DISPLAY segment is executed.

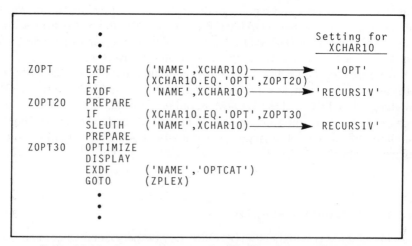

Figure 10.11   Program segment simplified by the use of SLEUTH.

Besides the architectural simplifications it achieves within the existing PLATOFORM system framework, SLEUTH offers a powerful, general mathematical programming tool. Its language elements provide access to the model at optimization as well as during all other phases from data input through reporting. The applications opportunities are limited more by ingenuity than by the lack of system capability. It is possible to tackle business problems that would have been impossible within the more traditional PCL environment, either because of unusual model characteristics or the number of suboptimizations involved. The DATAFORM operations have been resolved into four logical steps:

1. Managing the overall optimization.
2. Preparing the optimization strategy.
3. Executing the strategy.
4. Cataloging the solutions.

The binding among these programs is much less severe than before for two reasons. First, we have moved several key procedures into DATAFORM. Second, we can use stored DATAFORM tables to provide the links that remain. The PCL is almost independent of the particular optimization strategy needed for an application. We need only to include the necessary DATAFORM subprogram on the program library without disturbing the existing program.

We have limited this discussion to the provision of a subset of optimization strategies. This was intentional and was done partly because clarity was needed and partly because these were the major influences on the evolution of OPT. However, with the powerful, high-level DATAFORM language at our disposal, it becomes possible to explore a number of other interesting areas. For example, a networking capability can be incorporated into a DATAFORM subprogram by exploiting DATAFORM's ability to link directly to FORTRAN and PL/I. Several powerful, commercial network software packages are available that lend themselves to this approach, and a DATAFORM program can be written that employs such a system, with data access via PLATOFORM tables.

10.4.2 Parametric Extension

Before leaving the subject of the algorithmic extension of OPT, we briefly review the parametric algorithms. EMPS has access to separate PCL-callable procedures for processing parametric problems involving any row or column (including the objective row and the right-hand side). A "rim" parametric is also available for simultaneous changes to both objective row and right-hand side.

Two issues need to be resolved before these facilities can be incorporated into PLATOFORM. First, we have to decide where in the PCL framework to include the procedures. A natural place is immediately after the optimization phase, which will have created an optimal base point from which to begin the parametric study. The base matrix must be solved first; in the case of a previously optimized model, the overhead involved in reoptimizing from the final basis is small. Some peripheral PCL is also needed to support the procedures, both to select the desired algorithm and to support any off-normal parametric demands.

The second issue concerns the user's control of the parametric study. Relatively little data needs to be provided, but how should it be structured? Suppose we wish to investigate the effect of incremental cost changes to a given column.

Figure 10.12 illustrates a simple example drawn from the blending problem introduced in Chapter 9 (Figure 9.5). We assume that in the base-case solution the level of product CD at location UF is 1.333, (i.e., the column is in the basis). What is the effect of reducing the input cost on this column from 100 to 0? The PARAOBJ algorithm handles this request via a change objective row (CHOBJ), whose contribution to the objective increases as the parametric proceeds. At selected parametric break points, an intermediate solution may be printed, and the delta case may be permanently stored for later reports analysis. In summary, then, we need the names of the parametric algorithm and change row and also the unique names for the intermediate-solution cases.

The control stream illustrated in Figure 10.13 summarizes this information. The =PARAOBJ record defines the procedure name, solution increment, and parametric maximum values. Subsequent data records supply the name of the change row and the case names. Defaults may be assigned to essentially all these variables. (Case names may be built by the program by incremental operations on the base-case name.) —PSOLUTN requests that intermediate solutions be printed as well as stored. By careful use of defaults, the above stream can be simplified, as shown in Figure 10.14.

10.5 MANAGING THE SOLUTION CASES

We have reviewed in some detail the overall design of an optimization module that is both easy to use and sufficiently modular to allow extension over time.

	Availability at location UF
	ECD...UF
FAT1..J.	100
CHOBJ	-100
M1BOUND UP	NONE
LO	0

Figure 10.12 Matrix segment for a parametric run. CHOBJ is the change row to be applied to the objective function row FAT1..J.

```
+OPT
 -ACTPROB        RSIDE
 -PERIOD         M1
 -CASE           RSIDEM1
 -SOLUTION
 =PARAOBJ        0.1         1.0
       *ROW      CHOBJ
       *CASE
       RSIDEM1A
       RSIDEM1B
       RSIDEM1C
       RSIDEM1D
       RSIDEM1E
       RSIDEM1F
       RSIDEM1G
       RSIDEM1H
       RSIDEM1I
       RSIDEM1J
 -PSOLUTN
         •
         •
         •
```

Figure 10.13 Syntax for a parametric definition with no defaults.

```
+OPT
 -ACTPROB        RSIDE
 -PERIOD         M1
 -CASE           RSIDEM1
 -SOLUTION
 =PARAOBJ
 -PSOLUTN
         •
         •
         •
```

Figure 10.14 Parametric definition with defaults.

During this discussion, we made tacit assumptions about the cataloging and storing of the solution cases and the availability of bases. Now, let us look a little more closely at these issues from a systems design viewpoint.

The ultimate result from OPT, of course, is a set of preferred solution values for the given matrix and selected objective function. In most situations, these values need to be stored permanently in order to run reports without reoptimizing. We have seen how this physical storage is accomplished in PLATOFORM via the RECORD procedure. RECORD requires an assigned solution case name and the name of the ACTPROB that is to receive it. The TILLER tree is designed to allow any number of cases to be stored on a given ACTPROB.

Figure 10.15 reviews the basic ACTPROB structure. Any combination of the items illustrated is allowed. Thus, for instance, an ACTPROB may have two solution cases, no matrix, and one T-table. The cases do not have to be associated with their original host matrix. In some situations, the matrix may no longer be required and can be deleted. The case is still available for reports or as a starting basis. Alternatively, a case may have been copied from another ACTFILE. This implies the need for some kind of catalog structure that associates the case names with their physical residence.

Before we pursue further the organization of the cases on the file, we need to clarify the reasons for storing them. There are several uses for the information stored in a case. The first, and most obvious, is as input to the reporting programs. These may follow immediately after +OPT, or they may be delayed to a subsequent run. We may also wish to use historical cases for "side-by-side" or delta analysis against a more current version or to report a sensitivity case in terms of its base-case values. These issues are discussed in detail in Chapter 11.

Another major function is as input to successive passes through a nonlinear recursion, or a branch-and-bound search. The results of each suboptimization are subjected to the convergence or termination tests, and they ultimately drive the course of the overall optimization.

We see a third major use of the stored case in constructing an advanced starting basis. For efficiency, it is attractive to develop a starting basis routinely for any reasonably large model. But where is the basis to come from, and how should it be stored and managed? Presumably, it must be stored at the same time the solution is RECORDed, that is, following the previous optimization phase. We need to ensure that the basis is physically kept on the ACTFILE and not on a secondary basis file, if we are to retain the advantages of an integrated data base.

A little reflection offers a simple, practical technique. To construct a basis, we need to know the names of the matrix identifiers that are currently basic or are set to their bound limits. This information is routinely stored as part of the solution case. For each matrix identifier in the solution, its "status" (logical or structural, in basis or at bound) is filed along with the preferred activities, marginal values, and so on. We can, therefore, build a simple extract routine that creates a starting basis with the correct format, given a previously stored solution case. Since this routine can be entirely system driven, it is transparent to the analyst, who is now able to think of the solution case and basis as being the same, which at the level of the ACTFILE they are.

ACTPROB Structure		
Data Tables	LP Matrix	Solution Cases
Z-tables T-tables	One only	Any number, with assigned names

Figure 10.15   The elements of an ACTPROB.

10.5.1 The Solution Catalog

Having established the need for storing the cases, how can we organize the names into a catalog? We have seen that two sorts of catalog can be found within PLATOFORM. Catalogs that describe ACTFILE organization are kept on DATABASE and include the Problem Catalog PROB. . .C and the Dictionary Catalog DICT. . .C. Elements described in these catalogs occur only once on the ACTFILE (i.e., there is only one dictionary called PRODUCTD). In contrast, one of the catalogs (CONT. . .C) describes ACTPROB-specific data. CONT. . .C occurs on every ACTPROB and describes information that is uniquely associated with the ACTPROB itself (i.e., the named data tables). Multiple copies of these tables can occur within the ACTFILE as a whole, however. Should we catalog the case names in the same way as data tables, or as we catalog dictionaries? A single catalog on DATABASE gives us the advantage of an unambiguous case-name reference, but limits the names we can choose for each new solution. A separate catalog on each ACTPROB allows for duplicates on the ACTFILE, but it makes it harder to separate these duplicates if we need to distinguish them (i.e., for side-by-side reports). In this latter case, the analyst must define both the case name and the ACTPROB name to avoid ambiguity.

If we accept the limitation of unique case names (which has not been a problem in PLATOFORM), we can adopt the simpler solution of a single catalog. In fact, we see a way to turn this apparent limitation to advantage by the adoption of an optional case-naming convention. The catalog, which we call SOL. . . .C, should reside centrally on the ACTPROB DATABASE and should be maintained by the OPT DATAFORM program. The user should not need access to this catalog (via DBM), although certain user requests (e.g., to delete an entire ACTPROB) may result in system updates to SOL. . . .C by other PLATOFORM modules.

What information needs to be stored within the body of this table? Clearly, the ACTPROB name must be included. The date on which the case was last stored may also prove important. As in the example of the dictionary catalog DICT. . .C, a textual description of the table entries may also be useful.

The catalog illustrated in Figure 10.16 describes two cases: RBASEM1 and RSIDEM1. These are resident, respectively, on the ACTPROBs RBASE and RSIDE, and the last date each case was stored is also recorded. In a typical

Z:SOL....C	PROB	DATE	C1 - C3
RBASEM1	RBASE	82061001	BASECASE 1ST QUARTR PLAN
RSIDEM1	RSIDE	82072202	SULFUR MX @ 2.5 SIDECASE

Figure 10.16 An example Solution Catalog.

operating environment, the same case may be optimized a number of times before it is satisfactory. The program must allow the case to be overwritten (and the date entry modified) as long as the ACTPROB entry (PROB in the figure) is unchanged.

10.5.2 Case Mnemonic Structure

The case names in this example also suggest a relationship with the ACTPROB names. Since a given case can reside only in one place, the catalog gives us a way of unambiguously defining the ACTPROB, knowing only the case name. These names do not have to look like their host ACTPROB names for this purpose, since SOL. . . .C provides the lookup. However, it may be more convenient for the analyst to choose the names with their residence in mind. Why not take this memory-aid process one logical step further and encode at the same time the other pieces of data that characterize this solution?

We can provide an optional facility that takes advantage of these ideas to lessen the memory burden for the analyst. For a given matrix, two variables give rise to multiple cases. The first is the time period which, as we have seen, affects the choice of objective function. Second, we may wish to distinguish intermediate stages within a given optimization phase. These stages may represent intermediate nodes in a mixed-integer search or steps in a parametric study for which we wish to keep solution values. We can construct the eight-character case name to encode all of this data as follows:

Case name =   xxxxxyyz
where: xxxxx   = the associated ACTPROB name;
                      yy   = the time-period code;
                       z   = an arbitrary case counter.

If we adopt this convention, we can redefine the control syntax (see Figure 10.6) using the —SCASE (standard case) record. With this single record we define the ACTPROB RSIDE, period M1, and case-name RSIDEM1.

In the example given by Figure 10.17, OPT recognizes any eighth character as a valid case counter, including a blank. The SCASE convention limits the ACTPROB names to five characters and is therefore not mandatory. This is a recurrent theme throughout PLATOFORM: adoption of the conventions provides increased flexibility, but the rules may be broken where appropriate without serious loss of capability.

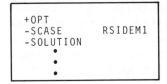

```
+OPT
 -SCASE        RSIDEM1
 -SOLUTION
    •
    •
    •
```

Figure 10.17   "Standard" case definition.

10.5.3 Case-Storage Considerations

To complete the discussion of case management, we need to understand what factors influence the decision to store a given solution on the file. Knowing that it may be used both for reports and for starting bases, are there any situations where we do *not* want to keep it? Consider first the basis. It is possible that *any* basis may be better than none as a starting point for a particular optimization. It is hard to anticipate whether the final basis for the current run will prove to be worth keeping. The conservative approach, therefore, is to store all cases as potential starting bases. But what about reporting requirements? Consider the typical situation presented in Figure 10.18.

The analyst wishes to receive a set of reports following a successful optimization. But suppose the model proves to be infeasible, has an unbounded column, or fails to complete for some other reason? In most cases, the reports will be meaningless, and the user may well wish to suppress them and print instead some solution-analysis and error-tracking output. How is OPT to control this situation? It is relatively easy to switch on any tracking routines automatically, since these are properly part of the OPT process. But should OPT be responsible for bypassing reports? This implies inspection of the overall run control and some sophisticated logic to determine which modules to bypass. What about the analyst who decides that some reports are useful even here? The decision is clearly the responsibility of the report programs themselves, and OPT should limit itself to signaling the status of the case. The subsequent program modules can then choose to ignore this status if required. So, how are we to provide a clear signal to dependent programs?

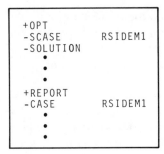

```
+OPT
 -SCASE        RSIDEM1
 -SOLUTION
    •
    •
    •
+REPORT
 -CASE         RSIDEM1
    •
    •
    •
```

Figure 10.18   Linking optimization and reports.

The Solution Catalog offers a solution. If OPT catalogs only solution cases that terminated successfully (i.e., were optimal), we have a mechanism for controlling the report programs. The reports can be designed to work from this catalog under normal conditions and will, therefore, fail to report a case deliberately omitted from SOL. . . .C. An optional control record included within the report-input stream can be employed to bypass the catalog lookup if we need to report on infeasible solutions. One subtle implication for OPT is in the cataloging of a previously optimized case. If an optimization is unsuccessful (not optimal), but this case was previously entered in the catalog, the old entry must also be deleted. This is so because the physically RECORDed solution has overwritten the previous version, and SOL. . . .C must reflect its status. Otherwise, any report programs will find the catalog entry and report the (nonoptimal) case.

However, we can still physically store *all* cases, regardless of status, as potential basis sources. The basis-building logic within OPT can simply ignore the catalog and search for the solutions directly on the ACTPROB.

10.6   THE ADVANCED STARTING BASIS

A number of issues surround the need for a starting basis. For example, many of Exxon's routine models are large and thus may consume an appreciable amount of machine time to solve. The choice of a suitably advanced starting basis can significantly reduce costs and improve turnaround time.

Along with the need to reduce overall machine time, the basis provides a security device in the event of computer failure. If appropriate, the OPT framework may be enhanced to take automatic snapshot bases at intervals during a long optimization so that a run restart can take advantage of the most recent basis. What, then, determines which starting basis, if any, to use in the current study? In most situations, the analyst can be relieved of the decision, and the program can be left to work it out.

10.6.1   The Search for a Basis

Given no other control information, OPT can search the ACTPROB for a solution with the same name as the case to be optimized. If one is present, the current run is probably a rerun of the updated model, a situation that occurs frequently in practice. From this stored case, a basis can then be constructed directly.

What if the case is for a new time period, and no previous basis exists? A brief look at the analyst's operating environment suggests another basis source. For many users, a new study is very similar to the previous one, differing only in its time period and modified economic and volume data. In basis terms, this new case may be very close to its predecessor. However, many of the logical and structural symbolic names will be different because of the new time-period code embedded in the matrix identifier names, as demonstrated in Figure 10.19.

We can immediately see that the mnemonic discipline of the matrix names offers a solution. Within our basis extract program, we can simply replace any matrix identifiers of the general form ***1*** with corresponding names ***2***. OPT can be designed to provide this facility automatically without involving the analyst, since the time-period codes for both periods can be deduced.

10.6.2 Alternative Basis Sources

Finally, we can offer the analyst explicit basis definition. The best basis may not be the obvious one; the most recent basis may be worse than no basis at all, and so forth. For these situations, a —BASIS record inserted within the control stream can be used to name an existing case/basis explicitly or to request an "all-slack" starting point. Figure 10.20 illustrates how the user might request basis RBASEM1 instead of the default RSIDEM1.

In summary, the program should be structured to search automatically for the most appropriate basis, if one exists. This is the normal situation, and requires no input from the user. The BASIS "prompt" record is needed only in unusual situations, where the knowledgeable user is aware of a better starting basis than the one that the program can derive.

Further extensions to this process can be perceived. In some instances (e.g., certain Markov formulation types), it may be possible to predict an appreciable

Basis for Previous Period	Status of Element	Current Matrix Element
LGL FAT1..J.	Basic Logical	FAT2..J.
SAB ERS.FLUF	Bounded Structural	ERS.FLUF
STR QFL1WTUF	Basic Structural	QFL2WTUF
•	•	•
•	•	•
•	•	•

Figure 10.19   Matrix identifiers for two different time periods.

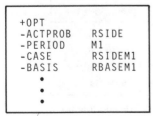

```
+OPT
-ACTPROB     RSIDE
-PERIOD      M1
-CASE        RSIDEM1
-BASIS       RBASEM1
     •
     •
     •
```

Figure 10.20   Adding a user-defined starting basis.

fraction of the final basis, and it can be advantageous to force in these logicals and structurals at the start. How can this be achieved? One way is to allow the user to input generic sets of basis candidates within +OPT, as in Figure 10.21.

This technique is satisfactory if the candidates happen to have names that can be submitted via a small number of such masks. This is rarely the case, however, and an obvious additional problem is the analytical burden on the user.

A more attractive route would be to generate the chosen basis names at the same time as the matrix is created. This requires an ability to build the symbolic basis from within DATAFORM and a clear set of rules for identifying the candidates. The approach is not really appropriate for the rather general +GENER module, but it may be important for more specialized matrix-generator codes.

10.7 DISPLAYING THE RESULTS

Let us turn, now, to a different issue that was introduced briefly at the beginning of this chapter: the decoding and text translation of matrix-identifier names. Why should we wish to do this? In philosophical terms, OPT has an unusual relationship with the user/analyst. A fundamental tenet of PLATOFORM has been to orient the analyst as much as possible to think of the business problem in terms of the data-base tables that represent it. OPT, however, is directly concerned with the mathematical programming matrix and only indirectly concerned with the data itself. Furthermore, this module may be viewed as the focus of the modeling process. In order to appreciate the results from the model, both the actual input and output must be directly accessible for inspection.

Figure 10.22 presents one way of looking at the four major phases of model

```
+OPT
-CMASK        DFL...UF   B*****UF   QFL*WTUF   • • •
     •
     •
     •
```

Figure 10.21   Identifying starting basis candidates.

Module	Tabular View for	Matrix View for
Data-Base Management	Input + Output	
Matrix Generation	Input only	Output only
Optimization		Input + Output
Report Writing	Output only	Input only

Figure 10.22   The four major phases of model development.

use. Data is input to and output from each phase of the process, but its actual form is different in each case. Data-base management is concerned with only a "tabular view" of the data (i.e., the two-dimensional PLATOFORM tables). These tables are also input to the matrix-generation phase, but the output is a "matrix view" of this same data. We can see that optimization is unique in this context, since it alone is concerned only with the latter "matrix view"—the one least suited to direct user interpretation.

This leads us to a potential conflict in the user's interface with OPT. In some ways, we would prefer to shield the analyst from the mathematical programming matrix, but this is clearly impractical. To control OPT properly, the user will have to understand the matrix formulation that was produced by the matrix generator. The interpretation of the results from OPT also demands a working knowledge of the matrix-mapping process, with its inherent limitations and idiosyncracies.

This overriding need to think in traditional LP terms during critical phases of any study places practical restrictions on the degree of matrix shielding that a support system should offer. The analyst who suppresses the output from the generator and optimizer and looks only to the data-base and management reports could completely misunderstand the results. This problem can be alleviated, to some extent, if powerful and comprehensive reports can be made available (see Chapter 11).

10.7.1 The SOLUTION Display

Alongside the need for management reports, there is a need to provide matrix-oriented reporting tools that allow the analyst to review the solution output of OPT. These tabular displays include the raw solution results (activities, bound-set values, and marginal costs). More specialized printout may include a TRANCOL (a display of the updated matrix following optimization) or a RANGE report (valid activity range for the printed marginal values, etc.).

Part of a typical MPS SOLUTION printout is illustrated in Figure 10.23. The output lists the sequence number and name of each matrix identifier in turn. The status of the element (Basic—BS, at Upper Limit—UL, etc.) is also given, along

NUMBER	COLUMN NAME	AT	ACTIVITY LEVEL	INPUT COST	LOWER LIMIT	UPPER LIMIT	REDUCED COST
7	DFL...UF	BS	5.333	-90.00	.	10.00	.
8	ECD...UF	BS	1.333	100.00	.	NONE	.
9	ERS...UF	UL	4.000	70.00	.	4.00	16.667-
10	BCD.FLUF	BS	1.333	.	.	NONE	.
11	BRS.FLUF	BS	4.000	.	.	NONE	.
12	QFL1WTUF	BS	5.333	.	.	NONE	.

Figure 10.23 A SOLUTION display.

with the input costs and the column-activity values obtained from the optimization phase. Although rather terse, this report contains a great deal of data about the solution and is a very important aid to the user. It is completely general, regardless of the kind of model being studied, but it is not very readable. How can the presentation be improved without loss of information? Clearly, the names of the rows and columns represent a considerable handicap to the rapid assimilation of the display. The eight-character names were constructed from the original data-table-attribute codes by the matrix generator, but their context and meaning are not at all obvious from a casual inspection. To understand the function of any matrix-identifier name, the user must not only know the meanings of the constituent codes but also the way they are assembled.

The solution, of course, is to extract the codes and translate them using the dictionaries at our disposal. The situation is analogous to the printing of data tables using the TPRINT submodule of DBM (Chapter 7), and the system requirements are similar. The rules for the generic construction of the rows and columns must be cataloged in the same way that tables are cataloged via GENTAB.C. Let us look at this process more closely.

10.7.2 Decoding the Matrix Identifiers

Consider a representative column name from the last example. In the previous chapter (Section 9.2.1), we saw how some illustrative names were constructed, using the generic dictionary codes for the various attributes. The BRS.FLUF vector in Figure 10.23 is a member of the general class for product blending.

Bpr.prlo	Content
B	Variable class (blend vectors)
pr	Product (stream into blend)
.	Invariant
pr	Product (finished grade)
lo	Location

Substituting the codes from the example, we obtain:

BRS.FLUF	Content
B	Blend
RS	Product RS
.	Invariant
FL	Product FL
UF	Location UF

Using the Dictionary Catalog (DICT. . .C), the dictionaries represented by codes "pr" and "lo" can be discovered. The codes can then be translated via a lookup in these dictionaries. What about the first character position in the name? In PLATOFORM, this character defines the generic vector set. Its translation should therefore come from the catalog, in the same way that GENTAB.C translates the generic table sets.

BRS.FLUF	Translation	Source
B	BLEND	From the catalog
RS	RESIDUE	Found in the product dictionary
.	.	Not translated
FL	FUEL OIL	Found in the product dictionary
UF	FAWLEY	Found in the location dictionary

The generic definition via a single character limits the total number of sets to fewer than 40. This restriction can be eased considerably by defining separate catalogs for the rows and columns. (Figure 9.4 illustrates character B, meaning a balance row and a blending column.)

Several ways of constructing these catalogs are possible. The general table catalog GENTAB.C illustrates one highly effective method that could be modified to serve the new requirements.

Figure 10.24 illustrates a simple, elegant way of defining the matrix column set via a new catalog COL. . . .C. The columns are defined generically within the stub, and they can be decoded using the mask in column M1 of the table, exactly as in GENTAB.C. The final three columns provide the translation for the set. No additional information is needed to translate all the columns in a matrix, assuming that this catalog has entries for each set and that each set is uniquely defined.

Although appealing, this design suffers from some practical limitations. The major one is the physical amount of processing required to translate each name. The structure of the name must first be unraveled via the mask, followed by multiple references to DICT. . .C. Unlike the analogous TPRINT processing for tables, a potentially large number of names needs to be decoded (routinely

COL....C	M1	D1 - D3
BPR.PRLO	XAAXBBCC	BLENDING VECTORS
EPR...LO	XAAXXXCC	STREAM AVAILABILITY
•		
•		
•		

Figure 10.24   One possible structure for a matrix column catalog.

2,000 to 5,000), and the time taken to do this could seriously degrade the performance of the printing routine.

10.7.3 The SOLPRINT Display

To solve the efficiency problem, a rather different catalog structure is needed (see Figure 10.25). This version is not so easily understood, but in practice it can be processed much faster, since it eliminates the need for the DICT. . .C lookup. The dictionary names are incorporated directly within the catalog itself. The mask is also eliminated by providing eight columns, one for each character position.

It is not necessary to describe the detailed syntax embedded in this table in order to understand the issues it addresses. The main advantage of defining the matrix elements on a character-by-character basis is that it permits the mapping of variable field lengths within the element names.

Occasionally, it has been necessary to translate names according to different "masks" or mapping rules. For instance, a given matrix may have two quite different kinds of rows that both begin with character B:

> Bpr. . .lo   Balance row for product pr at location lo
> B ...ML..   Audit row for following manufacturing losses

Both names are B rows, but the dictionary breakdown required is different for each. The catalog we develop must be able to identify easily the set to which any candidate matrix element belongs. So, our table embodies not only the two sets of mapping rules but also a method for distinguishing them. In practice, this can be handled by allocating special syntax characters along with the dictionary names within the catalog.

Although less satisfying visually, this structure is extremely efficient from a processing viewpoint. Since the two catalogs (ROW. . . .C is similar) are built only once at the beginning of an application, the exact format of the table is of

COL....C	F1	F2	F3	F4	F5	F6	F7	F8
BLEND	B	PRODUCTD	&	*	PRODUCTD	&	LOCATN.D	&
AVAILBLE	E	PRODUCTD	&	*	*	&	LOCATN.D	&
•								
•								
•								

Figure 10.25   The Column Catalog as used in PLATOFORM.

```
COLUMN     ACTIVITY            TYPE       ------------DESCRIPTION------------
 NAME      LEVEL

DFL...UF    5.333  • • •  DEMAND     FUEL OIL    .        ..        FAWLEY
ECD...UF    1.333  • • •  AVAILABL   CAT DIST    .        ..        FAWLEY
ERS...UF    4.000  • • •  AVAILABL   RESIDUE     .        ..        FAWLEY
BCD.FLUF    1.333  • • •  BLEND      CAT DIST    .      FUEL OIL    FAWLEY
BRS.FLUF    4.000  • • •  BLEND      RESIDUE     .      FUEL OIL    FAWLEY
QFL1WTUF    5.333  • • •  SPEC       FUEL OIL  PERIOD1  (WEIGHT)    FAWLEY
```

Figure 10.26   Improved display results achieved through use of a SOLPRINT module.

secondary importance here. If we now include the necessary logic within a routine to print the solution results, a greatly improved display results. In PLATO-FORM, this was achieved by a SOLPRINT module that combined the effect of SOLUTION with the translation logic described.

Part of the resulting output from SOLPRINT is illustrated in Figure 10.26. The entire solution is printed as before (it is abbreviated in the example, for clarity), and five additional fields are added to the right of each line. These describe the set to which the matrix identifier belongs and translate the constituent codes. In principle, the same technique can be applied to other kinds of displays involving the matrix-identifier names. To achieve this, the decoding logic needs to be decoupled from the solution display print.

10.8 SUMMARY

This chapter has described the development of a flexible and extensible frame-work for solving mathematical programming models starting from a simple skeleton. This embryo system is first designed to support a range of linear models in a generic fashion. The user is expected to provide an absolute minimum of control information by default, but has the option to supply additional data. The program itself is structured generally, and it makes few assumptions about the internal layout of the matrix. This, then, provides a basis for more radical extension and permits the evolution of a powerful matrix-solving tool, as additional techniques are incorporated under the control of the central OPT manager. A key ingredient in this evolutionary process is the planned extension of the underlying EMPS software. Such extensions necessarily precede any new capabilities within the PLATOFORM system, and they require foresight and a detailed understanding of future priorities.

In the latter part of the chapter, we described the need for three new catalogs: one for the solution cases and two for the matrix identifiers. These catalogs provide additional capability (standard case name, SOLPRINT) and greatly simplify the user's task. The row and column catalogs also introduce the idea of presenting more readable output. SOLPRINT is the first step in the process of creating reports that can be understood, not just by the working analyst, but by managers who may be unfamiliar with the internal details of the model. This topic, the design of a powerful reports capability, is the subject of the next chapter.

KENNETH H. PALMER

11. REPORT WRITING

11.1 INTRODUCTION

It is perhaps appropriate that we should leave this aspect of mathematical programming—report writing—until last, not only because report writing often naturally occurs at the end of a series of activities, but also because it represents an area that for many years defied any form of generalization or structure.

What we mean by report writing is the ability to restructure the basic solution values produced by the optimizing algorithms into a form that relates back to the real-world problem being modeled. The difficulty is that the form this restructuring takes tends to be rather personal, strongly linked to individual applications. Sometimes, it must match rigid reporting formats required by some other department of the company. Financial values might be required in local currency, other variables in units of measurement not necessarily the same as those used in the model. Translation of the mnemonic codes that make up the matrix row and column identifiers might be required in a local language, values might require totaling (or subtotaling), the number of decimal places required to express the values might depend on the units of those values, and so forth.

All of these and many other factors combine to make the development of even a very simple report—such as "crude oil processed by refinery"—a formidable task if the resulting report is to be usable by any application.

When PLATOFORM was first introduced in Exxon and began to spread to applications outside the original Esso Europe supply/refining model, in almost every case a new set of reports, written in DATAFORM, needed to be developed. This left a large amount of application-specific code (what we called Type-3 in Chapter 5) to be maintained locally. Because they were so specific, these reports were often time-consuming to write, required considerable maintenance, and executed rather slowly. The time required to set up a new application was very much influenced by the time required to develop a set of reports. Most other mathematical programming activities—data-base management, matrix generation, optimization—had already been generalized to the point at which a developer of a new application could select appropriate program modules "off the shelf." But, with one or two exceptions, no ready-made packages for report writing were available.

In 1976, Exxon embarked on an ambitious project to solve the problem of report writing. Four years later, a single DATAFORM module was handling more than 95% of all Exxon's mathematical programming report requirements. Not only had the development of new application reports written in DATAFORM virtually ceased, but almost all of those that had already been written for earlier applications had been converted to the new report-writing module. As a result, new applications could now be totally developed very quickly, and only rarely did they require new programming to support them. This new report-writing module (named ADHOC in recognition of its fundamental nature) had finally allowed all aspects of mathematical programming to be generalized.

In this chapter we trace the underlying concepts of this approach to report writing, and in Chapter 12 we give examples of some other report writers developed for specific functions outside the scope of ADHOC, but equally general in their design.

The original ADHOC module was very simple, once the basic ideas had been established. It was designed one evening and operating the next day. From that point, ADHOC developed in a natural evolution in which progressively more features were included, allowing an ever-expanding number of reports to be supported.

Two important features of this evolutionary phase are worth noting. The module remained fully operational throughout the four-year development period, and each new capability was added in such a way that all previously defined reports continued to operate as before. These aspects are vital to the proper evolution of a major system. The evolutionary approach is based on introducing a new major module as early as possible—as soon as it can perform a useful function. Its future development is then determined entirely by the demands of the system users, with no reliance on inspired estimates of what these demands might be. This is particularly important in the case of a general report writer such as ADHOC.

The success of a generalized approach to report writing is very much dependent on the mnemonic discipline imposed throughout the system, particularly in the naming of the matrix rows and columns. Here, the important point is not so much the sequence in which entities occur within a name, but the preservation of a constant field size for each entity. This enables classes of vectors (rows or columns) to be described in a general way by use of masks. We have already seen enough of PLATOFORM to realize that this mnemonic discipline based on fixed fields is a fundamental feature of the design (e.g., data structure as discussed in Chapter 4 and matrix structure in Chapter 9). The environment, then, is well-suited to the development of a general report writer.

In order to understand the thought processes that led to the development of the ADHOC report writer, we first take an example of a specific report and examine the basic steps required to produce it from a linear programming (LP) solution. Figure 11.1 illustrates part of a solution from a PLATOFORM-style

model, and Figure 11.2 a "crude run by location" report required from that
solution. We want to see how we can go from one to the other.

11.2 BASIC STEPS OF REPORT WRITING

To follow the basic steps, we must clarify the system context in which the
example shown in Figure 11.1 occurs. First, it is important to understand the
structure of the eight-character column names. In this example, character 1
defines the column function:

P   Process vector
B   Blend vector, and so on.

Of all the process vectors, only those that process crude oil through an
atmospheric distillation unit are relevant to the crude run report. These are
identified by a D in character position 5 (as well as a P in position 1). They can
be specified by means of the mask:

P***D***

	Column Name	Activity (X)	Reduced Cost (D)
1	PAH.D.UF	2	-
2	PAH.D2UF	1.5	-
3	PAL.D.UF	-	5.6
4	PAL.D2UF	0.2	-
5	PAM.D.UM	3.0	-
6	PKU.D3FF	1.5	-
7	PKU.V.FF	0.6	-
8	BXL.GPIA	0.7	-

Figure 11.1 Part of optimal solution (case—MPC11M1, ACTPROB—MPC11).

```
CRUDE RUNS BY REFINERY - KT/PERIØD

                       FRANCE   U.K.     U.K.
                       FØS      FAWLEY   MILFØRD
                                         HAVEN

                       FF       UF       UM

ARABIAN LIGHT    AL       .       0.2       .
ARABIAN MEDIUM   AM       .        .       3.0
ARABIAN HEAVY    AH       .       3.5       .
KUWAIT           KU     1.5        .        .
```

Figure 11.2 Required report. To prevent ambiguities, Ø is used in all the figures of this
chapter to identify an alphabetic character, which thus distinguishes it from a zero.

and this mask will select just the first six columns given in Figure 11.1. The full structure of these column names is as follows:

Pcr.unlo	Content
P	Process vector
cr	Name of crude
.	Invariant
un	Distillation unit
lo	Location

There are thus three entity classes involved:

cr	Crudes
un	Units
lo	Locations

Each entity class has an associated dictionary on ACTPROB DATABASE (see Figure 11.3). Figure 11.4 shows how these are cataloged in table DICT. . .C.

The entries in column C1 of DICT. . .C are the codes for the dictionaries and have the same number of characters as the stubs of the dictionaries (two in this example; see Chapter 4).

One other catalog, the Solution Catalog, is relevant to the report, since it defines the relationship between case name and ACTPROB on which the case is recorded (see Chapter 10). This is shown in Figure 11.5.

Since there are three entity classes involved (crudes, units, and locations), there are many ways in which the report could be laid out. In our example, we have chosen to report crudes down the side (stub) by location along the top (head) and not to be concerned with the distillation unit (i.e., report total crude run by type of crude at each location). The required report format was illustrated in Figure 11.2.

11.2.1   Step 1—Extracting the Subset of Solution Values

The first step in producing this report, then, is to extract from the solution recorded on the data-base ACTFILE a subset of column activities that refer to processing crude oils at refineries and to store these in a DATAFORM T-table.

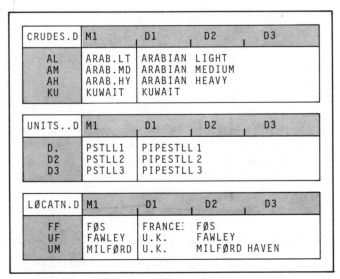

Figure 11.3   Relevant PLATOFORM dictionaries.

DICT...C	D1	D2	D3	C1
CRUDES.D	CRUDE	ØILS		CR
LØCATN.D	LØCATIØN	S		LØ
UNITS..D	PRØCESS	UNITS		UN

Figure 11.4   Part of the associated Dictionary Catalog.

SØL....C	PRØB	DATE
MPC11M1	MPC11	80060101

Figure 11.5   Part of Solution Catalog.

In DATAFORM this is achieved by use of a single verb SOLTAB (Solution to Table) to which the following parameters are passed:

1. Name of ACTFILE—via Communication Region (CR) cell.
2. Name of ACTPROB—MPC11 (via CR cell).
3. Name of CASE—MPC11M1.
4. Whether COLUMNS or ROWS—COLS.
5. Subset of columns—P***D*** (via mask).

6. Type of solution values—X (i.e., column activities).

7. Name of output table—SOLVALS.

We should note in passing that all of these parameters can be given indirectly to SOLTAB. This capability is essential for the later development of a more general program. The output from this step will be an internal DATAFORM table, as illustrated in Figure 11.6.

This table now contains all the information required for our final report, but note that it is structured essentially as a single column "list." To get to the required output report, we have to restructure this list as a two-dimensional array. This restructuring constitutes our second step and involves the decomposition of the eight-character column names.

11.2.2 Step 2—Decomposing the List to a Two-Dimensional Array

To carry out this decomposition, we have to consider which elements of the eight-character list names are to form the stub of the final report and which elements are to form the head. In our example, the stub of the report is to be crude oils. These are identified by two-character codes occurring in positions 2 and 3 of the list names. One way to extract these entities in DATAFORM is to make use of what is known as a DATAFORM mask such as:

0XX00000

In such a mask, the nonzeros specify the character positions to be selected. By successively applying this mask to all members of the list and removing duplicates, we can obtain a list of all two-character crude codes that are uniquely occurring (Figure 11.7).

SØLVALS	X
PAH.D.UF	2
PAH.D2UF	1.5
PAL.D.UF	0
PAL.D2UF	0.2
PAM.D.UM	3.0
PKU.D3FF	1.5

Figure 11.6 DATAFORM table created from a subset of solution values.

AH	
AL	
AM	
KU	

Figure 11.7 List of crude oil codes extracted from list names using DATAFORM mask 0XX00000.

Although we have used the same word "mask," it is important to notice the fundamental difference between this DATAFORM mask (0XX00000), used to identify character positions within an eight-character string, and the previous mask (P***D***) used to select a subset on the basis of matching characters; that is, all members of the set that have P in position 1 and D in position 5, asterisks being a universal match. We henceforth refer to this latter type of mask as a "matching mask."

The head of our report is to be composed of locations. These are identified by two-character codes occurring in positions 7 and 8 of the list names. By successively applying the DATAFORM mask 000000XX to each member of the list, removing duplicates, and attaching the resulting list of two-character entities as the head of our report table, we can produce the report skeleton shown in Figure 11.8.

Finally, we have to fill this skeleton table with values extracted from column X of the table SOLVALS (Figure 11.6). To do this, we process each of the list members, applying first the stub mask (0XX00000) and then the head mask (000000XX) to determine the stub/head intersection in which to place the value. For the first list member PAH.D.UF, this process will yield stub AH and head UF, and the value 2 is placed at this intersection (Figure 11.9).

The second list member, PAH.D2UF, yields the same stub/head intersection, AH/UF. The value 1.5 must be added to any nonzero value already at this intersection (Figure 11.10).

	UF	UM	FF
AH			
AL			
AM			
KU			

Figure 11.8   Report skeleton after applying stub mask 0XX00000 and head mask 000000XX to each list member.

	UF	UM	FF
AH	2		
AL			
AM			
KU			

Figure 11.9   Output report table filled with value from first list member.

	UF	UM	FF
AH	3.5		
AL			
AM			
KU			

Figure 11.10   Output report table after processing first two list members.

The list members PAH.D.UF and PAH.D2UF can be considered "logical duplicates" in the context of this report, since only character positions 2, 3 and 7, 8 are relevant to the final report. The decomposition step must ensure that the values associated with such logical duplicates are accumulated. Processing the remainder of the list members in a similar way will yield the report table shown in Figure 11.11.

This decomposition step is easily written in DATAFORM. Because of its key importance in the production of any two-dimensional solution report, there are obvious advantages in programming it as a DATAFORM subroutine to which any two DATAFORM masks can be passed as arguments. This second step can then, like the first, be executed via a single DATAFORM statement. The importance of this decomposition step had been recognized long before work started on the development of a general report-writing tool. In fact, the subroutine had been taken one stage further and rewritten as a FORTRAN function callable from DATAFORM with consequent gains in execution efficiency. The ease with which FORTRAN functions can be integrated into DATAFORM programs was to prove crucial to the development of an efficient general report writer. In these functions, FORTRAN is used as a vehicle to call the underlying data-handling language TILLER (see Chapter 3) directly. Not only do these functions significantly increase execution efficiency, but they also allow additional capabilities to be provided in the host language (DATAFORM).

11.2.3   Step 3—Dictionary Preparation

The report table shown in Figure 11.11 is now, basically, the right structure for the output report (Figure 11.2), but two things remain to be done before we print it. First, the two-character crude and location codes that form the stub and head need to have additional translation text added. This text must come from columns D1, D2, and D3 of the associated PLATOFORM dictionaries (CRUDES.D and LOCATN.D, respectively). Second, both the stubs and heads have to be reordered into the sequence given in these dictionaries (compare with TPRINT, Chapter 4).

As with TPRINT, we can use DATAFORM's TABULATE verb to achieve both of these requirements, but first we must prepare the dictionaries to ensure that all entities found in the list names have a corresponding entry in the dictionary. Missing entities can be added at the end of the dictionary, but without meaning-

	UF	UM	FF
AH	3.5		
AL	0.2		
AM		3.0	
KU			1.5

Figure 11.11   Output report table at completion of decomposition step.

ful translation text, and the dictionary can be reduced in size by eliminating all entities not found in the list names. The resulting smaller Z-table improves the efficiency of TABULATE (see Figure 11.12).

The dictionary-preparation step involves, essentially, three DATAFORM statements: the first to retrieve the original dictionaries from ACTPROB DATA-BASE and then two calls to a general FORTRAN function to prepare the small stub and head dictionaries to be used with TABULATE.

11.2.4 Step 4—Printing the Report

This final step in report writing consists primarily of a single TABULATE statement. But before we do this, we have to attend to two other matters. The first is to determine where on the page we want to start printing the report. To do this, we need to take into account the current line position; the size of the table; the tiers of head translation; titling and footnotes; and the number of strips required to accommodate the total number of columns to be printed. While TABULATE automatically divides the table into strips (see Figure 3.11), the report program must aim to avoid splitting one of these strips across two pages by "skipping" to a new page if necessary. Such an algorithm can be wrapped into a general subroutine to which are passed the dimensions of the report table.

The second thing we must do is to print a title for the report using a simple

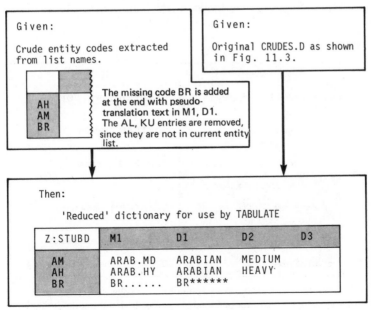

Figure 11.12 Producing a "tailor-made" dictionary for TABULATE.

PRINT statement. We are now ready to TABULATE the report itself. TABULATE works in conjunction with a FORMAT definition that specifies the number of columns across a page, column width, and number of decimal places to be printed for the values (Chapter 3). Here, because of its importance in developing a general report-writing program, we should note that a TABULATE statement can be indirectly linked to any FORMAT definition.

11.2.5 Summary of Steps

In describing these four steps, we have deliberately taken a close look at some of the internal workings of the DATAFORM language. This was partly to illustrate the type of support-language features that are helpful in developing solution reports and partly to provide sufficient technical background to understand the way in which the PLATOFORM ADHOC module developed.

Once the necessary tools (FORTRAN functions and DATAFORM subroutines) have been assembled, the development of a DATAFORM program to produce a specific report of the type illustrated in our example has been quite simplified (see Figure 11.13). Of course, this is a very simple report, but it does provide a framework on which more complex reports can be developed.

11.2.6 "Split" Reports

Before going on to see how these basic steps can be generalized, we should consider two more types of report, one of which introduces a further fundamental step.

	STEP	TOOL
1	Extract subset of solution values	DATAFORM verb – SØLTAB
2	Decompose this 'list' to two-dimensional array	FORTRAN function
3	Prepare stub/head translation dictionaries from PLATOFORM dictionaries	FORTRAN function
4	a) Determine position on page to start printing	DATAFORM subroutine
	b) Print title	DATAFORM verb – PRINT
	c) Print table	DATAFORM verb – TABULATE

Figure 11.13 Summary of basic steps required to produce simple report. The entire report can be developed with about fifteen DATAFORM statements (including calls to FORTRAN functions and DATAFORM subroutines).

We have already noted (Section 11.2) that the crude oil processing vectors contain three qualifying attributes: name of crude oil, name of distillation unit, and name of refinery (location). Our previous example showed a report that made use of only two of these attributes (crude oil and location). Suppose that we want to report all three; where can we put our third entity (distillation unit)?

There are, of course, a number of possibilities. We could, for instance, double up the reported entities in either the stub or the head of our output report. Figure 11.14 illustrates one such possibility where the distillation unit has been added as a second segment in the heads. Such "multisegment" reports do not alter the basic steps required, although they add complexity to the dictionary-preparation step (Step 3). In this example, the two head dictionaries (LOCATN.D and UNITS. .D) must first be merged into one, another function readily amenable for development as a general subroutine.

But this is not the only way we can accommodate our third entity. We could instead produce a multitable report: one table for each unique occurrence of a third entity. Basically, we could report crude runs (stub) by distillation unit (head) and have one such report for each location (see Figure 11.15). We refer to such multitable reports as "split" reports.

Conceptually, these split reports must be considered as a single report: one that produces multiple report tables. To achieve this, we must introduce one additional step into our program logic. This step will occur after Step 1, when we have extracted via SOLTAB the list of solution vectors and their values using the matching mask P***D*** (see Figure 11.6). Before decomposing this list according to a stub mask (0XX00000) and a head mask (this time, 0000XX00), we must first break up the list (split it) into sublists based on the occurrence of unique entities in positions 7, 8 of the list names (locations). To do this, we can again make use of a DATAFORM mask (000000XX) to select the appropriate characters. Figure 11.16 illustrates what we want this split function to achieve.

```
CRUDE RUNS BY REFINERY AND DISTILLATIØN UNIT - KT/PERIØD

                      FØS       FAWLEY             MILFØRD
                      PSTLL3    PSTLL1    PSTLL2   PSTLL1

                      FFD3      UFD.      UFD2     UMD.

ARABIAN LIGHT    AL      .         .        0.2      .
ARABIAN MEDIUM   AM      .         .         .      3.C
ARABIAN HEAVY    AH      .        2.0       1.5      .
KUWAIT           KU    1.5         .         .       .
```

Figure 11.14 A "multisegment head" report. Note that the short eight-character text from column M1 of the head dictionaries (see Figure 11.3) is used to avoid printing forty-eight characters of text. The dictionary sequence is preserved hierarchically, and the "topmost" head segment is printed only on its first occurrence.

```
CRUDE RUN BY DISTILLATIØN UNIT - KT/PERIØD  REFINERY:  FØS      FF

                    PIPESTLL
                    3

                    D3
KUWAIT              KU 1.5


CRUDE RUN BY DISTILLATIØN UNIT - KT/PERIØD  REFINERY:  FAWLEY   UF

                    PIPESTLL       PIPESTLL
                    1              2

                    D.             D2
ARABIAN LIGHT    AL .              0.2
ARABIAN HEAVY    AH 2.0            1.5


CRUDE RUN BY DISTILLATIØN UNIT - KT/PERIØD  REFINERY:  MILFØRD  UM

                    PIPESTLL
                    1

                    D.
ARABIAN MEDIUM   AM 3.0
```

Figure 11.15   A split report. The translation of split entity is added to the title line. Tables are printed in LOCATN.D sequence (FF, UF, UM).

Because of the fundamental nature of this splitting operation in report writing, it is again useful to develop a general-purpose FORTRAN function to which is passed:

- The name of the table to be split (e.g., SOLVALS).
- The split mask (e.g., 000000XX).
- A short key name for forming the names of the split tables (e.g., SOL).

Having split the master list into an unknown number of sublists, we can simply carry out Steps 2 through 4 on each of the sublists by looping through the list of tables (stub of table SOL in our example), first sorting them into split-dictionary (LOCATN.D) order. The basic steps to produce a split report are summarized in Figure 11.17.

By packaging all the major report-writing activities into subroutines or FOR-TRAN functions, we not only make the writing of subsequent reports much simpler, but we also ensure very fast execution, since each of the packages can be carefully tuned for maximum performance.

We are now ready to consider how we can generalize the report-writing process. How can we make the same program module, as developed here for

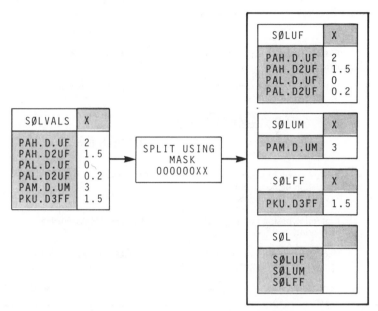

Figure 11.16 Splitting a "list" into sublists. Table SOL records the names of the sublist tables created. The name of each sublist table reflects the entity to which it refers.

the crude run report, serve to produce other, quite different reports; for example, a product distribution report or a capacity utilization report?

11.3 GENERALIZING THE BASIC STEPS

As we have gone through the basic steps to produce the crude run report (in any of its forms), we have noted that the arguments required to drive the major programming functions (DATAFORM verbs, subroutines, or FORTRAN functions) can invariably be provided indirectly. In other words, it is not necessary to "hard code" a particular matching mask (e.g., P***D***) whenever we call the DATAFORM verb SOLTAB. SOLTAB can be directed to look for this mask in an internal DATAFORM name cell or to extract it from the body of a Z-table.

To see how far we can take this generalization, we need to consider all the factors that make the particular report shown, for instance, in Figure 11.2, specific. These factors are summarized in Figure 11.18 and completely define the requirements for the crude run report. A different set of factors would give rise to a different report. As long as we can input these factors from an external source, then the same DATAFORM program can be used to produce either report or, indeed, any number of similar two-dimensional reports.

If we look carefully at the list of factors given in Figure 11.18, we can notice

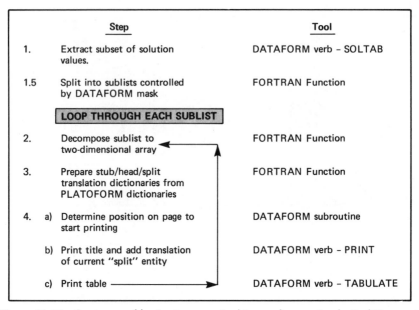

Figure 11.17   Summary of basic steps required to produce a simple "split" report.

that they fall into two types. There are those that represent a permanent definition of the report itself (factors 4 through 11) and those that represent a temporary definition of the solution case to which we want to apply the report (factors
1 through 3). In designing an input syntax for a general report writer, we must
be careful to distinguish these two. The permanent nature of the report-definition
factors suggests that we store them in a PLATOFORM table on the data base,
where, of course, they could still be modified by using the update facilities of
PLATOFORM's DBM. The case-specific factors could more conveniently be
supplied at run time through the regular PLATOFORM input syntax. In other
words, we can set up in advance a set of report definitions in a table and decide
what case we want to report on later. These same report definitions would be
expected to apply to any case.

Let us consider first, then, the more permanent report-definition factors and
see how we might represent them in tabular form.

11.3.1   Report-Definition Table

Most of the permanent factors given in Figure 11.18 can conveniently be identified by an eight-character string. They could thus be contained in a column of

	Factor	Example for Crude Run Report
1	Name of ACTFILE	CR-cell XACTFILE
2	Name of ACTPROB on which solution is recorded	MPC11
3	Name of solution case	MPC11M1
4	Report title	CRUDE RUNS BY REFINERY - KT/PERIØD
5	Subset of column vectors required for report	P***D***
6	Type of solution values	Column Activities (X)
7	Character position of entities to form stubs of report	0XX00000
8	PLATOFORM dictionary to translate stubs	CRUDES.D
9	Character position of entities to form heads of report	000000XX
10	PLATOFORM dictionary to translate heads	LØCATN.D
11	Number of decimal places to be printed	1

Figure 11.18 Factors specific to the simple crude run report illustrated in Figure 11.2.

a DATAFORM Z-table whose stubs contain appropriate keywords identifying the factors. Factors 5 through 11 are shown this way in Figure 11.19.

In fact, we can simplify this definition of the report by taking advantage of PLATOFORM's data structure. Instead of using XX as the nonzero characters in the DATAFORM stub and head masks, we could substitute the generic codes for the related dictionaries (see Figure 11.4). Thus, a STUBMASK of 0CR00000 would not only indicate the character positions within the column vector names

ZR.AH..Z	CRUN
MASK	P***D***
SØLTYPE	X
STUBMASK	0XX00000
STUBDICT	CRUDES.D
HEADMASK	000000XX
HEADDICT	LØCATN.D
PØINT	1

Figure 11.19 Definition of a simple crude run report arranged as a column of a data table. The table name conforms to PLATOFORM naming conventions, and the head name identifies the particular report.

containing the stub entities but would, at the same time, indicate which dictionary is to be used to translate them. By this means we do not need the separate STUBDICT/HEADDICT definitions (unless, of course, the generic dictionary code contains a "zero" character).

The one factor that we have not shown in this table is the report title (factor 4). Here, we have a problem, since report titles will generally require many more than eight characters. Once again, however, we can take advantage of the PLATOFORM data structure and utilize the associated "comments table" to store up to fifty-six characters of text for each title line. The appropriate title line can then be cross-referenced by use of the line number. Figure 11.20 illustrates these features and includes definitions for the two other reports discussed earlier in this chapter: the multisegment head report shown in Figure 11.14 (CRUNLO) and the split report shown in Figure 11.15 (CRSPLT).

It is important to realize that one simple DATAFORM program can be written to interpret any (or all) of these three reports, if it is written in a general way with indirect references to each of the factors defining the report. These references are then resolved by extracting the appropriate values from the control table (ZR.AH..Z). This simple concept formed the basic idea of the PLATOFORM ADHOC report-writing module. Note the ease with which a complete report can be defined. A few minutes effort is all that is required to define and add, say, a product distribution report as a fourth column of the table (see Figure 11.21). No additional DATAFORM programming is involved.

Note, also, the ease with which a report format can be changed. If we want to see our simple crude run report (CRUN) with crude oils along the top and locations down the side, we need only interchange the STUBMASK and HEADMASK, using the UPDATE submodule of DBM (Figure 11.22).

This approach allows report definition and maintenance to come under the control of the users. Reports can be written as and when they are needed, on

ZR.AH..Z	CRUN	CRUNLØ	CRSPLT	Titles are stored in the associated comments table and cross-referenced via line numbers.
MASK	P***D***	P***D***	P***D***	Nonblank SPLTMASK indicates a multi-table report required for CRSPLT; one for each location (LØ).
SPLTMASK			000000LØ	
STUBMASK	0CR00000	0CR00000	0CR00000	
HEADMASK	000000LØ	000000LØ	0000UN00	HEADMSK1 defines a second head segment for report CRUNLØ.
HEADMSK1		0000UN00		
PØINT	1	1	1	SØLTYPE is omitted since it is convenient to make column activities (X) the default type of solution values.
TITLE	L01	L02	L03	

ZR.AH..X	
L01	CRUDE RUNS BY REFINERY - KT/PERIØD
L02	CRUDE RUNS BY REFINERY AND DISTILLATION UNIT - KT/PERIØD
L03	CRUDE RUN BY DISTILLATIØN UNIT - KT/PERIØD REFINERY:

Figure 11.20   Definition of three crude run reports.

ZR.AH..Z	CRUN	CRUNLØ	CRSPLT	PRDIST	Structure of Transportation Column Names	
MASK				T*******		
SPLTMASK				000000LØ	Tpr.exim	Content
STUBMASK				0PR00000	T	Transportation vector
HEADMASK				0000LØ00	pr	Product
PØINT				3	.	Invariant
TITLE				L04	ex	Exporting location (LØ)
TITLE2				L05	im	Importing location (LØ)

ZR.AH..X				
L01				
L02				
L03				
L04	PRØDUCT MØVES BY EXPØRTER - KT/PERIØD		IMPØRTER	
L05	=======================================			

Figure 11.21 Addition of a fourth report to show product moves, one table for each importing location (only new data shown).

an ad hoc basis. The role of the systems analyst is to maintain and extend the single generalized DATAFORM module that interprets these reports.

Before going on to see how these basic concepts can be extended to more complex reports, we must look at the input syntax required to call the ADHOC module.

11.3.2 Supplying Case-Specific Data

The control table (ZR . AH. .Z) contains a "permanent" definition of each report in terms of its structure. This structure is independent of the particular optimal solution to be analyzed, but it does, of course, assume a constant matrix structure. This structural definition of each report remains permanently on the data ACTFILE. Only when we run one of these reports is it relevant to supply data relating to the solution case we want to analyze. This requires a PLATO-FORM call to the ADHOC module, followed by information defining the case and an indication of which reports we want to run. Our factors 1 through 3 must be included in this information.

In general, there is no need to pass the name of the ACTFILE to a PLATO-FORM module since each PLATOFORM run relates to one ACTFILE. The control

```
+DBM              DATABASE
-UPDATE
  U ZR.AH..Z      CRUN
    STUBMASK      000000LØ
    HEADMASK      0CR00000
```

Figure 11.22 Switching stubs/heads in a report.

module PLEX (see Chapter 5) can establish this name at the start of a run. If we are using a Solution Catalog (see Figure 11.5), then we have no need to pass the name of the ACTPROB on which our solution case has been recorded since this is already available in catalog SOL. . . .C. Hence, it is necessary to indicate only the name of the case, and the ADHOC module can do the rest. A suitable input syntax for ADHOC is illustrated in Figure 11.23.

Here, —CASE acts as a keyword to define the case name. Other "minus" requests refer to individual reports in the sequence in which they are required. They cross-reference head names for the reports in the control table (ZR.AH..Z).

This syntax is suitable so long as only a few reports are required in a run. It quickly becomes tedious if a standard set of twenty or thirty reports is required each time. One solution is to permit the definition of "sets" of reports. A set can conveniently be defined within the control table (see Figure 11.24).

A set can then be requested at input time, as shown in Figure 11.25.

As we go on to look at extending these basic concepts, we encounter other factors that are strictly case-specific and hence require definition at input time via a suitable keyword.

```
+ADHØC
-CASE       MPC11M1
-CRUN
-PRDIST
```

Figure 11.23   ADHOC input syntax. Reports CRUN and PRDIST are required for solution case MPC11M1.

ZR.AH..Z	CRUN	CRUNLØ	CRSPLT	PRDIST	• • •
MASK					
		•			
		•			
		•			
★SET1		1		2	• • •
★SET2	2		3	1	• • •

Figure 11.24   Definition of report sets. *SET1 gives report CRUNLO followed by PRDIST. *SET2 gives reports PRDIST, CRUN, and CRSPLT in that order. Any number of such sets can be defined.

```
+ADHØC
-CASE       MPC11M1
-★SET2
```

Figure 11.25   Requesting a set of reports.

11.4 EXTENDING THE BASIC IDEAS

What we have seen, so far, is a generalized reporting tool, driven by data stored in a PLATOFORM data table with case-specific information supplied at input time. While such a tool can produce many different reports from a basic solution to an LP problem, it is nevertheless restricted to a narrow range of report types—simple two-dimensional arrays or multitable reports involving two or three entity classes. Values are reported in the same "units" in which they occur in the model solution and are derived from a set of vectors that can be specified via a single matching mask. Very limited totaling capabilities are included.

Some of these capabilities are easily extendable without need to alter the basic structure of the underlying ADHOC module. Thus it is easy to allow any number of title lines for a report (we have already seen an example of two in Figure 11.21) and to extend this idea to footnotes, or "trailers," to be printed at the bottom of a report.

Solution types need not be confined to column activities (the default—X). We could just as easily report on dual activities of rows (P = PI values), reduced costs of columns (D = D/J values), upper or lower bounds, input costs—indeed, any solution type forming part of the recorded solution. Figure 11.26 shows the definition of a report giving marginal values of product streams by location based on the dual activities of the balance rows. Figure 11.27 illustrates the resulting output report.

We can extend the multisegment-head concept, already illustrated in report CRUNLO (Figure 11.20), to allow up to three entities in both the stub and the head of a report. This allows a maximum of six entities to be reported. Note that

ZR.AH..Z	PRVALS
TITLE	L01
MASK	B**...**
SØLTYPE	P
STUBMASK	0PR00000
HEADMASK	000000LØ
PØINT	2
TRAIL	L02

Structure of row names

Bpr . . . lo	Content
B	Balance row
pr	Product
. . .	Invariant
lo	Location

ZR.AH..X	
L01	MARGINAL VALUES ØF PRØDUCTS
L02	UNITS - $/T

Figure 11.26 Report definition for product marginal values. A footnote (TRAIL) is included to show units of marginal values. ADHOC knows that the mask B**. . .** refers to rows and not columns because of the SOLTYPE P (PI value of row).

```
┌─────────────────────────────────────────────────────────────────┐
│  MARGINAL VALUE OF PRØDUCTS                                        │
│                                                                   │
│                           FRANCE      U.K.       U.K.             │
│                           FØS         FAWLEY     MILFØRD    • • •  │
│                                                  HAVEN            │
│                           FF          UF         UM              │
│                                                                   │
│   VIRGIN NAPHTHA     NL   101.22      100.86      99.51          │
│   PREMIUM GASØLINE   GP   130.60      129.81     128.07          │
│   REGULAR GASØLINE   GR   125.90      124.87     123.11          │
│                      •                                           │
│                      •                                           │
│                      •                                           │
│                                                                   │
│                                                                   │
│                           UNITS - $/T                            │
│                                                                   │
└─────────────────────────────────────────────────────────────────┘
```

Figure 11.27   Type of output from report PRVALS.

this is two more than is handled by TPRINT in DBM, PLATOFORM tables being limited to a maximum of two entities in stub and/or head.

A totaling capability is conveniently provided by a minor extension to the decomposition function (Step 2, Section 11.2.2), independent row and column totals being controlled by inserting appropriate matching stub entries in the reduced dictionaries (Step 3, Section 11.2.3). This leads to a control table syntax of the form:

TØTAL	YES	Both column and row totals.
TØTAL	STUB	Just column totals (i.e. a stub of the report named TOTALS).
TØTAL	HEAD	Just row totals (i.e. a head of the report named TOTALS).
TØTAL	NØ	No totals – also the default if not specified.

Subtotals in a report require a more major extension to the supporting code and are discussed in Section 11.6. Three other areas in which additional capabilities can be provided with only minor changes to the underlying code deserve special consideration.

11.4.1 Reporting Zero Values

Typical LP problems have many more columns than rows, frequently by a factor of three. This implies that the majority of column activities have a value of zero in the optimal solution (although some of these nonbasic columns may be at a nonzero bound). Some reports, then, will consume a considerable amount of paper with very little to show for it, so that a general report writer must provide an option to allow these zero values to be suppressed or retained when the report is defined. A suitable control table syntax can be built on the lines already discussed for TOTALS:

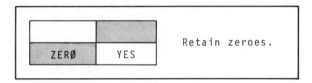

with zero suppression being the default. The zero values can be conveniently removed (by a DATAFORM subroutine) after the initial list has been extracted in Step 1.

If we look at the list for our original crude run report example (Figure 11.6), we can see that this would involve the removal of the element PAL.D.UF from this list. Note that, in this instance, this would have no impact on our final report (Figure 11.2), since there remains some nonzero activity for crude AL and for location UF after the list is decomposed. If the column PAL.D.UF had been the only instance of processing crude AL, then the AL row would not have appeared in the final report.

This leads to another problem with such two-dimensional reports. If we look carefully at the multisegment head report of Figure 11.14 in which the "dots" imply zero values, there is no way to see that the Arabian Light crude run on PSTLL1 at Fawley came out with a zero value in the optimal solution. The "dot" in the report looks no different from the "dot" referring to the Arabian Light crude run on PSTLL1 at Milford. But this did not have a zero value in the solution: the option to process Arabian Light at Milford did not even exist in the matrix. It is often useful to distinguish these two cases by "marking" the genuine zero values so that they print with a "0.0" rather than a ".". This can be achieved conveniently by substituting a very small number for zero-valued list elements. A suitable control-table syntax is:

ZR.AH..Z	CRUNLØ
ZERØ	MARK

11.4.2 Multiple Masks

To be useful, a general report writer must allow the user to extract values from any subset of solution vectors. This implies provision in the syntax to specify any number of matching masks to represent both selection and exclusion. Figure 11.28 illustrates a catalytic reforming process report in which the required vector names have the structure shown in Figure 11.29.

In this example, the four-character keyword DMSK defines a "mask-specific function," since it is linked by further matching characters to one particular MASK. We see further examples of mask-specific functions that are a necessary consequence of the introduction of multiple masks. This extension requires a simple LOOP (over all MASK**** stubs) to be introduced in Step 1, calling SOLTAB each time and appending the resulting lists together.

11.4.3 Scaling

Our examples thus far have reported numerical values in the same units in which they occur in the model solution. For the crude run reports, we assumed that these units were thousands of tons of crude oil per planning period (KT/P). To report them in some other units, a scaling capability must be provided. In its simplest form, this capability must permit each value in the final list to be multiplied or divided (but not necessarily both) by a constant value. Thus, if we want to see our crude runs reported in tons/period instead of K tons/period, we need to multiply each solution value by 1,000 (or divide it by 0.001). A suitable syntax, based on the more frequently occurring "division" scaling, is illustrated in Figure 11.30.

In the program, this scaling can easily be accommodated in Step 1 by a simple scaling of all values extracted by the SOLTABs. Note, however, that we must first convert the alphanumeric representation of the scale factor (as stored in a DATAFORM Z-table) to its numeric equivalent. A simple FORTRAN function can be used to achieve this.

The keyword SCALE (five characters) implies that this scaling function is to apply to values defined by all masks. Let us now consider an example that illustrates a need to provide, in addition, a mask-specific scaling function whose keyword would need to be four characters.

A report that is required very frequently is one that will immediately highlight any instances where market sales requirements are not being fully satisfied by

ZR.AH..Z	REFØRM	
TITLE	L06	Note that the 'delete mask'
TRAIL	L07	DMSKC is linked to the
MASKA	P***P***	specific MASKC and, hence,
MASKB	P***Q***	will not delete any PA, QA
MASKC	P***R***	units.
DMSKC	*****A**	
SPLTMASK	000000LØ	Column totals will be
STUBMASK	0PRO0000	calculated and zero-valued
HEADMASK	0000UN00	solution vectors 'marked'.
ZERØ	MARK	
TØTAL	STUB	

ZR.AH..X		
•		
•		
•		
L06	FEED TØ REFØRMING UNITS - KT	
L07	EXCLUDES ARØMATICS UNIT (RA)	

Figure 11.28   Catalytic reforming process report that excludes a specific process unit.

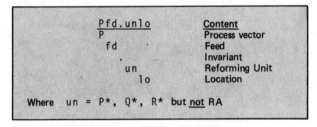

Pfd.unlo	Content
P	Process vector
fd	Feed
.	Invariant
un	Reforming Unit
lo	Location

Where un = P*, Q*, R* but **not** RA

Figure 11.29   Structure of catalytic reforming process vectors.

refinery production. In a large model where there can be many such market "demands," this can involve a very tedious search through the printed solution, with the strong possibility that a particular "shortfall" may be overlooked. Based on a typical PLATOFORM matrix structure (see Chapter 9), such shortfalls are characterized by a nonzero difference between the upper bound and the optimal activity on a demand column vector (D*******).

Figure 11.31 illustrates how mask-specific scaling (as well as mask-specific, solution-type definition) can provide a very simple syntax for specifying such a report.

These examples have been based on scaling all members of a list (or any

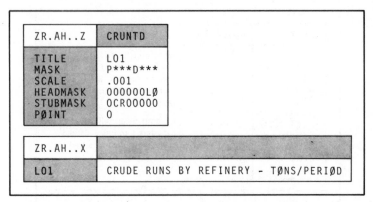

Figure 11.30   Crude run report using scaling to convert values to tons/period. The POINT 0 will cause the values to be printed as rounded integers.

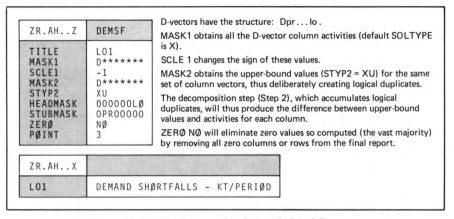

Figure 11.31   Definition of a demand shortfall report.

mask-specific subset) by a constant, known, scaling factor. Suppose, now, that we would like to print our crude runs not in tons per period, but in tons per day. Clearly we need to divide the values by a second factor, namely the number of days in the period. But how can we know this number ahead of time? It is a function of the particular case we want to report on at run time. CASE MPC11M1, as used in previous examples, might refer to a planning period of one month, say 31 days, whereas another case, MPC11Y2, might refer to a full year of 365 days. What number do we put in our report definition table? The answer, of course, is that we put no number. This is an example of data that is case specific (just as is the case name) and hence must be given at input time. But in our report definition we must indicate that such a scaling is required for a particular report. This can be done conveniently by using any nonnumeric string of characters in place of the numeric value. This same character string can be cross-referenced at input time to provide the appropriate numeric value (see Figure 11.32).

A more serious difficulty arises if we want to report our crude runs in volumetric units (e.g., barrels of crude oil per day) instead of the weight units used in the model. The problem, here, is that we cannot specify a single common scale factor, since each column activity must be scaled by the specific gravity (tons per cubic meter) of the particular crude oil represented by that column (i.e., as indicated by the contents of character positions 2, 3 of the column name). We refer to this type of scaling as "variable scaling," since the factors used vary in accordance with a particular entity occurring in a known position of the vector name.

To avoid having to provide all possible scale factors [e.g., crude oil specific gravities (SG)] either at run time or in the body of the report-definition table, we must allow our generalized report writer to access any numeric PLATOFORM data table in which the required scale factors might be stored as a column. In most cases, these factors will already exist as part of the data base, since they have probably been required as input to the matrix generator. We have already noted in previous chapters how DATAFORM's fundamental data-storage

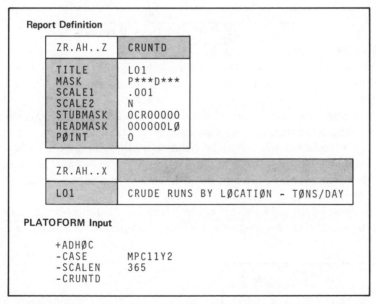

Figure 11.32   Crude run report in tons per day. SCALE2 N causes the scale value to be picked up from input keyword SCALEN (number of days in period associated with CASE MPC11Y2). SCALE1 converts values from KT to T.

structure enables the input data tables and output solution values to be readily integrated, since both are components of the same file.

Let us return to our crude run example and assume that crude quality data is already available in a data-base table as illustrated in Figure 11.33.

Three pieces of information are required to define how data from this table is to be used as scale factors:

1. Character positions in the vector name   0 XX00000 (DATAFORM mask)
to be used as a lookup
2. Name of the table   QC . . . . . T
3. Column of the table   S G

It remains to design some syntax that will allow multiple uses of such variable scaling within a report. We must also, this time, allow the factors to be used in both a multiplication and division mode, since we are aiming to use data that already exists (and is thus unlikely to exist in its reciprocal form). Such a syntax is illustrated in Figure 11.34, which shows the definition of a crude run report converting the solution values from kilotons per period (KT/P) to barrels per day (B/D) according to the following scaling:

$$\frac{B}{D} = \frac{KT}{P} \times \frac{1000}{SG \times m^3/B \times \text{days in period}}$$

where m^3/B = cubic meters per barrel.

Incorporation of variable scaling can again be done in Step 1, but requires a rather more complex routine than is required for simple scaling by a known constant.

At this point, we have provided the three basic scaling modes necessary to allow most solution reports to be defined in a general, data-driven manner. But many report requirements cannot be handled by these methods. To be useful, many extensions to these basic scaling methods are necessary, particularly in connection with variable scaling, but it is outside the scope of this book to discuss these in detail.

All of these additional features provide tools that the report designer can use in various ways to produce reports of varying complexity. In building a general

QC . . . . . T	SU	SG
AL	1.8	.858
AH	2.1	.888
AM	1.9	.872
KU	1.9	.865

Figure 11.33   Crude quality data: SU = wt% sulfur, SG = specific gravity (metric tons/cubic meter).

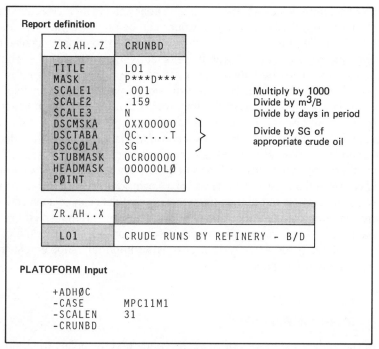

Figure 11.34   Crude run report in barrels per day. Note that the stubs DSCMSKA, DSCTABA, and DSCCOLA form a linked triplet to define the variable scaling. Mnemonic linking is provided by characters 1, 7, and 8 (here D, A, and blank). Character 1 is also used to indicate dividing scales rather than multiplying scales. The characters SCMSK, SCTAB, and SCCOL are keywords defining the three pieces of information required. The table QC. . . . .T is located via the Contents Catalog (CONT. . .C) of the ACTPROB on which the solution case is recorded. Notice the three types of scale factors: 1,000— constant; m³/B—constant (0.159 cubic meters per barrel); days in period—case-specific (31); SG—entity-specific (specific gravity of each crude oil).

report writer, it is important that each new enhancement not be tied to a specific application, but increase the general capability of the overall tool.

Thus far, we have considered extensions that can be included within the framework of the basic steps described in Section 11.2. We want, now, to go on to consider some fundamental extensions in scope.

11.5 EXTENDING THE SCOPE OF ADHOC

By building on the basic steps, we have seen how to provide some of the fundamental capabilities required to design solution reports. Addition and subtraction of solution values is provided by the ability to create logical duplicates; multiplication and division of solution values can be provided via extensions to the variable scaling function. Any part of a recorded solution is available for reporting, and rows or columns can be selected with any degree of sophistication. Scaling of values by known constants can be provided on an absolute, case-specific or an entity-specific basis. Provision is made for handling totals and zeros, and the printed format of the final report can be controlled. Space has not allowed us to discuss all of these enhancements, for example, the provision of a capability to allow incrementing by a constant so that values such as degrees Celsius can be converted to degrees Fahrenheit.

All of these enhancements enable us to design reports of increasing complexity, but we are still bound to the analysis of a single solution case. What we want to look at, now, is how to broaden the scope of what data can be analyzed, such as more than one concurrent solution case or items that are not solution cases. This leads us into new areas, where we see that our ADHOC module can be used not only as a report writer but also as an important supplement to the data-management function.

11.5.1 Multicase Reporting

One of the earliest extensions of ADHOC concerned the ability to combine in a single report solution values extracted from more than one solution case. This need arose from the practice of producing plans for several time periods during a single planning cycle and reporting the optimal values "side-by-side." (Alternatively, it might be useful to compare the January plan as predicted this month with the January plan as predicted last month, etc.)

Through the Solution Catalog (SOL. . . .C), PLATOFORM allows solution cases to be retained on the data-base ACTFILE for as long as they are needed and so long as they are named uniquely. What is needed, then, is the ability to allow ADHOC to access data from any number of solution cases in such a way as to be able to identify the case source of the data by inspection of the eight-character row or column names.

To illustrate this problem, let us return to our simple crude run example and assume that a second case MPC11M2 exists on the same ACTPROB MPC11. Figure 11.35 illustrates part of this optimal solution.

Now, let us suppose that we want to take values from this case and combine them with values from our first case, MPC11M1 (see Figure 11.1), to produce the report illustrated in Figure 11.36.

It is not difficult to develop an input syntax that provides the necessary data for such a multicase report. Figure 11.37 illustrates such a syntax.

The problems begin when we consider how we might integrate these ideas into our basic report-writing steps. The obvious approach would be to enclose

	Column Name	Activity (X)	Reduced Cost (D)
1	PAH.D.UF	4.5	-
2	PAH.D2UF	-	3.0
3	PAL.D.UF	2.7	-
4	PAL.D2UF	-	4.2
5	PAM.D.UM	3.0	-
6	PKU.D3FF	1.8	-
7	PAL.D3FF	-	1.5

Figure 11.35 Part of optimal solution for second case MPC11M2 on ACTPROB MPC11.

```
CRUDE RUNS AT EACH REFINERY BY CASE        UNITS   T/DAY


                                 CASE         CASE
                                 MPC11M1      MPC11M2
                                 JAN.82       FEB.82
FØS        ARAB.LT   FFAL           .            0.
           KUWAIT    FFKU         48.           64.
FAWLEY     ARAB.LT   UFAL          6.           96.
           ARAB.HY   UFAH        113.          161.
MILFØRD    ARAB.MD   UMAM         98.          107.
*TØTALS*                         265.          429.
```

    Solution values are scaled by
        a) .001
        b) number of days in period (31 for 1st case,
           28 for 2nd case)
    to convert to units of T/day.

Figure 11.36 A multicase report set up as a multisegment stub report (LOCATIONS/ CRUDES).

Step 1 (extraction of solution values) in an outer loop that would address each case in turn and append the resulting "lists" together. But this, of course, would lead to duplicate names for list members whose values would be accumulated in Step 2 (decomposition of the list). We must somehow tag the eight-character names to indicate from which case they were extracted. A single character tag (e.g., A for the first case, and B for the second) would suffice, but where can we put this tag?

The problem is that any of the eight character positions in a row or column name may be significant for any particular report, so we cannot preempt any one of them. The solution, however, lies in a corollary of the above statement, namely, that it is very unlikely that all eight characters will be significant for any

```
+ADHØC                                          Up to six cases can be specified.
-CASE       MPC11M1      MPC11M2               -TEXT provides 8-character translation
-SCALEN     31           28                     of the CASE name.
-TEXT       JAN.82       FEB.82                 -SCALE provides appropriate case-
-CRUNMC                                          specific scale values.
                                                The report name is CRUNMC.
```

Figure 11.37   A possible input syntax for specifying a multicase report.

one report. In our crude run report (Figure 11.36), for example, only character positions 2, 3 (crudes) and 7, 8 (locations) are significant—leaving any of the character positions 1, 4, 5, or 6 available for a case tag. Thus we can solve the problem by allowing the character position of the case tag to be report-specific. This implies that the report designer must select an appropriate free character position in which to put the case tag for a multicase report and that syntax must be provided to enable this to be done when the report is defined in table ZR.AH..Z.

Such a syntax is illustrated in Figure 11.38, which shows a complete definition of the report CRUNMC.

By introducing the entity "case" into the list member names, cumulative reports can be designed by deliberately ignoring this entity (and hence creating logical duplicates), and delta reports can be defined by again ignoring the entity and using a negative case-specific scale factor for one of the cases (see Figure 11.39).

ZR.AH..Z	CRUNMC	
TITLE	L01	
MASK	P***D***	The nonblank CASEMASK entry defines
CASEMASK	OXXXXXXX	the report as multicase.
SCALE	.001	The zero in the case mask gives the
SCALE2	N	character position available for a case tag.
STUBMASK	000000LØ	The @ in the HEADMASK references an
STUBMSK2	OCR00000	internally constructed CASE dictionary
HEADMASK	@0000000	to translate the heads.
TØTAL	STUB	
ZERØ	MARK	
PØINT	0	

Figure 11.38   Definition of a multicase report.

```
ZR.AH..Z      CRUDELT

TITLE         L01
MASK          P***D***
CASEMASK      OXXXXXXX
SCALE         N
HEADMASK      000000LØ
STUBMASK      OCR00000
ZERØ          NØ


ZR.AH..X

L01           DELTA CRUDE RUNS BY LØCATIØN - KT/D


+ADHØC
-CASE         MPC10M1       MPC11M1
-SCALEN         -31            31
-CRUDELT
```

Figure 11.39   Crude run report comparing January as forecast this month with January as forecast last month. Note negative scale factor for case MPC10M1 (last month). Entity "case" (@) not referenced in HEAD or STUB segments.

11.5.2 Accessing Data Tables

The next major step in expanding the scope of the ADHOC report writer resulted in some reorientation of the prime purpose of this PLATOFORM module. Since PLATOFORM, through its powerful and instantly available data-management functions (DBM), offered a convenient repository for all kinds of data in tabular form, a number of applications took advantage of this to store data, not all of which was destined to find its way into a mathematical programming matrix. Frequently, this data would be precursors to further data that would then be accessible to a matrix generator. This implied some manipulation of the primary data, such as aggregation into predefined categories, restructuring of the tabular arrays, and combining data from different tables.

Could ADHOC, with its already comprehensive entity-manipulating capabilities, assist in this process? There seemed to be two major problems to solve: first, how to extract data from data tables in a form compatible with ADHOC's eight-character list and, second, how to structure the printed report into a physical table that could be filed back on the ACTFILE. The second of these problems presented few difficulties, since the table used by TABULATE to print the report was already structured in a form compatible with the PLATOFORM data structure. Basically, all that was required was a syntax mechanism for providing a name (or names in the case of a "split" report) for the output table(s).

The first problem was much more fundamental. Solution values are associated with a single matrix vector and, hence, can be fully described by eight characters. Only when we combine solution values for more than one case does this eight-character limitation become restricting, and in the last section (11.5.1) we saw how this required a special technique in order to handle multicase reports. When we consider an element in a data table, we can see that the problem is far more severe. Each such element can be qualified by attributes that could require a maximum of twenty-three characters for their full description: table name, seven characters (remember that the last character of a numeric PLATOFORM table (T) can never be used to describe an entity); plus stub name, eight characters; plus head name, eight characters. If we are to extend these ideas to multi-ACTPROBs, then a further character will be required to identify the ACTPROB source of the data, in the same way as for multicase solution reports.

Accordingly, we are faced with a possible twenty-four-character description of a data table item that must somehow be condensed into eight characters. Failure to do this would imply a complete redesign of Steps 2 through 4 and their associated packaged tools (FORTRAN functions, subroutines, etc.).

Here, there seemed to be no answer other than to permit data-table processing only when the required qualifying entities could be described in a maximum of eight characters. Subsequent experience showed that this restriction was not limiting in the vast majority of data-table manipulations. As with case tagging, syntax had to be provided to allow the user to specify which characters, out of the potential twenty-four, were to be selected to identify any particular set of data items. A natural choice was to use "positional indicators," but to be usable, these would need to have some mnemonic significance. Fortunately, a natural choice of characters also proved to have mnemonic significance, as illustrated in Figure 11.40.

To see how these KEY characters can be used to define a report, let us see

1, 2, 3, 4, 5, 6, 7, 8	—	Character positions within table names
A, B, C, D, E, F, G, H	—	Character positions within head names
S, T, U, V, W, X, Y, Z	—	Character positions within stub names

Figure 11.40 Positional indicators for selection of characters to describe data items comprised of: the first eight integers, the first eight characters of the alphabet, and the last eight characters of the alphabet. Note that the letters of the name HEAD are totally contained within the first eight characters of the alphabet, and the name STUB starts with the same first three characters as the last eight characters of the alphabet.

how we might use ADHOC to produce the XPRINT illustrated in Figure 7.10. Here, we are extracting the sulfur qualities for each product stream from a set of QB tables relating to various locations. Figure 11.41 illustrates the definition of a report that would achieve essentially the same result as the standard DBM XPRINT. The output report will resemble the one shown in Figure 7.10.

Unlike its DBM counterpart, ADHOC can file back on the data base the tables formed by such reports. Thus if we want to restructure our component quality data so as to have one table for each quality (rather than one table for each location), this can be achieved by the report definition shown in Figure 11.42. Here, the newly structured tables are in a set defined as QQ tables, which must have been previously cataloged in table GENTAB.C.

It is important to notice that, in this context, ADHOC is being used to supplement the data-management function and in no way is it being used as a

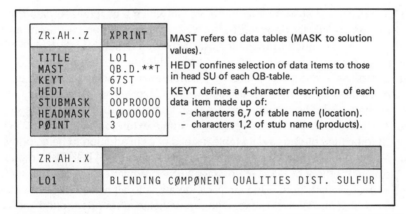

ZR.AH..Z	XPRINT	MAST refers to data tables (MASK to solution values).
TITLE	L01	
MAST	QB.D.**T	HEDT confines selection of data items to those in head SU of each QB-table.
KEYT	67ST	
HEDT	SU	KEYT defines a 4-character description of each data item made up of:
STUBMASK	00PR0000	
HEADMASK	LØ000000	– characters 6,7 of table name (location).
PØINT	3	– characters 1,2 of stub name (products).

ZR.AH..X	
L01	BLENDING CØMPØNENT QUALITIES DIST. SULFUR

Figure 11.41   Definition of a report to parallel the function of XPRINT in PLATOFORM's DBM.

ZR.AH..Z	REFØRMAT	SPLTMASK produces one table for each quality.
MAST	QB.D.**T	
KEYT	67STAB	FILETAB gives table name for filed tables with SPLIT code being substituted for **s.
SPLTMASK	0000QU00	
STUBMASK	00PR0000	
HEADMASK	LØ000000	PØINT NØ suppresses all printing and dictionary processing for the report.
FILETAB	QQ.D.**T	
PØINT	NØ	

Figure 11.42   Definition of report to restructure quality data by quality type from data organized by location.

report writer. The output tables that are filed back on the ACTFILE as a result of executing "report" REFORMAT, based on the input tables illustrated in Figure 7.10, are shown in Figure 11.43.

This type of mechanism provides a very powerful tool for "bridging" data between quite different data bases—even if only one of these is a PLATOFORM data base. If, for instance, we want to pass data from an IMS data base to a PLATOFORM data base, the data can be communicated via a sequential file readable by PLATOFORM (e.g., formatted either as input to PLATORD or as output from PLATORD). The structure of the data on this file can match the IMS data-base structure and can then be restructured via a set of ADHOC reports. This greatly simplifies the bridging programs.

11.5.3 Other Extensions

We have discussed at some length the way in which a generalized data-driven report writer, originally designed to analyze a single solution case from an LP model, can be extended to multiple cases and even data tables. We have seen how these extensions can be incorporated within the basic framework of the original module without reconstituting this framework. In each case, the extensions result in an eight-character list at the end of Step 1. The sources from which the members of this list are derived are then immaterial to the remaining steps.

At the other end, we have seen how a simple extension to Step 4, allowing a report to be filed back as a data table instead of or in addition to being printed, adds a new dimension to the fundamental uses of this PLATOFORM module.

Pursuing this thought, we can envisage further, similar extensions that we do not discuss at length here. One clear candidate is to allow ADHOC to access matrix coefficients. With this capability, mass-balance reports around sections of the LP model (for example, process units) become possible, even when the

QQ.D.SUT	UF	UM
VH	1	1.1
VD	1.7	1.72
CD	1.5	
CF		1.55

QQ.D.V1T	UF	UM
VH	2.874	2.8
VD	7.42	7
CD	5.63	
CF		5.9

Figure 11.43 Output tables filed by report REFORMAT, based on input tables shown in Figure 7.10.

corresponding data in the data tables is no longer valid. An example of this use of ADHOC is given in Chapter 12.

A similar extension in Step 4 can allow a report to be filed out as a piece of a matrix, thus effectively providing an ad hoc matrix generator. Within the Enhanced Mathematical Programming System (EMPS)/DATAFORM environment, these extensions present no implementation problems.

Before leaving the PLATOFORM ADHOC module, we should consider one more important aspect of the report-writing process that applies equally to data derived from LP solutions or from data-base tables. This concerns the aggregation of individual entities into categories.

11.6 MAPPING, AGGREGATION, AND SUBTOTALING

No general-purpose report writer would be complete without some means of rearranging mnemonics in order to build a consistent naming structure. Such rearrangement must include the substitution of literal characters into a name as well as the shifting of existing characters into new positions within the name. We refer to such operations as a mapping process. As an example, we can refer to a "mapping" of the eight-character string PAH.D.UF into the eight-character string AAH.UFD., a process that involves substitution of the literal character A into the first position of the name and the interchange of characters 5, 6 and 7, 8. This same mapping process could be applied to all members of a list of names.

A syntax with which to specify such a mapping process can be designed by using a mixture of "positional indicators" and literal characters. Thus if we use the characters 1 through 8 to indicate position within the name and any other character to indicate a literal, then the "maps key" A2347856 would define the mapping process described above. Here, the characters 2347856 are positional indicators specifying a rearrangement of existing characters, and A is a literal to be inserted in position 1 of the name. Of course, to be fully general, we must allow any of the characters 1 through 8 to be used as a literal, but this can be done by allowing an optional redefinition of the eight positional indicators.

As an illustration of how we might use this feature, let us suppose that our LP model contains a single vector XA..DDQS representing the total refining operations at refinery QS, and that this operation includes running a fixed crude oil mix of 80% Light Arabian crude (AL) and 20% Heavy Arabian crude (AH). We want to incorporate these crude runs into our simple crude run report. Figure 11.44 illustrates how we could use the MAPS feature to accommodate such a vector, which lacks the normal structure of a crude oil processing vector.

There are, of course, many other contexts in which such a mapping process can be useful. We could, for instance, use MAPS to calculate our own totals as an alternative to using the TOTAL feature. (See Figure 11.45.)

ZR.AH..Z	CRUNX	The single matrix column XA..DDQS is split into two list elements via MASK2, MASK3.
TITLE	L01	
MASK1	P***D***	
MASK2	XA..DDQS	MAPS2, MAPS3 restructure the name to conform to that of a normal crude-processing vector.
MAPS2	PAH.D.78	
SCLE2	5	
MASK3	XA..DDQS	SCLE2, SCLE3 apportion the value to 20% AH crude and 80% AL crude, respectively.
MAPS3	PAL.D.78	
SCLE3	1.25	
STUBMASK	0CR00000	Also, notice that 'MAPS' is a mask-specific function.
HEADMASK	000000LØ	
PØINT	3	

Figure 11.44   Simple crude run report including special "total process" vector.

ZR.AH..Z	CRUNT	Crude vectors are retrieved twice.
TITLE	L01	In the second set, 'logical duplicates' are created by changing all crude codes to '..' (these are accumulated on decomposition to give totals).
MASK1	P***D***	
MASK2	P***D***	
MAPS2	1..45678	
STUBMASK	0CR00000	Dictionary CRUDES.D requires the entry '..'.
HEADMASK	000000LØ	

Figure 11.45   Crude run report with calculated total crude run.

In this example, the literal ".." is substituted for every crude code occurring in positions 2, 3 of the vector names. Such syntax, however, proves inadequate if we want to apply this mapping process to produce subtotals. In this case, the literal we want to insert is a function of the particular crude code being substituted. If we want to structure our report so as to print subtotals by crude type (e.g., total Arabian crude, total Iranian crude, etc.), then we must select an appropriate literal linked to the actual crude represented by the column vector. Thus AL, AH, AM crude oils might "map" to A. (total Arabian crude oil), and IL, IH to I. (total Iranian crude oil).

These concepts have a very strong analogy to our previous discussion on scaling (Section 11.4.3). While scaling referred to operations on the numerical values, mapping refers to analogous operations on the names of the list members. With scaling, we saw that there was little syntactical difficulty so long as we were dealing with constants, just as mapping of literals presents few problems. But once we needed to handle entity-specific scale factors (i.e., scale values linked to entities within the list member names), then we needed to introduce a new syntax based on "variable scaling." The scale values had to be looked up in a data table.

In the same way, entity-specific mapping, as required, for instance, in subtotaling, calls for a new syntax. Here, in place of the data table, we can conveniently use the PLATOFORM dictionaries as a source from which to extract appropriate "tags" linked to entities.

Let us illustrate these ideas by returning to our crude run report in which we want to report subtotals by crude type as well as individual crudes. We first have to expand our crudes dictionary (CRUDES..D) in order to define the aggregation rules. The way this can be done is illustrated in Figure 11.46. (The original dictionary was illustrated in Figure 11.3.)

Figure 11.47 shows our simple crude run report with subtotals by crude type, based on the "tagged" dictionary of Figure 11.46.

The output from this report definition, based on the solution data given in Figure 11.1, is illustrated in Figure 11.48.

While this example illustrates aggregation on a single stub segment of a report,

CRUDES..D	M1	D1	D2	D3	T1
AL	ARAB.LT	ARABIAN LIGHT			A.
AM	ARAB.MD	ARABIAN MEDIUM			A.
AH	ARAB.HY	ARABIAN HEAVY			A.
A.	TØT.ARAB	**TØTAL ARABIAN			
KU	KUWAIT	KUWAIT			K.
K.	TØT.KUW	**TØTAL KUWAIT			

Figure 11.46 Crudes Dictionary expanded to define subtotals. The additional head T1 is a permissible "tag" column in a PLATOFORM dictionary (any number of T* heads allowed). Stubs A. and K. have been added to represent aggregate sets whose members are defined by matching entries in column T1. A. and K. are positioned as required in an output report.

ZR.AH..Z	CRUNST
TITLE	L01
MASK	P***D***
STUBMASK	0CR00000
STUBAGGR	11S
HEADMASK	000000LØ
PØINT	3
TØTALS	STUB

Code 11S implies:

1 — Aggregation tags are in column T1 of dictionary.

1 — The tags start in character position 1 of this column.

S — Subtotals required (i.e., also report individual entities).

Figure 11.47 Definition of simple crude run report with subtotals by crude type. The keyword STUBAGGR links aggregation to entities defined by STUBMASK, and hence to dictionary CR.

```
CRUDE RUNS BY REFINERY - KT/PERIØD


                              FRANCE     U.K.       U.K.
                              FØS        FAWLEY     MILFØRD
                                                    HAVEN
                              FF         UF         UM

ARABIAN LIGHT     AL     .         0.2         .
ARABIAN MEDIUM    AM     .          .         3.0
ARABIAN HEAVY     AH     .         3.5         .
**TØTAL ARABIAN   A.     .         3.7        3.0
KUWAIT            KU    1.5          .          .
**TØTAL KUWAIT    K.    1.5          .          .
*TØTALS*               1.5         3.7        3.0
```

Figure 11.48   Crude run report with subtotals.

aggregation can, of course, be extended by use of appropriate keywords to any (or all) of the three stub segments, three head segments, or to the split segment.

Incorporation of these aggregation techniques into the module involves a mapping process on the final list produced at the conclusion of Step 1. For subtotaling, the list is duplicated, and the appropriate aggregation code is "mapped" into the copy. Aggregation then occurs naturally on decomposition as a result of the deliberate creation of logical duplicates. Figure 11.49 illustrates our subtotaled crude run report enhanced with underlines.

11.7 SUMMARY

In this chapter, we have tried to show how a comprehensive, data-driven report writer can be built around four or five simple basic steps to provide a single module that will produce an unlimited variety of reports. This one DATAFORM program of about 2,000 program statements currently supports more than 500 separate reports throughout Exxon—reports that, if programmed individually, might require some 50,000 to 100,000 statements. Program maintenance is thereby reduced by a factor of at least 25, a figure that is grossly conservative since specially written reports require much more maintenance than a single general report program.

Apart from the obvious advantages of program maintenance, this approach to report writing allows new reports to be developed very quickly, frequently by the end users. In one instance, a set of ten complex reports was written by a skilled DATAFORM programmer in ten weeks. Today, using ADHOC, these same reports can be defined in just two days; execution time will also be improved.

To be useful, a general report writer must be comprehensive. Early versions

```
CRUDES.D   M1            D1            D2          D3        T1         T2

AL         ARAB.LT       ARABIAN LIGHT                      A.
AM         ARAB.MD       ARABIAN MEDIUM                     A.
AH         ARAB.HY       ARABIAN HEAVY                      A.
A.         TØT.ARAB      **TØTAL ARABIAN                                #
KU         KUWAIT        KUWAIT                             K.
K.         TOT.KUW       **TØTAL KUWAIT                                 #
                         CRUDES DICTIONARY
Position 1 of tag column T2 is used for underline markers


ZR.AH..Z   CRUNST

TITLE      L01                        Report definition:
MASK       P***D***                      ULINE 21 requests underlining as
$TUBMASK   0CR00000                      determined by markers in tag col.
STUBAGGR   11S                           T2, position 1 of the stub
HEADMASK   000000LØ                      dictionary (CR = CRUDES.D)
PØINT      3
TØTALS     STUB
ULINE      21


           CRUDE RUNS BY REFINERY - KT/PERIØD

                           FRANCE      U.K.         U.K.
                           FØS         FAWLEY       MILFØRD
                                                    HAVEN
                           FF          UF           UM

ARABIAN LIGHT      AL        .          0.2          .
ARABIAN MEDIUM     AM        .           .          3.0
ARABIAN HEAVY      AH        .          3.5          .
**TØTAL ARABIAN    A.        .          3.7         3.0
**********************************************************
KUWAIT             KU      1.5           .           .
**TØTAL KUWAIT     K.      1.5  .        .           .
**********************************************************
*TØTALS*                   1.5         3.7         3.0
```

Figure 11.49   Use of underlines in a report.

of the ADHOC module received minimal acceptance from users, since at that stage the module could not handle all of the required reports. Using a mixture of ADHOC and specifically developed reports resulted in an awkward calling syntax and untidy, discontinuous output. On the other hand, making a general report writer fully comprehensive necessarily involves introducing more complex syntax, which, in turn, increases the level of expertise required to define certain reports, although the level of expertise is far below that required to specially write the reports in DATAFORM. Such expertise can readily be imparted to key analysts through regularly held, two-day training seminars. Such seminars can be organized to enable attendees to adjust their attendance to the degree of sophistication they require. Simple reports require very little expertise.

Throughout this chapter, we have tried to stress the fundamental nature of this module. Basically, it provides a set of capabilities or tools that the report designer can then use in any combination to produce the report required. Often, an expert in the use of ADHOC will surprise the designer with a report that the designer had never considered a possible outcome of the ADHOC module. One example is a bar chart showing major European process-unit shutdowns on a weekly basis, derived from numerical data used as input to the capacity sub-module of the matrix generator (see Figure 11.50).

In retrospect, we should ask: What is the minimum set of capabilities required that, when used in combination, will yield any foreseen reporting requirement? Inevitably, ADHOC, today, contains some redundancies. There is often more than one way to define a particular report requirement. Some of these redundancies are justified to balance execution efficiency for large reports against easier definition for smaller reports. However, the minimum set of capabilities must encompass the following:

- Ability to access any part of the total model environment (data base, matrix, solution values, etc.).
- Selection of any subset of this environment via character matching.
- Selection on the basis of numerical values.
- Basic arithmetic functions on these values.
- Comprehensive character manipulation of the names.
- Output format control, including filing back on the data base.

Throughout, we see a parallelism between operations on names (mapping) and operations on values (scaling).

The real challenge in producing a general report writer of this type is not so much in adapting it to the language in which it will be programmed, but, rather, in designing a succinct and self-consistent syntax in which report designers can express their requirements. With ADHOC, we have found that whenever a new feature is requested, how to express this feature syntactically in the control table, not how to program it in DATAFORM, is the overriding consideration in deciding whether (or how) to implement it. The problems of character length, as discussed, for example, under case tagging, are in no way germane to DATA-FORM. Had we used a ten-character implementation language, we would then require eleven characters to include the case tag.

We said at the outset that ADHOC currently handles about 95% of all Exxon's mathematical programming reporting needs. What of the other 5%? There remain, of course, some unsolved difficulties. Perhaps the most severe is the eight-character limitation imposed by ADHOC's fundamental design of the list names. This prohibits certain data-table manipulations. Also, in handling data items the lists can frequently become rather long. Once a list exceeds 2,000

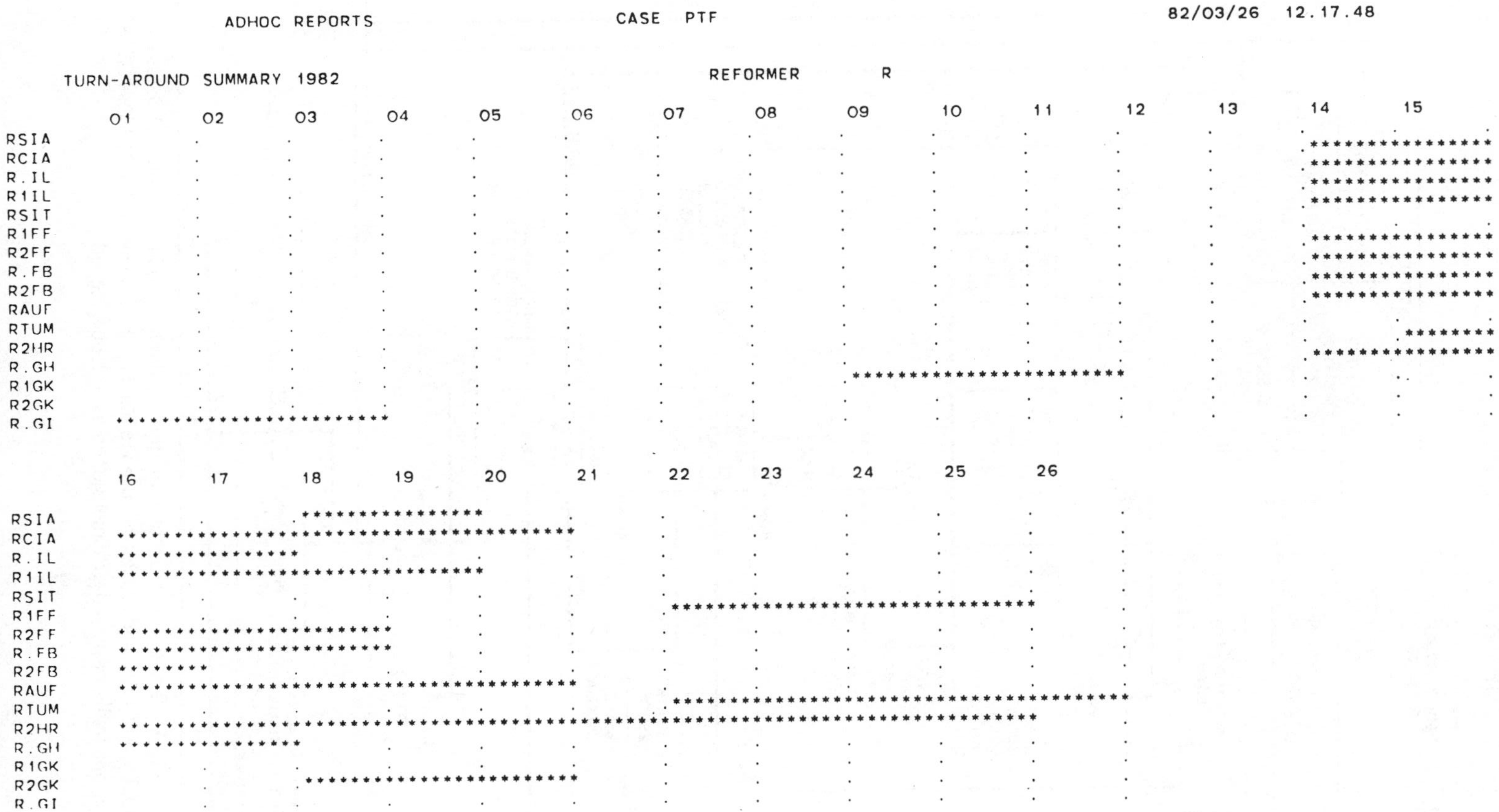

Figure 11.50 Bar chart showing process unit shutdowns over a twenty-six-week period.

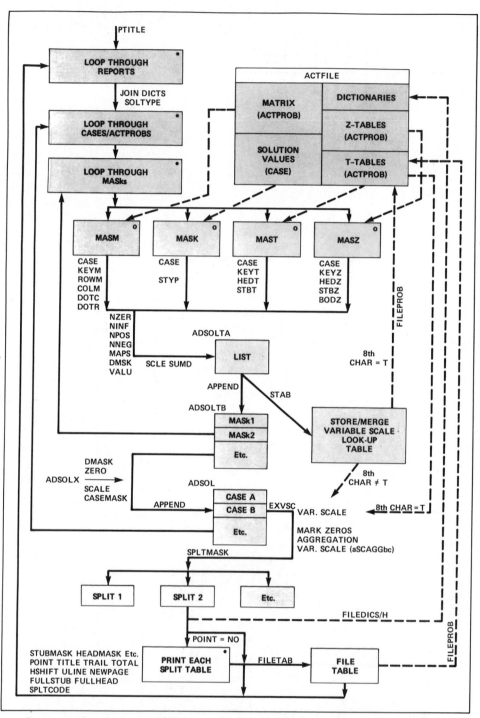

Figure 11.51 System architecture of the ADHOC module. Tables ADSOLTA, ADSOLTB, ADSOLX, and ADSOL are used to build up the final list in Step 1. These tables can be optionally displayed or dimensioned as a debugging aid.

members, the execution efficiency of ADHOC decreases when using aggregation and variable scaling, and more specific programming techniques become preferable. Yet other reports require complex nonlinear conversions in which model units (e.g., blending values) require conversion to more fundamental units, as in pool-blending reports.

Problems such as these can often be resolved by developing other general-purpose modules similar to ADHOC. An example of PLATOFORM's pool-analysis report writer (PANREP) is given in Chapter 12, as well as another general module designed to analyze values from the updated solution basis (TRANREP report). Such a report allows a detailed analysis of the marginal mechanisms, costed out to show their contribution to any selected dual value.

It is possible that the future evolution of ADHOC will see the incorporation of the capabilities represented by such modules as PANREP and TRANREP in the same way that it has already absorbed the capabilities of several earlier modules of a similar nature through carefully controlled evolution.

In following the development of ADHOC, we have seen how a module originally intended for analysis of solution cases has been extended to incorporate the analysis of data stored in tables and matrix coefficients. We have seen how the printed report can be filed as a table on the data base, thus endowing ADHOC with a powerful capability to supplement the functions of the Data-Base Manager (DBM) and, potentially, the matrix generator. All of these extensions have been incorporated without destroying the simple underlying structure of the code or disabling any existing report definitions. We have noted the importance of packaging the basic reporting functions into DATAFORM subroutines or FORTRAN functions, a task that is an essential preliminary to the development of such a general module.

Space has allowed us to discuss only some of the more important capabilities required, but we hope that these have been sufficiently representative to show the required scope of such a module. To complete the picture, Figure 11.51 illustrates the total architecture of ADHOC.

N. Kenneth Boudwin

12. AN EXAMPLE

12.1 INTRODUCTION

The purpose of this chapter is to tie the previous chapters together and to give an idea of how a user would see the PLATOFORM system. This affords a comparison of the design objectives with the characteristics of the present system. As discussed in Chapter 2, these design objectives were:

- A single integrated system.
- A simple, basically nonprocedural, user input.
- A comprehensive facility for model management.
- A modular system that can be easily tailored to support individual users.

Recall that the PLATOFORM system can be thought of as:

- Data-base management.
- Matrix generation.
- Matrix optimization.
- Report writing.

These functions are linked together by a central data base to support the wide range of capabilities desired in using mathematical programming models.

PLATOFORM has greatly expanded the range of linear programming (LP) applications that can be supported within Exxon, basically, for two reasons:

- The amount of time to develop a new application has been greatly reduced.
- The number of LP experts needed to support regular use of an LP model has been reduced.

As a result, we have come a long way from using LP only in transportation models, small blending models, and refinery models, as discussed in Chapter 1.

The evolution of PLATOFORM began with its use in the large supply model in Esso Europe, Inc. From there, the number of applications rapidly increased, and today we support a wide range of applications with PLATOFORM. These include:

- Single refinery and chemical plant models.
- Regional models for the coordination of several refineries or chemical plants.
- Product distribution through intermediate facilities.
- Distribution facility investment.
- Drilling rig scheduling.
- Portfolio investment.
- Blocked operations scheduling.
- Refinery energy optimization.
- Truck fleet sizing.
- Vessel scheduling.

One of the key features of PLATOFORM is the open-ended structure that has enabled it to evolve to support a wide range of applications of mathematical programming techniques. While there is a wide range of LP applications to choose from, we intentionally select a simple abstraction to illustrate PLATOFORM concepts. The selected example is a simplified illustration of the use of PLATOFORM for supply/refinery planning.

12.2   THE SUPPLY PLANNING BUSINESS

To set the stage for our example, it is helpful to briefly describe the supply planning business in Esso Europe at the time when PLATOFORM was originally designed. Supply planning is characterized by data coming from many different sources within the corporation:

- Crude oil supply data for nonlocal crude oils is provided by Exxon Corporation Headquarters in New York City.
- Crude oil properties and assays are determined by our research laboratories and are distributed as part of a library of crude oil assays.
- Product sales data is provided by the marketing organizations in the affiliates throughout Europe.
- Refinery process yields and process unit capacities are provided by the refineries throughout Europe.
- Interaffiliate movements of products are determined by Esso Europe Headquarters in London.

Esso Europe was then responsible for coordinating supply planning through-out Europe. In this context, it was responsible for allocating some fifty different crude oils between seventeen refineries in a number of countries. It was also responsible for coordinating moves of intermediate or final products between the different countries. While Esso Europe was not responsible for detailed operating plans of the refineries, it had to have sufficient data on refinery operating capabilities to allow optimal decisions to be made on crude allocations and intercountry product movements.

Figure 12.1 presents a greatly simplified view of a typical European medium-term planning cycle as it was at the time of PLATOFORM's original develop-ment:

1. The refineries communicated their processing capabilities and capacities to their local supply department.
2. The supply departments transmitted their data to Esso Europe in London, including refinery capabilities and local product demands.
3. Exxon Corporation advised Esso Europe of the crude oils that will be available from outside Europe, the amounts, and the prices.
4. All of this data became part of the PLATOFORM data base and was used to build a European regional supply-coordination model. Based on the results of the model runs, Esso Europe advised the affiliates on forecast of:

   Affiliate crude allocations.

   Interaffiliate product movements.

   Gross operating guidelines.

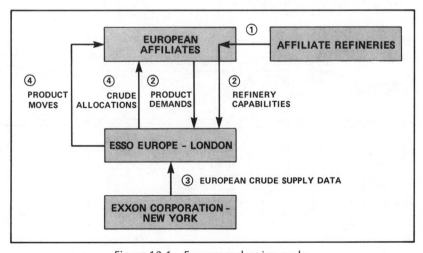

Figure 12.1   European planning cycle.

The PLATOFORM system must perform several important functions to support such a planning cycle:

- It must provide mechanisms for data input from many different sources.
- It must map the many types of data into a formal mathematical model (the LP).
- It must provide capability to optimize the LP, including capabilities to do side studies as well as parametric analysis.
- It must provide the user with report-writing capabilities.

In the next section, we take a closer look at the composition of this data, and in the remaining sections we consider how a PLATOFORM model would be built using this data.

12.3 SUPPLY PLANNING DATA

To demonstrate the use of PLATOFORM, we have selected a simple refinery representation. This representation does not show all the detail of either a full refinery model or a regional supply model, but it will be sufficiently complex to illustrate the basic capabilities and use of PLATOFORM. A flow diagram of the refinery we are going to model is given in Figure 12.2.

The data elements necessary to model this simple refinery are: crude oil supply data, product demand data, and refinery capabilities data.

There are three important aspects of supply-planning data with regard to crude oils. The first is, very simply, what crude oils are available. Following that, it is necessary to know how much of each crude is available and its price. Example data is shown in Figure 12.3.

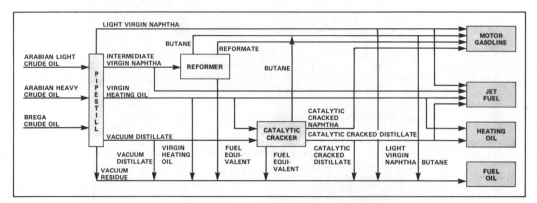

Figure 12.2 Example refinery flow diagram. The reformer and catalytic cracker are specifically designed to improve the refinery's ability to produce motor gasoline.

	Crude Availability			Crude FOB Cost	Cost of Transportation to Refinery
	Thousands of Tons			$ Per Barrel	$ Per Ton
	January	February	March	All Months Under Study	All Months Under Study
Arabian light	110	105	115	35	24.15
Arabian heavy	165	155	170	34	24.15
Brega	80	70	80	42	10.05

Figure 12.3 Crude supply data.

Figure 12.3 raises an immediate question. Crude availabilities and transportation costs are given on a weight basis (tons), but crude oil FOB (Free on Board) prices are given in $/barrel, barrel being a volumetric unit. Therefore, it becomes necessary to convert crude oil FOB prices into $/ton. To do this, crude-oil specific gravities are needed. These are provided in Figure 12.4.

Product demand data is very similar to crude supply data in that the important items are the products to be sold, the amounts of each product to be sold, and the revenues realized from product sales. Figure 12.5 shows example product demand data.

Data specifying refinery capabilities includes refinery throughput capacities, refinery operating costs, and products produced by the refinery. The data used

Arabian light	.858
Arabian heavy	.886
Brega	.823

Figure 12.4 Crude oil specific gravities.

	Product Demands Thousands of Tons			Product Revenues $ Per Ton
	Jan.	Feb.	Mar.	All Months Under Study
Motor gasoline	40	35	45	430
Jet fuel	20	20	20	300
Heating oil	50	50	40	315
Fuel oil	145	130	145	250

Fuel oil can also be imported from an affiliated refinery at a cost of $245 per ton in all time periods. It is assumed to satisfy fuel oil specifications.

All demands for products must be satisfied fully.

Figure 12.5 Product demand data.

here (as elsewhere) is entirely fictitious, but it is representative of the data needed to model a refinery.

We consider first the case of refinery throughput capacities and operating costs (see Figure 12.6), then turn our attention to the issue of products produced by the refinery.

The refinery data concerning products is more complicated. Our discussion centers, in turn, on each of the major process units (the pipestill, the reformer, and the catalytic cracker) and on the blending of finished products from the streams produced by the process units.

The pipestill splits crude oil into various streams via a distillation process. Ordinarily, a refinery has both an atmospheric pipestill (which distills the crude oil) and a vacuum pipestill (which distills the residue from the atmospheric pipestill). In our simplified case, these are combined into one unit, which means that the refinery does not have the option of allowing the atmospheric pipestill residue to bypass the vacuum pipestill. The pipestill data indicates the relative amounts of each stream produced from each of the crude oils. Example data is given in Figure 12.7.

The reformer is a process unit that takes streams ordinarily blended to make

	Throughput Capacity KB/D	Operating Cost $/B of Feed	Capacities and costs are given on a per-barrel basis.
Pipestill	65.0	0.15	Effective daily capacity will be decreased in January due to the planned shutdown of the catalytic cracker and reformer.
Reformer	7.5	0.60	
Catalytic cracker	8.0	0.65	

Figure 12.6   Refinery capacity and cost data.

	Arabian Light Crude Oil	Arabian Heavy Crude Oil	Brega Crude Oil	
Light virgin naphtha	3.5	3.0	4.5	Ordinarily, these yields are derived from basic crude-oil-assay data using PLATOFORM's capability to bridge to other data bases.
Intermediate virgin naphtha	10.0	7.5	13.5	
Virgin heating oil	39.0	30.0	43.0	The user then simply specifies the boiling ranges and desired properties for each stream.
Vacuum distillate	28.5	23.0	28.0	
Vacuum residue	16.5	33.5	10.0	The bridge calculates the yields.
Fuel consumption and loss	2.5	3.0	1.0	

Figure 12.7   Pipestill yields as a percentage of crude oil feed.

jet fuel or motor gasoline and, under the influence of a catalyst, partially converts them to a stream that is more suited for gasoline blending. This increases the refinery's ability to manufacture gasoline suitable for use in automobiles. Figure 12.8 shows the effect of the reformer.

The catalytic cracker is similar in purpose to the reformer. However, it accepts heavier feed streams that are generally blended into heating oil and fuel oil. Again, a catalyst is used to produce a chemical reaction, yielding a stream that is suitable for gasoline blending. Our modeling of the catalytic cracker is somewhat more complicated since, as shown in Figure 12.2, two different streams can be used as feeds to the unit. Additional complexity is introduced by allowing the unit to be operated in two modes:

- A low-severity mode, producing a stream that is less blendable to gasoline.
- A high-severity mode, producing a stream that is more blendable to gasoline.

Figure 12.9 shows the results of catalytic cracker operations.

The next area to consider is the blending of finished products. In our example, heating oil is the simplest product to blend. Either virgin heating oil or catalytic-cracked distillate can be used directly to produce heating oil suitable for sale.

Jet fuel has a more complex manufacturing representation in our example. Because of its use as an aviation fuel, it must meet special performance specifications. In our simple refinery example, we assume that there are several basic

Butane	2
Reformate	90
Fuel equivalent	8

Figure 12.8   Reformer yields as a percent of intermediate virgin naphtha feed.

	Virgin Heating Oil Feed		Vacuum Distillate Feed	
	Low Severity	High Severity	Low Severity	High Severity
Butane	2.0	3.25	5.0	6
Catalytic-cracked naphtha	27.5	37.75	32.5	45
Catalytic-cracked distillate	68.0	55.50	58.5	44
Fuel equivalent	2.5	3.50	4.0	5

Figure 12.9   Catalytic cracker yields as a percent of feed.

recipes for blending jet fuel that meet these specifications. The recipes are given in Figure 12.10.

Gasoline and fuel oil blending are more complex than heating oil and jet fuel blending in our example. Because of the wide variety of blends that can make these products, it is difficult to specify a comprehensive set of recipes to span the alternatives. In addition, the product specifications may change over time, necessitating corresponding adjustments to the recipes. Therefore, instead of a simple recipe approach, a specification approach is used, in which the blend qualities are directly calculated to assure that they meet specifications. This requires that the data describe not only blend specifications, but also how each component of the blend affects the blend's quality. In our example, gasoline is blended to three quality specifications:

- A Research Octane Number, which is a measure of its antiknock properties in a laboratory engine.
- A Reid vapor pressure (RVP) specification, which is a measure of its tendency to produce evaporative emissions.
- A percent distilled at 100°C (212°F), which is a measure of its volatility at engine temperatures.

Figure 12.11 shows the gasoline blend specifications and the properties of the blend components for the qualities specified.

Since gasoline qualities are volume dependent, it is also necessary to know the specific gravities of gasoline components in order to calculate the volumetric contribution of a component to the blend. The pool volume is also an important factor, because the cost of adding lead depends on the pool volume. Thus a blend with a lower specific gravity will have a greater volume, and the cost of lead will be higher. The component specific gravities are given in Figure 12.12.

For fuel oil blending, the important product specifications are:

- Sulfur content of the fuel, because of environmental concerns.
- Viscosity of the fuel, which is a measure of how well the fuel flows.

	Recipe 1	Recipe 2	Recipe 3
Light virgin naphtha		20%	
Intermediate virgin naphtha	30%		10%
Virgin heating oil	70%	80%	10%
Catalytic-cracked distillate			80%

Figure 12.10   Recipes for producing jet fuel. Any combination of these recipes will also produce jet fuel that meets the specifications.

	RVP	% Distilled at 100°C	Research Octane	
Butane	75	100	101.61	Gasoline is assumed to be blended with 0.4 gram per liter lead content and all octanes are given for this lead level.
Light virgin naphtha	12	95	86.3	The lead used in gasolines has a cost of $7.80 per kilogram.
Reformate	6	35	102.43	RVP and % distilled at 100°C are linear functions of the volume of the components used.
Catalytic-cracked naphtha:				
Low severity	7	60	94.94	Research octane cannot be modeled directly, but requires
High severity	9	64	99.05	conversion to a research blending number (using Exxon proprietary
Gasoline-blend specifications:				technology). The research blending
All months under study	10 Minimum	60 Maximum	99.37 Minimum	number is a linear function of the volume of components used.

Figure 12.11   Specifications and component qualities for gasoline blending.

Butane	0.570
Virgin naphtha	0.650
Reformate	0.737
Catalytic-cracked naphtha:	
Low severity	0.730
High severity	0.750

Figure 12.12   Specific gravities of gasoline blend components.

Viscosity is similar to octane in that the typical units in which it is measured (centistokes) are not suitable for linear-blending methods. Another Exxon proprietary correlation formula is needed to calculate a viscosity-blending number (VBN). Figure 12.13 provides the fuel oil specifications and blend component properties.

To conclude our consideration of fuel oil blending, we must ask how plant fuel consumption affects this blend. The fuel-equivalent streams produced by the reformer and catalytic cracker can be burned as plant fuel, thus releasing plant fuel from other sources for use in the fuel oil blend pool. Butane and light virgin naphtha can also be burned in the plant, releasing heavier plant fuel for use in the fuel oil blend pool. Figure 12.14 shows the relevant data for plant fuel considerations.

Inventory change requirements complete our description of the data. Suppose that we have a three-day shutdown of the reformer and the catalytic cracker planned for the end of January. During December, in anticipation of this, inventories of catalytic-cracked naphtha were built above typical levels for use in January. During January, reformate inventories will be built to protect against an unforeseen extension of the shutdown. These will be drawn down to normal levels in February. Figure 12.15 shows the inventory change requirements.

	Weight %	Viscosity		
	Sulfur Content	Centistokes	VBN	
Vacuum residues:				Sulfur and VBN are linear functions of the **weight** of the components used.
Arabian light	4.0		48.0	Vacuum residues from the various crude oils are significantly different in their properties and require modeling on a segregated basis.
Arabian heavy	5.0		51.4	
Brega	0.6		44.6	
Virgin heating oil	1.0	2.874		
Vacuum distillate	1.7	7.420		Direct viscosity-blending numbers (VBN) are available for the vacuum residues.
Catalytic-cracked distillate	1.5	5.630		
Fuel-oil-blend specifications	3.0 Maximum	500.000 Maximum		

Figure 12.13   Specifications and component qualities for fuel oil blending.

	Sulfur Weight %	VBN	Fuel Value
Fuel equivalent	3.5	44	1.000
Butane			1.110
Light virgin naphtha			1.107

Figure 12.14   Plant fuel considerations. Butane and light virgin naphtha have a higher heating value (BTU content) than other fuels. Thus, burning a ton of these releases more than a ton of plant fuel, as shown by the fuel value figures.

	January	February	March
Low-severity, catalytic-cracked naphtha	(580)	0	0
Reformate	1650	(1650)	0

Figure 12.15   Net buildup (drawdown) of inventories in tons due to scheduled shutdowns.

Figures 12.3 through 12.15 provide the basic data we need to build our refinery model. In Section 12.5 we construct the data base, but first we must consider the entity classes that are used to describe the data.

12.4   ENTITY CLASSES FOR SUPPLY PLANNING DATA

In designing a data base for an LP application using PLATOFORM, this design becomes, in effect, a user specification for the matrix generator that builds the

LP matrix. Chapter 9 discussed Exxon's standard supply/manufacturing generator at length. In Section 12.5 we build a data base to use this generator, but in this section we analyze some of the design considerations that led to this particular generator.

Section 9.3 discussed the concept of generator submodules and the advantages of such a design. The discussions in Section 12.3 were intentionally organized to suggest an approach based on submodules, the submodules being designed to support the various major categories of data used in supply planning. Thus reviewing the data discussed in Section 12.3, we see that we need generator submodules to handle the following basic aspects of supply and refinery planning:

- Crude oil supply submodule.
- Product supply submodules:
    Product demands.
    External import/export considerations.
- Refinery capacity submodule.
- Refinery process submodules:
    Crude oil distillation processes.
    Other processes.
- Refinery product blending submodules:
    Recipe blending.
    Gasoline blending.
    Fuel oil blending.

Each of these aspects of supply and refinery planning will have its own particular organization of data, which is discussed in Section 12.5.

Each entity class we are attempting to model will have its own dictionary for translation, but, more importantly, the entity classes will affect how we organize our data tables with regard to table name structure, head structure, and stub structure. This aspect of data-base design, discussed in Chapter 4, is a central consideration in any new application area using PLATOFORM. A good database design will:

- Organize the data in a way that is meaningful and useful to the business analyst.
- Support a generator design that can be efficiently programmed and maintained.

Figure 12.16 gives a brief summary of the major entity classes used in describing the data in Section 12.3.

| | ACTIVITY | | | | | | | | |
ENTITY CLASS	Crude Oil Supply	Product Demands	External Export/ Import	Refinery Capacities	Crude Oil Distillation	Other Processes	Recipe Blending	Gasoline Blending	Fuel Oil Blending
Crude Oils	X				X				
Time Periods	X	X	X	X				X	X
Data Units	X								
Products		X	X		X	X	X	X	X
Process Units				X	X	X			
Capacity Types				X	X	X			
Qualities	X							X	X
Constraint Types	X	X	X	X				X	X

Figure 12.16   Major entity classes encountered in supply and refinery planning data.

The entity classes in Figure 12.16 are not, in fact, a complete list of those needed for supply planning, but they are the ones immediately suggested by the data. We identify the few remaining entity classes as we need them in the next section.

12.5 BUILDING THE DATA BASE

In this section we follow the step-by-step process required to build a PLATO-FORM data base. As we build, we format the data into tables as expected by Exxon's standard supply/manufacturing generator. While we do not fully address fundamental data-base design, we do discuss some of the thought processes that led to the particular design used. We also give some attention to building user dictionaries and catalogs, but we do not address every entry.

It is most useful to study the data in the order in which it was discussed in Section 12.3, since this logical grouping parallels the structure of the matrix generator. First, let us recall that the crude oil supply data consisted of:

- The types of crude oils that are available.
- The amount of crude oil available.
- The price of the crude oil.

Thinking about the basic design of the tables, we see several entities that we must classify using table names, heads, and stubs. These entities are:

- The type of data (e.g., price or availability).
- The crude oil.

- The time period.
- The basic units (e.g., tons or barrels).

While many types of organization are possible, we have found the following conventions to be quite useful throughout many PLATOFORM applications:

- Amounts of resources available are specified in tables whose first character is B (for Bounds).
- Prices of resources are specified in tables whose first character is F (for Financial).
- Selection of options to model are specified in tables whose first character is I (for Incidence).
- Insofar as possible, we let the second character of the table name designate the category for which the table supplies data (e.g., A for crude oil Availabilities).

Using these conventions, then:

- The crudes to be run will be given in IA tables.
- The crude oil amounts available will be given in BA tables.
- The crude oil prices will be given in FA tables.

The crudes to be run can simply be given in a list, that is, a table with stub only. Figure 12.17 shows a table to specify that Arabian Light (AL), Arabian Heavy (AH), and Brega (BR) crudes are to be used.

For crude oil availabilities, the BA table has an additional entity to describe the time period under consideration. We can use the head of the table for this entity as shown in Figure 12.18.

In specifying crude oil prices, we also have to specify the units ($/barrel),

IA.....Z	
AL	
AH	
BR	

Figure 12.17   Crude oils to be run.

BA.....T	M1	M2	M3
AL	110	105	115
AH	165	155	170
BR	80	70	80

Figure 12.18   Amounts of crude oil available.

since they differ from the basic financial units in which we are modeling ($/ton). We can use part of the table name for this information, as shown in Figure 12.19.

In Figure 12.19, note the use of "''.." as an indicator for default time-period data. It is quite common throughout the PLATOFORM system to allow the code ".." to specify default data across all time periods under consideration. This convention often results in considerable input savings and reduced data redundancy. It is but one example of the more general principle of data overlay discussed in Chapter 9. We discuss later how we define "all time periods under consideration."

Having chosen to give crude oil prices in $/barrel and to build our model in tons, it becomes necessary to provide crude oil specific gravity data in order to convert these prices into $/ton. A table giving this data (a QC table) is shown in Figure 12.20.

Crude oil prices in $/barrel can be converted into crude oil prices in $/ton using the formula:

$$\$/\text{ton} = \$/\text{barrel} \times 6.29 \frac{\text{barrels}}{\text{cubic meter}}/SG$$

where SG is the crude gravity expressed in tons per cubic meter.

This is but one illustration of the way in which the PLATOFORM system has come to provide maximum support for the analyst to use data in its original form. The option remains for the analyst to use a table FA. . . . .T to provide data directly in $/ton if desired, although this option has seen very little use.

It may also seem that crude oil transportation costs should be input with the crude oil supply data. In a single refinery model this is possible. For a multirefinery model it is not possible, since crude supply data represents crudes available to all refineries, while crude transportation data is specific to the refinery receiving the crude. We see, later, how we can account for crude transportation costs.

Figure 12.21 shows the PLATOFORM syntax needed to input the tables

FA.BL..T	..
AL	35
AH	34
BR	42

Figure 12.19   Crude oil prices in $/barrel.

QC.....T	SG
AL	.858
AH	.886
BR	.823

Figure 12.20   Crude oil specific gravities. QC is suggestive of Qualities of Crudes.

```
+DBM              BCASE      DUMMY
-UPDATE
  A IA.....Z
    AL
    AH
    BR
  A BA.....T   M1          M2          M3
    AL         110         105         115
    AH         165         155         170
    BR         80          70          80
  A FA.BL..T   ..
    AL         35
    AH         34
    BR         42
  A QC.....T   SG
    AL         .858
    AH         .886
    BR         .823
```

Figure 12.21   Input of crude supply data tables.

containing crude oil supply data. Several things should be noted from this input:

- A new ACTPROB BCASE is being set up from scratch (DUMMY used as the old ACTPROB).
- The UPDATE submodule of DBM is used.
- Since we are setting up a new ACTPROB, all table-update types are "A" (for add).

At this point, it is worth taking a step back to ask how we can catalog the crude supply data tables so that the submodule TPRINT of DBM can print the tables with dictionary translation. This cataloging is enforced by DBM, since no tables can be added if the set to which they belong has not been previously cataloged in the generic table catalog, GENTAB.C, that is resident on ACTPROB DATABASE.

Let us start by cataloging the incidence table of crude oils to be run. We notice several things in cataloging this set of tables:

- There will be only one IA table.
- The table is alphabetic (Z-table).
- The table stubs come from the crude oils entity class.

This suggests that we must have a dictionary describing the crude oils. Notice that we are using two-character codes for the crude oils and that such codes need to be translated. We can choose a suggestive name, say CRUDES.D, for this dictionary. This dictionary must also be cataloged in the Dictionary Catalog (DICT. . .C), and in particular we need a code to use in referring to this dictionary. We can use the code CR as a memory aid. Figure 12.22 shows the dictionaries and catalogs needed to catalog the IA table set.

Next, we can catalog the BA table set that gives limits on crude oil availabilities. The cataloging of this table set is similar to the IA table set, but we notice some differences:

- The BA table set is numeric.
- An additional entity class, time periods, is used in the head of the table.

As with cataloging the IA table set, we must create dictionaries for all entity classes. Thus we need a dictionary for time periods that we might call PERNAM2D (period names with two characters). We can use the code P2 to catalog this dictionary in DICT. . .C. (See Figures 12.23 and 12.24.)

Cataloging the FA table set requires another entity class. Recall that we said crude oil prices could be given in $/barrel or $/ton and that positions 4, 5 of the table name can be used to specify which price form is used. We also translate this entity class from a dictionary, say DAUNIT.D (for data units). This dictionary must be cataloged in DICT. . .C, and we can use the code DA.

In cataloging the FA table set, we now catalog a true generic set of tables (more than one table is possible in the set), and the table set is cataloged using the generic name.

Cataloging the QC table set will require yet another entity class, qualities, and

GENTAB.C	S1	H1	M1	F1	D1-D3
IA.....Z	CRUDES.D	*		A8	CRUDE OILS AVAILABLE

DICT...C	D1-D3	C1
CRUDES.D	CRUDE OILS	CR

CRUDES.D	M1	D1-D3
AL	ARAB.LT	ARABIAN LIGHT
AH	ARAB.HY	ARABIAN HEAVY
BR	BREGA	BREGA

Figure 12.22 Cataloging the IA table set. Note the use of "*" in GENTAB.C to stipulate that the heads (H1) of the IA table set are not to be translated.

its dictionary. Otherwise, cataloging the QC-table set introduces no new concepts and is not discussed in detail. Figure 12.23 shows a portion of the PLATO-FORM input needed to catalog the crude supply data tables.

Note particularly in Figure 12.23 the use of the STARTUP submodule of UTILITY to initiate a new PLATOFORM ACTFILE. The input in Figure 12.23 must, of course, come before the input in Figure 12.21 in order to start the new file and catalog the data tables to be added. Figure 12.24 shows the contents of the dictionaries and catalogs that would result from running the input listed

```
+UTILITY
-STARTUP
+DBM          DATABASE
-UPDICT
      GENTAB.C IA.....Z S1H1T1M1 CRUDES.D *            S
               F1D1D2D3 A8            CRUDE OI LS AVAIL ABLE
            BA.....T S1H1T1M1 CRUDES.D PERNAM2D S
               F1D1D2D3 D3            CRUDE OI L AVAILA BILITIES
            FA.DA..T S1H1T1M1 CRUDES.D PERNAM2D G        XXXAAXXX
               F1D1D2D3 D2            CRUDE OI L PRICES  $/
            QC.....T S1H1T1   CRUDES.D QUALITYD S
               F1D1D2D3 D3            CRUDE OI L PROPER TIES
      DICT...C CRUDES.D D1D2D3C1 CRUDE OI LS              CR
               PERNAM2D D1D2D3C1 2-CH TIM E PERIOD  CODES P2
               DAUNIT.D D1D2D3C1 DATA UNI TS              DA
               QUALITYD D1D2D3C1 QUALITIE S               QU
      CRUDES.D *ADD
            AL       M1D1D2D3 ARAB.LT  ARABIAN  LIGHT
            AH       M1D1D2D3 ARAB.HY  ARABIAN  HEAVY
            BR       M1D1D2D3 BREGA    BREGA
```

Figure 12.23   Portion of input necessary to catalog the table sets containing crude supply data.

GENTAB.C	S1	S2	H1	H2	T1	F1	M1	D1	D2	D3
IA.....Z	CRUDES.D		*		S	A8		CRUDE OI	LS AVAIL	ABLE
BA.....T	CRUDES.D		PERNAM2D		S	D3		CRUDE OI	L AVAILA	BILITIES
FA.DA..T	CRUDES.D		PERNAM2D		G	D2	XXXAAXXX	CRUDE OI	L PRICES	$/
QC.....T	CRUDES.D		QUALITYD		S	D3		CRUDE OI	L PROPER	TIES

DICT...C	D1	D2	D3	C1
CRUDES.D	CRUDE OI	LS		CR
PERNAM2D	2-CH TIM	E PERIOD	CODES	P2
DAUNIT.D	DATA UNI	TS		DA
QUALITYD	QUALITIE	S		QU

DICTIONARY	CRUDES.D		CRUDE OILS			
CODE	MNEMONIC		DESCRIPTOR		TAGS	
	M1	D1	D2	D3	T1	T2
AL	ARAB.LT	ARABIAN LIGHT				
AH	ARAB.HY	ARABIAN HEAVY				
BR	BREGA	BREGA				

Figure 12.24   Example print of catalogs and dictionaries.

in Figure 12.23. This list is produced using the DISPLAY and DICPRNT sub-modules of UTILITY.

The next major area of data to be input is product demand data. This requires us to input information on product demands and on revenues from product sales. At this point, we need to introduce the product entity class, since our demands are given by product. Description of product demand data requires at least two more entity classes:

- A class of constraint types to indicate whether demands are upper limits, lower limits, or fixed requirements.
- A class of locations. Whereas crude oils were made available to the entire model, demands can be given by location. In our particular model, this is not a significant distinction. However, in a regional supply coordination model, locations become an important entity class.

For product demand data, then, there are at least four entity classes:

- Locations.
- Products.
- Time periods.
- Constraint types.

For obvious reasons, it is helpful to consider the location as part of the table name. This allows each location to have exclusive responsibility for a well-defined group of tables. For consistency, it is also helpful to maintain as much similarity of design as possible between table sets. This means that time periods should be in the heads of product demand table sets as they are in the heads of crude supply table sets. The analog to crudes (products in this case) will occupy table stubs. Constraint types, being associated with bounds that are built on a time period basis, naturally associate themselves with time periods and thus give rise to a complex head in our product demand table. Figure 12.25 shows a table that can be used to represent product demands.

BS...UFT	M1FX	M2FX	M3FX	
GS	40	35	45	BS table set represents Bounds on Sales.
JF	20	20	20	FX in head shows that demands are FiXed requirements.
HØ	50	50	40	
FL	145	130	145	UF is location standing for Fawley in the United Kingdom.

Figure 12.25   Product demands in thousands of tons.

We have a corresponding table for product revenues. It is worth noting, here, that with fixed demand requirements, revenues are not particularly relevant. Nevertheless, for completeness and to anticipate the possibility of input of discretionary demand data, we can input sales revenues in a table, as shown in Figure 12.26.

We can handle the inventory changes and fuel oil imports as part of the external import/export considerations. These are similar to demands, but they can be used to represent interactions with locations that are not being explicitly modeled. Data table organization is similar to that for demand data, except that we use a character "E" (for external) in the table set names. Figure 12.27 shows one possible organization of this inventory and import data.

The data regarding refinery capabilities is rather extensive. To save space, we do not describe input of all this data in detail, but instead we concentrate on concepts in table design that have not previously been discussed. Other data tables are mentioned in passing, but are displayed in detail only in the OPRINT (ordered table print) in Appendix A.

FS.....T	..	Revenues are represented as negative numbers; costs as positive.
GS	-430	These are prices for all model locations (..). Overlay data could be given in tables FS...U.T (all U.K.) and FS...UFT (Fawley), if desired.
JF	-300	
HØ	-315	
FL	-250	

Figure 12.26   Product sales revenues in $/ton.

BE.UMUFT	M1FX	M2FX	M3FX
CL	.58	1E-5	1E-5
PF	1E-5	1.65	1E-5

BE.UFUMT	M1FX	M2FX	M3FX
PF	1.65	1E-5	1E-5

FE.UMUFT	..
FI	245

- We have used UM (Milford Haven, U.K.) as our arbitrary external location.
- 1E−5 is an artificial convention for a fixed limit of zero, since the default is a minimum of zero and a maximum of infinity.
- FI will have default limits on import of zero (minimum) and infinite (maximum).
- UMUF in table name shows moves *to* Fawley.
- UFUM in table name shows moves *from* Fawley.
- FI is assumed to meet fuel-oil specifications of 3wt% sulfur and 500 centistokes viscosity.

Figure 12.27   Imports and inventories shown as external import/export conditions.

Refinery-capacity data is given in an IU table (Incidence of Unit capacities), BU tables (Bounds on Unit capacities), and an FU table (Financial data for Units). These require the introduction of several new entity classes:

- Capacity types (e.g., total feed to a unit).
- Capacity data classes (e.g., existing capacity, grass roots investment, turn-around times).
- Cost types (e.g., operating cost).

The crude oil processing data is worth looking into in greater detail, since it introduces several new concepts. Our crude oil processing data must include:

- A description of what crude oils are to be processed and how to process them.
- A description of the product yields from crude oil processing.
- A description of the capacities consumed in crude oil processing.

As might be expected, we use an incidence table to describe what crude oils have to be processed and how to process them. Figure 12.28 shows the incidence table, which, in our example, is quite simple.

Product yields from crude oil processing are generally derived from the Exxon Assay System (an Exxon proprietary data base of crude assay information) via a bridge. This is the standard PLATOFORM approach. However, for our case we consider the product yields as user supplied. Specification of distillation yields requires naming the crude oil, the distillation option, and the product. One possible organization is shown in Figure 12.29.

We said earlier that we would regard the crude transportation cost as part of the data for crude oil distillation. We do this because the transportation cost depends on the refinery receiving the crude. Figure 12.30 shows the input data for crude oil transportation.

The only data remaining to be specified for crude oil distillation is the capaci-

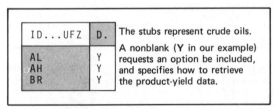

Figure 12.28   Crude oil processing incidence table.

YD...UFT	ALD.	AHD.	BRD.
LN	3.5	3.0	4.5
IN	10.0	7.5	13.5
VH	39.0	30.0	43.0
VD	28.5	23.0	28.0
YL	16.5		
YH		33.5	
YB			10.0
ML	2.5	3.0	1.0

YD stands for Yields from Distillation.

Vacuum Residues (YL, YH, and YB) from different crudes are handled separately in the processing which follows.

Figure 12.29 Crude oil distillation yields.

FC...UFT	..
AL	24.15
AH	24.15
BR	10.05

Figure 12.30 Crude oil transportation costs in $/ton. FC stands for Financial data for Crude transportation.

ties consumed. This requires a subsidiary incidence table, as shown in Figure 12.31. This table allows several different distillation options (e.g., options D1, D2, and D3) to consume capacity on the same unit (e.g., pipestill D.).

The data for nondistillation processes is organized similarly to crude distillation data, that is:

- An IP (Incidence of Processing) table is used to specify feeds and processing options.
- A set of YP (Yields from Processing) tables is used to specify process yields.

The YP tables are broken down by major class of processing (one table for catalytic cracking and one table for reforming). This improves efficiency by giving small, dense tables rather than large, sparse tables. In small data tables, data display is also more useful.

KD...UFZ	D.
TF	D.

Capacities are assumed consumed on a volume basis.

TF stands for Total Feed.

Figure 12.31 Consumption incidence of distillation capacity.

How can we input capacity requirements for noncrude processing, since these are not directly calculated by the system? We can use several methods. The one we choose is useful because it illustrates the use of exception structure, discussed in Section 9.3.4. Exception structure is given in an optional EB table that is specific to a generator submodule and location. The exception structure for noncrude processing (PR) at Fawley (UF) is shown in Figure 12.32.

Using exception structure in this way requires the user to be familiar with matrix row and column mnemonics and formulation style. While indiscriminate use of exception structure is not recommended, part of the generator-design philosophy is that the user must have a method for incorporating structure that is not built as part of the base generator. If many users are building the same type of exception structure, this structure obviously becomes a candidate for incorporation as a standard option within the base generator.

Recipe blending considerations are very similar to processing considerations:

- The blend options to use are specified in an IM (Incidence of Miscellaneous blending) table.
- The recipes to use in the blends are specified in a set of YM tables.

The plant fuel considerations illustrate how formulation style can be used to support reporting requirements. Recall that we said that butane, virgin heating oil, and the fuel-equivalent stream from the reformer and catalytic cracker can be burned, releasing plant fuel. Figure 12.33 shows how this can be modeled in our data base.

We complete our discussion of building the data base by considering specification blending. Gasoline blending and fuel oil blending tables are quite similar, so we confine our consideration to fuel oil blending.

For fuel oil blending, the following data must be given in our tables:

- Fuel products to be blended.
- The components to be used in the blends.
- Qualities to be controlled on each of the blends.
- The numeric values of the specifications that the fuel blends must satisfy.

EB.PRUFT	PVD.C*UF	PVH.C*UF	PI..F*UF	Capacity consumption coefficients are divided by specific gravity of feed. Assumed gravities: VD = .92, VH = .886, I. = .737.
CTF.C.UF	-1.087	-1.129		Note the use of '*' in table-head names
CTF.F.UF			-1.357	to specify a set of matrix columns.

Figure 12.32   Exception structure to show capacity consumption.

- The qualities of the components.
- The way in which component qualities are blended together.

We again use a set of incidence tables to specify the options (fuels to blend, qualities to control, and components to use) and a set of numeric data tables to provide the supporting data (blend specifications and component qualities).

Figure 12.34 exhibits the fuel oil blending incidence tables that are used to specify the options for fuel blending. An IQ (Incidence of Qualities) table and an IB (Incidence of Blend component) table are used to specify the options.

Figure 12.35 shows the data tables needed for fuel oil blending. Product specifications are given in an SB (Specifications for Blends) table, and component qualities are given in a QB (Qualities for Blend components) table.

The only task remaining for fuel oil blending is to define how the component

YM.R.UFT	C4RF	VHRF	RF is the Refinery Fuel equivalent stream.
C4	-1		C4 and VH are equivalent to more than 1 ton of fuel.
VH		-1	
RF	1.11	1.07	The ML (Material Loss) is an accounting entry to enable material-balance reports to be written.
ML	-.11	-.07	

Figure 12.33   Plant fuel considerations.

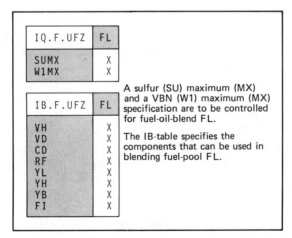

Figure 12.34   Incidence tables showing options to be used in fuel oil blending.

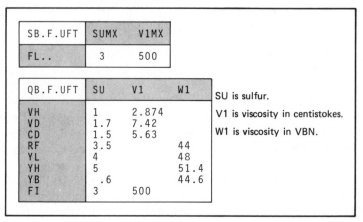

Figure 12.35   Quality data for fuel oil blending.

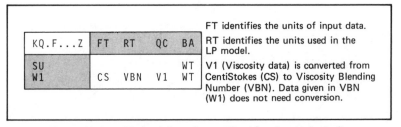

Figure 12.36   Methodology for quality blending in fuel oils.

qualities are blended together. In this table, we define the conversion routines needed and whether blending of qualities is on a volume or weight basis. Each quality has its technology displayed in the KQ (a secondary incidence table for qualities) table of Figure 12.36. The KQ table is extremely powerful, and it allows the specification blending submodule of the generator to be truly general. Otherwise, reserved quality mnemonics, each with its own hard-coded blending technology, would be needed for the blending submodule.

A full OPRINT of all data tables needed for the generator to build our simple model is available in Appendix A. Before looking at this printout, we need, briefly, to consider how we control the generator.

12.6   CONTROLLING THE GENERATOR

There remains the question of how we control the generator. This requires the definition of:

- The submodules of the generator to use.
- The locations for which matrix structure is to be built.
- The time periods for which matrix structure is to be built.

Section 9.3.2 discussed the concept of a generator-module dictionary to define the valid submodules for a particular application. For our example, the necessary generator-module dictionary (GENRMODD) is illustrated in Figure 12.37.

This table defines all of the submodules that are valid for any ACTPROB on the ACTFILE. It says nothing about locations for which they are valid. This is the job of a generator-incidence table (IG) on each ACTPROB, which gives the locations for which to build structure and tells what structure to build for each location. In our case, we have three locations under consideration:

- The arbitrary ".." location at which crude oils are made available.
- The location UM (Milford Haven) from which imports are available and to which exports are sent.
- The location UF (Fawley) representing the refinery.

The IG table for our ACTPROB is shown in Figure 12.38.

Finally, we must include a table defining the time periods under study, as shown in Figure 12.39.

GENRMODD	M1	D1-D3
CA	CRAVAIL	CRUDE AVAILABILITIES
BØ	EXTERN	EXTERNAL IMPORT/EXPORT
CI	CAPINV	CAPACITY/INVESTMENT
PE	PSTLL	PIPESTILL
PR	PROCESS	SPECIAL PROCESSES
GB	GBLEND	MOGAS BLENDING
FB	FBLEND	FUEL BLENDING
RE	RECIPE	MISC. BLENDS (RECIPES)
DM	DEMAND	DEMANDS

Figure 12.37   Generator-module dictionary for example model.

IG.....Z	..	UM	UF
CA	X		
BØ		X	
CI			X
PE			X
PR			X
GB			X
FB			X
RE			X
DM			X

Figure 12.38   Generator incidence table.

P2 . . . . . Z	T1
M1	1
M2	2
M3	3

Figure 12.39   A time period control table. T1 head gives one character equivalent for use in building row and column names.

Having built our data base, it is now a simple matter to build the matrix via a call to the generator. Since this is the first time we have built the matrix, we request that all submodules be generated for all locations. Figure 12.40a shows the input required to generate the matrix and display a picture of the matrix (Figure 12.40b) and a print of the bound sets (Figure 12.40c).

12.7   MODEL OPTIMIZATION

In this section, we arrive at the heart of our problem, determining an optimal solution to the supply/refinery planning problem. The optimization module of PLATOFORM, as described in Chapter 10, is quite complex and has a wide range of capabilities including:

- Conventional linear-continuous optimization.
- Parametric optimization.
- Nonlinear optimization.
- Mixed-integer optimization.

Here, we concentrate on straightforward linear optimization. A later section considers parametric optimization.

The syntax required to run an optimization for each of our three time periods is developed in this section. The information that must be passed to the optimization module is:

- The ACTPROB to be used.
- The time period to be used.

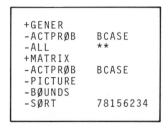

```
+GENER
 -ACTPRØB     BCASE
 -ALL         **
+MATRIX
 -ACTPRØB     BCASE
 -PICTURE
 -BØUNDS
 -SØRT        78156234
```

Figure 12.40   (a) PLATOFORM input that generates and displays, on the following pages, (b) the matrix and (c) the bound-set printout.

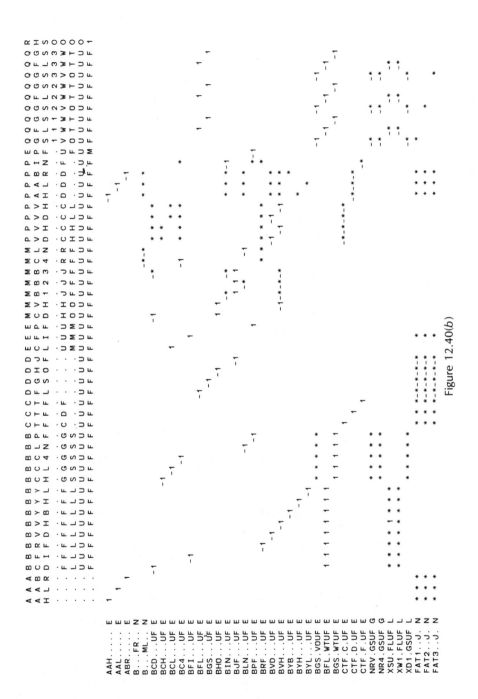

Figure 12.40(b)

VECTOR NAME	BOUNDSET ========		BOUNDSET ========		BOUNDSET ========	
	M1BOUND		M2BOUND		M3BOUND	
	---LOWER--	---UPPER--	---LOWER--	---UPPER--	---LOWER--	---UPPER--
AAH.....	.	165.00000	.	155.00000	.	170.00000
AAL.....	.	110.00000	.	105.00000	.	115.00000
ABR.....	.	80.000000	.	70.000000	.	80.000000
CTF.C.UF	.	35.612083	.	35.612083	.	39.427663
CTF.D.UF	.	320.34976	.	289.34817	.	320.34976
CTF.F.UF	.	33.386328	.	33.386328	.	36.963434
DFL...UF	145.00000	145.00000	130.00000	130.00000	145.00000	145.00000
DGS...UF	40.000000	40.000000	35.000000	35.000000	45.000000	45.000000
DHO...UF	50.000000	50.000000	50.000000	50.000000	40.000000	40.000000
DJF...UF	20.000000	20.000000	20.000000	20.000000	20.000000	20.000000
ECL.UMUF	.5800000	.5800000	.	.	.	.
EPF.UMUF	.	.	1.6500000	1.6500000	.	.
EPF.UFUM	1.6500000	1.6500000	.	.	.	.

Figure 12.40(c)

- The case name to be given to the results.
- The printed output desired.

Figure 12.41 shows the input needed to request optimization of three cases and to print a translated solution (SOLPRINT). SOLPRINT was discussed in Chapter 10 and, of course, requires definition of mnemonic structure via a ROW. . . .C and a COL. . . .C, which are shown in Appendix C with the printout of ACTPROB DATABASE. Figure 12.42 shows the result of the SOLPRINT for the M1 (January) time period.

To complete our discussion of optimization, we turn to the types of operating

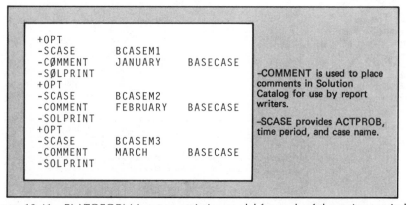

Figure 12.41   PLATOFORM input to optimize model for each of three time periods.

guidelines that would typically be derived from an optimal solution. As an example, the January SOLPRINT shows:

- Fuel oil has no sulfur "giveaway" (that portion of the product quality that is better than the specification).
- Motor gasoline has no octane giveaway.
- All of the available Arabian Light crude is used.
- The catalytic cracker capacity is filled with vacuum distillate at high severity.
- All available butane is used in gasoline.
- Jet fuel is produced strictly from blend 2.
- Marginal fuel oil demand is satisfied from imports.

12.8 REPORT WRITERS

The report-writing step is crucial to the successful application of linear programming. The application reports translate the results of the LP optimization into a format and organization that are meaningful and useful to the business analyst. As pointed out in Chapter 11, Exxon has made significant progress in this area with PLATOFORM. ADHOC, discussed in Chapter 11, now handles about 95% of all our report writing requirements. Through a series of evolutionary steps, it has become a powerful tool. In this section, we illustrate the versatility of ADHOC and other report-writing capabilities in PLATOFORM with some specific examples.

12.8.1 Solution Printing

We have already seen one example of a report made possible by PLATO-FORM's data structure and mnemonic discipline. This is the SOLPRINT (translated solution print) shown in Figure 12.42. It is important to notice that the report depends on two features of PLATOFORM:

- All row and column names adhere to strict mnemonic disciplines that can be described completely in the ROW. . . .C and COL. . . .C residing on ACTPROB DATABASE.
- The PLATOFORM dictionaries are used to translate the row and column entities as described in the ROW. . . .C and COL. . . .C.

SOLPRINT is already a significant step forward from the standard solution-print procedure available in any mathematical programming system (MPS). Basically, the user is freed from having to memorize the full mnemonic structure of

SECTION 1 - ROWS
================

ROW	AT	ACTIVITY	SLCK ACT	LO LIMIT	UP LIMIT	DUAL ACT	NUMBER	TYPE	F1	F2	F3	F4
AAH....	EQ	.	.	.	.	241.37698-	1	AVAILBTY	ARAB.HY	.	.	EUROPE
AAL....	EQ	.	.	.	.	276.50053-	2	AVAILBTY	ARAB.LT	.	.	EUROPE
ABR....	EQ	.	.	.	.	320.99635-	3	AVAILBTY	BREGA	.	.	EUROPE
B...FR..	BS	.	6.9021848-	NONE	NONE	.	4	ACCNTING	.	.	REFFUEL	EUROPE
B...ML..	BS	6.9021848	6.9021848-	NONE	NONE	.	5	ACCNTING	.	.	LOSS	EUROPE
FAT1..J..	BS	2883.9638-	2883.9638	NONE	NONE	1.0000000	6	FINANCL	AFTR TX	JAN	.	EXXON
FAT2..J..	BS	3046.2156	3046.2156	NONE	NONE	.	7	FINANCL	AFTR TX	FEB	.	EXXON
FAT3..J..	BS	3046.2156-	3046.2156	NONE	NONE	.	8	FINANCL	AFTR TX	MAR	.	EXXON
BCD...UF	EQ	.	.			316.50642-	9	BALANCE	CATDIST			FAWLEY
BCH...UF	EQ	.	.			487.36371-	10	BALANCE	CATNAPH			FAWLEY
BCL...UF	EQ	.	.			407.20074-	11	BALANCE	CATNAPL			FAWLEY
BC4...UF	EQ	.	.			555.53435-	12	BALANCE	BUTANE			FAWLEY
BFI...UF	EQ	.	.			245.00000-	13	BALANCE	FO IMP			FAWLEY
BFL...UF	EQ	.	.			245.00000	14	BALANCE	FUEL			FAWLEY
BGS...UF	EQ	.	.			498.29977-	15	BALANCE	MOGAS			FAWLEY
BHO...UF	EQ	.	.			316.50642-	16	BALANCE	HEATOIL			FAWLEY
BIN...UF	EQ	.	.			519.15494-	17	BALANCE	IVN			FAWLEY
BJF...UF	EQ	.	.			303.49026-	18	BALANCE	JETFUEL			FAWLEY
BLN...UF	EQ	.	.			251.42560-	19	BALANCE	LVN			FAWLEY
BPF...UF	EQ	.	.			549.99459-	20	BALANCE	REFMT			FAWLEY
BRF...UF	EQ	.	.			227.12339-	21	BALANCE	REFFUEL			FAWLEY
BVD...UF	EQ	.	.			291.47918-	22	BALANCE	VACDIST			FAWLEY
BVH...UF	EQ	.	.			316.50642-	23	BALANCE	VIRG HO			FAWLEY
BYB...UF	EQ	.	.			330.80771-	24	BALANCE	BR RESID			FAWLEY
BYH...UF	EQ	.	.			173.49358-	25	BALANCE	AH RESID			FAWLEY
BYL...UF	EQ	.	.			209.24679-	26	BALANCE	AL RESID			FAWLEY
BGS.VOUF	EQ	.	.			1195.3156	27	BALANCE	MOGAS		VOLUME	FAWLEY
BFL.WTUF	EQ	.	.			352.25964-	28	BALANCE	FUEL		WEIGHT	FAWLEY
BGS.WTUF	EQ	.	.			498.29977-	29	BALANCE	MOGAS		WEIGHT	FAWLEY
CTF.C.UF	EQ	.	.			102.83859-	30	CAPACITY	TOTFEED		CATCRAK	FAWLEY
CTF.D.UF	EQ	.	.			.9435000-	31	CAPACITY	TOTFEED		PIPSTIL	FAWLEY
CTF.F.UF	EQ	.	.			3.7740000-	32	CAPACITY	TOTFEED		REFORM	FAWLEY
NRV.GSUF	BS	181.15938	181.15938		NONE	17.104406-	33	MIN®SPEC	RVP		MOGAS	FAWLEY
NR4.GSUF	BS	.	.		NONE	35.753212	34	MIN®SPEC	RON .4G		MOGAS	FAWLEY
XSU.FLUF	UL	.	.	NONE	.	.	35	MAX®SPEC	SULFUR		FUEL	FAWLEY
XW1.FLUF	BS	431.50281-	431.50281	NONE	.	.	36	MAX®SPEC	VBN		FUEL	FAWLEY
XD1.GSUF	BS	.5339642-	.5339642	NONE	.	.	37	MAX®SPEC	% 212F		MOGAS	FAWLEY

SECTION 2 - COLUMNS
===================

COLUMN	AT	ACTIVITY	INP COST	LO LIMIT	UP LIMIT	RED COST	NUMBER	TYPE	F1	F2	F3	F4
AAH.....	BS	136.81632	241.37698	.	165.00000	.	38	AVAILBTY	ARAB.HY	.	.	EUROPE
AAL.....	UL	110.00000	256.58508	.	110.00000	19.915448-	39	AVAILBTY	ARAB.LT	.	.	EUROPE
ABR.....	BS	5.2931594	320.99635	.	80.000000	.	40	AVAILBTY	BREGA	.	.	EUROPE
BCD.FLUF	LL	.	.	.	.	17.876606	41	BLEND	CATDIST		FUEL	FAWLEY
BFI.FLUF	BS	10.862701	.	.	NONE	.	42	BLEND	FO IMP		FUEL	FAWLEY
BRF.FLUF	BS	3.4503304	.	.	NONE	.	43	BLEND	REFFUEL		FUEL	FAWLEY
BVD.FLUF	BS	31.538034	.	.	NONE	.	44	BLEND	VACDIST		FUEL	FAWLEY
BVH.FLUF	BS	34.636150	.	.	NONE	.	45	BLEND	VIRG HO		FUEL	FAWLEY
BYB.FLUF	BS	.5293159	.	.	NONE	.	46	BLEND	BR RESID		FUEL	FAWLEY
BYH.FLUF	BS	45.833469	.	.	NONE	.	47	BLEND	AH RESID		FUEL	FAWLEY
BYL.FLUF	BS	18.150000	.	.	NONE	.	48	BLEND	AL RESID		FUEL	FAWLEY
BCH.GSUF	BS	14.742813	.	.	NONE	.	49	BLEND	CATNAPH		MOGAS	FAWLEY
BCL.GSUF	BS	.5800000	.	.	NONE	.	50	BLEND	CATNAPL		MOGAS	FAWLEY
BC4.GSUF	BS	2.4052244	.	.	NONE	.	51	BLEND	BUTANE		MOGAS	FAWLEY
BLN.GSUF	BS	4.1437424	.	.	NONE	.	52	BLEND	LVN		MOGAS	FAWLEY
BPF.GSUF	BS	18.128221	.	.	NONE	.	53	BLEND	REFMT		MOGAS	FAWLEY
CTF.C.UF	UL	35.612083	4.0885000	.	35.612083	98.750094-	54	CAPACITY	TOTFEED		CATCRAK	FAWLEY
CTF.D.UF	BS	289.05690	.9435000	.	320.34976	.	55	CAPACITY	TOTFEED		PIPSTIL	FAWLEY
CTF.F.UF	BS	29.817865	3.7740000	.	33.386328	.	56	CAPACITY	TOTFEED		REFORM	FAWLEY
DFL...UF	EQ	145.00000	250.00000-	145.00000	145.00000	5.0000000-	57	DEMAND	FUEL			FAWLEY
DGS...UF	EQ	40.000000	430.00000-	40.000000	40.000000	68.299773	58	DEMAND	MOGAS			FAWLEY
DHO...UF	EQ	50.000000	315.00000-	50.000000	50.000000	1.5064238	59	DEMAND	HEATOIL			FAWLEY
DJF...UF	EQ	20.000000	300.00000-	20.000000	20.000000	3.4902585	60	DEMAND	JETFUEL			FAWLEY
ECL.UMUF	EQ	.5800000	.	.5800000	.5800000	407.20074-	61	EXTERNAL	CATNAPL		MILFORD	FAWLEY
EFI.UMUF	BS	10.862701	.	.	NONE	.	62	EXTERNAL	FO IMP		MILFORD	FAWLEY
EPF.UMUF	EQ	.	245.00000	.5800000	NONE	549.99459-	63	EXTERNAL	REFMT		MILFORD	FAWLEY
MCD.HOUF	BS	14.415194	.		NONE	.	64	RECIPE	CATDIST		HEATOIL	FAWLEY
MVH.HOUF	BS	35.584806	.		NONE	.	65	RECIPE	VIRG HO		HEATOIL	FAWLEY
MB1.JFUF	LL	.	.		NONE	73.810719	66	RECIPE	JF BL 1		JETFUEL	FAWLEY
MB2.JFUF	LL	.	.		NONE	33.281016	67	RECIPE	JF BL 2		JETFUEL	FAWLEY
MB3.JFUF	BS	20.000000	.		NONE	.	68	RECIPE	JF BL 3		JETFUEL	FAWLEY
MC4.RFUF	LL	.	.		NONE	303.42738	69	RECIPE	BUTANE		REFFUEL	FAWLEY
MLN.RFUF	LL	.0489395	.		NONE	.	70	RECIPE	LVN		REFFUEL	FAWLEY
PVD.CHUF	LL	.	.		NONE	46.932208	71	PROCESS	VACDIST		CC---HS	FAWLEY
PVH.CHUF	BS	32.761806	.		NONE	.	72	PROCESS	VIRG HO		CC---HS	FAWLEY
PVD.CLUF	LL	.	.		NONE	48.906576	73	PROCESS	VACDIST		CC---LS	FAWLEY
PVH.CLUF	LL	.	.		NONE	88.583916	74	PROCESS	VIRG HO		CC---LS	FAWLEY
PAH.D.UF	BS	136.81632	24.150000		NONE	.	75	DISTILL	ARAB.HY		PIPSTIL	FAWLEY
PAL.D.UF	BS	110.00000	24.150000		NONE	.	76	DISTILL	ARAB.LT		PIPSTIL	FAWLEY
PBR.D.UF	BS	5.2931594	10.050000		NONE	.	77	DISTILL	BREGA		PIPSTIL	FAWLEY
PIN.F.UF	BS	21.975801	3.1200000		NONE	.	78	PROCESS	IVN		REFORM	FAWLEY
QGS1VOUF	BS	52.003765	.		NONE	.	79	SPECIFY	MOGAS	JAN	VOLUME	FAWLEY
QFL1WTUF	BS	145.000000	.		NONE	.	82	SPECIFY	FUEL	JAN	WEIGHT	FAWLEY
QGS1WTUF	BS	40.000000	.		NONE	.	85	SPECIFY	MOGAS	JAN	WEIGHT	FAWLEY
EPF.UFUM	EQ	1.6500000	.	1.6500000	1.6500000	549.99459	88	EXTERNAL	REFMT		FAWLEY	MILFORD

Figure 12.42 SOLPRINT results for the M1 period.

317

the model and the entity codes in order to use a standard solution print. The basic organization of a SOLPRINT is, however, no different from that in the SOLUTION procedure. This means that the business analyst still needs to extract data from the solution (and possibly the matrix or tables), make calculations with it, and rearrange it for further analysis and management reporting. In this area, report writers are critical to the successful use of LP.

12.8.2 Application-Specific Reports

The first major step was the development of application-specific report writers. These tools handled the manual analysis and reorganization of data typically done by business analysts. They are, of necessity, developed to operate with specific matrix generators, since they depend on an assumed mnemonic and matrix structure. While they are specific in that respect, they still have at least two useful, general-purpose features:

- They use dictionary information to translate entity codes.
- They are data driven, meaning that if the scope of the LP model expands or decreases, they will also expand or decrease to match the entities of the model.

Perhaps the best way to illustrate these types of reports is by a specific example. In analyzing refinery supply planning models, it is often necessary to look closely at the pool blending area of the model. For each pool it is important to know:

- The relative composition by weight and possibly by volume.
- The blended-pool properties in relation to specifications.
- The component properties.
- The incentive to relax pool specifications.

The PLATOFORM PANREP (pool analysis report) is designed to produce this type of information for each pool blended in a standard PLATOFORM refinery model. If there are changes in the pools blended, the qualities followed, or the components used in the blend, then the report automatically changes, since it is based on an analysis of all model rows and columns with a specific mnemonic structure. The PANREP report writer is written specifically for use with the pool blending submodule of the PLATOFORM supply/refinery matrix generator.

The user syntax for requesting a PANREP is extremely simple, requiring only that the solution case to be analyzed be named. Optionally, the user can limit the analysis to specific pools. Figure 12.43 shows an example of a PANREP report for the fuel oil pool in our example.

POOL FL FUEL OIL   (FAWLEY)   PERIOD M1
===

		WEIGHT	WEIGHT PERCENT	WEIGHT PERCENT SULFUR	VISCSTY BLENDNG NUMBER
		WT	PW	SU	W1(VBN)
VIRGIN HEATING OIL	VH	34.636	23.887	1.000	14.803
VACUUM DISTILAT	VD	31.538	21.750	1.700	21.804
BREGA VACUUM RESID	YB	0.529	0.365	0.600	44.600
ARAB.LT VACUUM RESID	YL	18.150	12.517	4.000	48.000
ARAB.HY VACUUM RESID	YH	45.833	31.609	5.000	51.400
FUEL OIL IMPORT 3.0% SU	FI	10.863	7.492	3.000	37.531
REFINRY FUEL EQUIV	RF	3.450	2.380	3.500	44.000
TOTAL POOL		145.000	100.000	3.000	34.555

SPECIFICATION SUMMARY

		ACTUAL	MINIMUM	MAXIMUM	INCENTIVE TO RELAX SPECIFICATION
WEIGHT PERCENT SULFUR	SU	3.000	.	3.000	35.75 DOLLAR/ SU *TONNE
VISCSTY CENTSTKS	V1	157.623	.	500.000	

Figure 12.43   PANREP report for the fuel oil pool.

12.8.3 ADHOC Reports

Application-specific reports, such as PANREP, suffer from the limitations discussed in Chapter 11—high development effort and occasionally high maintenance effort. They also, of course, are not general purpose in terms of being able to support a wide variety of applications. Here, the strength of ADHOC begins to show.

ADHOC was discussed in detail in Chapter 11, so we do not repeat the arguments made for its usefulness. Rather, we show how this module can be used to rapidly develop user-oriented reports to support a particular application.

In our example, one interesting element is that it covers multiple cases. Previously, a business analyst often had to flip pages of output from one case to another to compare or add across the months of a quarter. Thus, in our discussion of ADHOC we particularly look at ADHOC's multicase facilities.

In defining ADHOC reports, we must ask what type of information the user will want to extract from a solution. Defining ADHOC reports, then, is user intensive, at least in terms of defining requirements. For many reports, it is also entirely possible for the user to do the development work. The specific reports required are, of course, highly dependent on the users of a model. For our case, however, we can define some typical reports that would be useful. Some examples might be:

- Crude runs reported in KT/period and in KB/day.
- Incentives to increase (or decrease) the demand for finished products.
- A summary of the utilization of process unit capacities.
- A net material balance around the entire refinery.
- Material balances around each process unit.
- Economic balances around each process unit. (Just as material flows must balance in a feasible solution to an LP, the economic flows must balance in an optimal solution.)

Each of these is highlighted in one of the ADHOC reports featured in Appendix B. The full definitions of the various reports are shown in the ZR.AH. .Z table in Appendix C. While we do not discuss each definition in detail, we do look at the capacity summary and the material balance around process units to show how reports can be developed readily using ADHOC.

First, let us study the report entitled "Process Unit Capacity Summary," on the second page of Appendix B. In this report, we take a detailed look at process unit capacities and, in particular, recognize that model units are different from the units in which we want to analyze the data. We want the report to display the following:

- Total feed to the unit. This can be in tons (as in the model), but it must be per day rather than per period.

- The capacity used in barrels per day, since our input data is in barrels per day.
- The spare capacity in barrels per day.
- The maximum capacity in barrels per day.
- The incentive to increase capacity (if any) in $/barrel.

This, in fact, is a fairly complex report that gives us a good opportunity to demonstrate how to develop an ADHOC report. First, we consider the total feed to the unit. The feed can be retrieved from the activities of the process vectors (first character P).

    MASK1   P**.**UF

Since we want to distinguish these as feed, we map an F into the first position of the list names:

    MAPS1   F2345678

Next, we consider the actual capacity used. For this we need the activity of the capacity vectors (first character C):

    MASK2   C**.**UF

To designate these as capacity used, map a U into the first position of the list names:

    MAPS2   U2345678

We must also convert these activities from cubic meters to barrels (1 barrel = 0.159 cubic meter):

    SCLE2   .159

We also need to include maximum unit capacities in the report. These are retrieved from the upper limit (XU) of the capacity vectors. Again, we must convert cubic meters to barrels:

    MASK3   C**.**UF
    STYP3   XU
    SCLE3   .159

To designate these as maximum capacities, we map an M into the first position of the list names:

    MAPS3   M2345678

Spare capacity is a bit more complex. This requires subtracting actual capacity used from maximum capacity. This subtraction can be caused by:

- Scaling capacity used by −1.
- Creating logical duplicates using MAPs.

We can create the upper limit for this subtraction by a second MAP applied to the vectors retrieved by MASK3. Multiple MAPs can be used to expand a list:

    MAPT3   S2345678

From these elements in the list, we must subtract actual capacity used. We scale by 0.159 to convert to barrels and by −1 to subtract:

    MASK4   C**.**UF
    MAPS4   S2345678
    SCLE4   −.159

The incentive to increase capacity is found on the reduced cost (D) of the capacity vectors. To identify this as an incentive, we map an I into the first position of the list names. Also we must convert $/cubic meter into $/barrel (6.29 barrels = 1 cubic meter). Since negative reduced costs, by convention, represent an incentive to increase capacity, we also change negative numbers to positive with our scaling:

    MASK5   C**.**UF
    STYP5   D
    MAPS5   I2345678
    SCLE5   −6.29.

At this point, it is advantageous to look at the content of the list that has been built by ADHOC.

- Ffd.unUF   KT/period of feed to unit   (MASK1/MAPS1)
- UTF.unUF   KB/period of capacity utilized   (MASK2/MAPS2)
- MTF.unUF   KB/period maximum capacity   (MASK3/MAPS3)
- STF.unUF   KB/period maximum capacity   (MASK3/MAPT3)
- STF.unUF   −KB/period of capacity utilized   (MASK4/MAPS4)
- ITF.unUF $/B of incentive to increase capacity   (MASK5/MAPS5)

where: fd = feed to unit; un = unit code.

Now, we wish to make this report a multicase report. As explained in Section 11.5.1, this requires us to specify (via CASEMASK) a position that can be used for "tagging" the members of the list to indicate which case they correspond to. In our list, we can see that position 4 (always a dot) is available for this purpose:

    CASEMASK ***0****

This will result in a tag (A for first case, B for second case, etc.) in position 4 of our list names.

Next, we need to convert all the KB/period figures into KB/day. This will involve scaling all elements in our list (except the ITF.unUF) by a case-specific scale, the number of days in the period for the case. This information can be provided at input time via a —SCALExxx card. The scaling is based on position 4 of our list names (the case tag created by CASEMASK).

VSCMSK1	000*0000	
VSCTAB1	*SCALE	Scales from —SCALExxx given at input
VSCCOL1	DP	Scales will be on —SCALEDP card
VSCDSL1	I*******	Do not scale $/B elements of list

What remains is to specify how the output report is to be printed:

STUBMASK	0000UN00	Stubs are process units
HEADMASK	C0000000	Heads are total feed, used capacity, and so on
SPLTMASK	000@0000	One report for each case
POINT	3	3 decimal places
TITLE	LO4	Title line
TITLE1	LO1	Title line
NEWPAGE	X	Always begin this report on a new page

This report would also require definition of a dictionary (say, CAPREP.D for Capacity Report Dictionary) defining the codes F, U, M, S, and I used in the head of the report.

It is also instructive to look at how, using ADHOC, we could write a material-balance report around all the process units. Figure 12.44 shows a skeleton of this report.

	PIPESTILL	REFORMER	CATALYTIC CRACKER	TOTAL	
Feeds	-	-	-	-	Feeds will be by total crude for pipestill, and by individual feed for other units.
Yields	+	+	+	+	Yields will be by individual product.
TOTAL	0	0	0	0	TOTAL stub will always show zero if unit yields are in material balance.

Figure 12.44 Skeleton of material balance report.

For this report we need a product of matrix coefficients on balance rows (both crude and product) with process vector (P**.****) activities. Figure 12.45 shows this report definition with appropriate annotation.

An economic-balance report around each unit could be similarly defined, using the ADHOC capability to multiply matrix coefficients by dual activities (row marginal values). Multiplying by both primal and dual activities will produce a total dollar balance.

We have talked about a set of six useful reports for our model. We may not want all reports in all cases. For example, we may want the crude run, capacity summary, demand incentive, and net material balance every time we run a plan. However, we may want our detailed balance reports only when starting a plan, to make certain that our balances are correct. ADHOC can define groups of reports. A possible group definition is shown in Figure 12.46.

Finally, we look at the syntax for calling ADHOC. We must specify to ADHOC:

- The cases to report.
- The reports (or sets of reports) to run.

MASM1	MATRIX	retrieve matrix coefficients,
COLM1	P**.****	for process columns,
ROWM1	B**...**	for product balance rows,
DOTC1	X	multiply coefficients by activities,
KEYM1	TUE.	list name will be product from balance row (TU) and unit class from process vector (E.),
MASM2	MATRIX	retrieve matrix coefficients
COLM2	P**.****	for crude distillation vectors,
ROWM2	A**.....	for crude balance rows,
DOTC2	X	multiply coefficients by activities,
KEYM2	TUE.	list name will be crude from balance row (TU) and unit class from process vector (E.),
MAPS2	XX345	retag all crudes to single code XX for total crude feed,
CASEMASK	****0***	multicase report using position 5 in list names for case tag,
STUBMASK	PR000000	stubs are products,
HEADMASK	00UN0000	heads are process units,
SPLTMASK	0000@000	one report for each case,
POINT	3	3 decimal places,
TOTAL	X	print row and column totals,
NEWPAGE	X	begin this report on a new page,
TITLE	L07	title line,
TITLE1	L01	title line.

Figure 12.45   Material balance report definition.

ZR.AH..Z	CRUDES CAPSUM PRODVALS MATBAL UNITMB UNITDB	Two groups of reports are defined.
*PLAN *BALANCE	1 3 2 4 1 2	Order of sequence within a group is also defined.

Figure 12.46   Defining groups of reports.

- Any case-specific scale.
- Textual information for case translation (optional).
- A title for the top of each page (optional).

The syntax to run the sets of reports, defined in Figure 12.46 for our three cases in the plan period, is shown in Figure 12.47. The total definition of the reports is shown in the ZR.AH..Z table in the display of ACTPROB DATABASE in Appendix C.

12.8.4 Marginal Analysis Reports

One of the strongest capabilities of LP is its ability to perform detailed analysis on the marginal (dual) values produced. This shows the marginal mechanisms behind various changes and justifies in detail the marginal economics. Such analysis can often be quite useful when the model gives results we find difficult to understand.

An example should help to illustrate this. If we look at the ADHOC report entitled "Incentive to Increase Product Demand—$/Ton" in Appendix B, we notice an extraordinarily high disincentive ($663.32 per ton) to increase the motor-gasoline demand in CASE BCASEM3. This number contrasts sharply with other marginal economics. It would be quite useful to find out exactly how additional motor gasoline would be made and why it is so expensive.

The TRANREP report writer of PLATOFORM is designed to answer this type

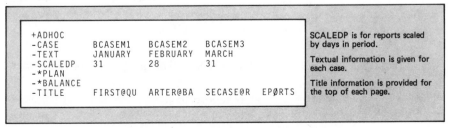

Figure 12.47   Input syntax for ADHOC reports.

```
+TRANREP
 -CASE          BCASEM3
 -CØLUMN        DGS...UF
```
Figure 12.48   Input for TRANREP report.

of query. It analyzes the changes in basic activities that would result from a change in a specified nonbasic activity. The changes are multiplied by objective function coefficients to determine the economic effect of each change.

In our case, the nonbasic activity we wish to analyze is the column named DGS. . .UF (demand for motor gasoline at Fawley), which is out of the basis at a fixed bound. TRANREP will show what basic variable changes are necessary to sell one additional ton of gasoline. At the same time, it will show the economic impact of each of the changes. The syntax for TRANREP is quite simple, requiring us to specify only the case and the nonbasic activity to be analyzed. The input syntax is shown in Figure 12.48.

Figure 12.49 shows the results of the TRANREP report. Examination of this report quickly reveals the reason for the high cost of making additional gasoline. To make one more ton of gasoline, Brega crude runs must be increased by 13.48 tons while decreasing Arabian heavy crude runs by 12.73 tons. The additional cost of Brega crude over Arabian heavy ($331.05 − 265.53 = $65.52 per ton, including transportation and processing cost) and the large crude switch needed account for the increase. The business analyst needs to evaluate whether this is a reasonable crude switch or whether there is some problem in the model data.

12.9   AN EXAMPLE SIDE STUDY

One of the fundamental objectives of PLATOFORM design is to make it very easy to run side cases. A typical side case would involve the following steps:

- Use DBM to update tables.
- Use GENER to regenerate relevant portion of the matrix.
- Use OPT to reoptimize the case.
- Run reports.

We illustrate how easily we can run side cases in PLATOFORM by looking at a change in the January base case.

- The maximum sulfur specification on fuel oil in January is to be decreased from 3.0% to 2.5%.
- We wish to run our *PLAN set of reports to show the differences between the 3% case and the 2.5% case.

TRANREP REPORT FOR NON-BASIC COLUMN: DGS...UF   DEMAND   MOGAS   FAWLEY

ON SOLUTION CASE BCASEM3   MARCH   BASECASE 3% SU   STORED ON ACTPROB BCASE

BASIC VECTOR			CODE	SUBSTITUTION RATE *	ORIGINAL COST	DJ VALUE CONTRIBUTION
AVAILBTY ARAB.HY	.	EUROPE	AAH......	-12.73	241.38	-3071.54
AVAILBTY BREGA	.	EUROPE	ABR......	13.48	321.00	4326.42
CAPACITY TOTFEED	.	PIPSTIL FAWLEY	CTF.D.UF	2.01	0.94	1.90
CAPACITY TOTFEED	.	REFORM FAWLEY	CTF.F.UF	1.20	3.77	4.53
DISTILL ARAB.HY	.	PIPSTIL FAWLEY	PAH.D.UF	-12.73	24.15	-307.31
DISTILL BREGA	.	PIPSTIL FAWLEY	PBR.D.UF	13.48	10.05	135.45
SPECIFY MOGAS	MAR	VOLUME FAWLEY	QGS3VOUF	1.24	3.12	3.86

DJ VALUE CONTRIBUTION FROM BASIC STRUCTURALS   1093.32
ORIGINAL COST IN OBJ FUNCTION   -430.00
DJ   663.32
VALID FOR NON-BASIC VARIABLE CHANGE OF (THETA)   4.18

DGS...UF IS OUT OF THE BASIS AT ITS FIXED LIMIT OF   45.00

* SUBSTITUTION RATE = -1 TIMES TRANCOL COEFFICIENT AFTER OPTIMIZATION.
  IT REPRESENTS THE CHANGE IN THE BASIC VECTOR RESULTING FROM A UNIT INCREASE IN THE NON-BASIC VECTOR.

Figure 12.49   Output from TRANREP report. Note that a large crude switch from Arabian
Heavy to Brega is required to produce more gasoline.

Running a side case requires the user to:

- Determine which data tables must be changed.
- Determine which generator submodules must be generated.

For this side study, after consulting our OPRINT, we find that the data to be changed is in table SB.F.UFT (specifications for blending fuels at Fawley). The only generator submodule affected is FB (fuel oil blending) at UF (Fawley).

In this side study, we also take the opportunity to make some very simple changes to our *PLAN set of ADHOC reports to make them more useful in comparing our side case and our base case. First, we wish to be able to report explicitly the differences between the two cases. If we multiply the side case by $+1$ and the base case by -1, then the TOTAL option of ADHOC will show the differences between cases any time cases are in the stub or head of the report:

ZR.AH..Z	CRUDES	CAPSUM	PRODVALS	MATBAL
SCALEDLT	DLT	DLT	DLT	DLT

We use a $-$SCALEDLT card in ADHOC to supply the scales used for delta reporting. Second, since we wish to report differences between cases, it is useful to put cases into the stub of the CAPSUM report. We reverse the role of cases and units in this report:

ZR.AH..Z	CRUDES	CAPSUM	PRODVALS	MATBAL
SPLTMASK		0000UN00		
STUBMASK		000@0000		
SCALEDLT	DLT	DLT	DLT	DLT

Finally, since we report differences, certain totals that were not previously meaningful now become meaningful, so we adjust our total requests accordingly, as follows:

ZR.AH..Z	CRUDES	CAPSUM	PRODVALS	MATBAL
TOTAL	X	STUB	HEAD	X
SPLTMASK		0000UN00		
STUBMASK		000@0000		
SCALEDLT	DLT	DLT	DLT	DLT

(c)

The entire PLATOFORM input to run this side study (including changes to the ADHOC report definitions) is shown in Figure 12.50. The ADHOC reports resulting from this side-case run are shown in Figure 12.51a and b.

```
+DBM          SCASE      BCASE
-UPDATE
 U SB.F.UFT   SUMX
   FLM1       2.5
+GENER
-ACTPRØB      SCASE
-FB           UF
+ØPT
-SCASE        SCASEM1
-SØLPRINT
-CØMMENT      JANUARY    SIDECASE   @2.5%@SU
+UNLØAD
+DBM          DATABASE
-UPDATE
 U ZR.AH..Z   CRUDES     CAPSUM     PRØDVALS   MATBAL
   TØTAL      X          STUB       HEAD       X
   SPLTMASK              0000UN00
   STUBMASK              000@0000
   SCALEDLT   DLT        DLT        DLT        DLT
+ADHOC
-CASE         SCASEM1    BCASEM1
-SCALEDP      31         31
-SCALEDLT     1          -1
-*PLAN
-TITLE        JANUARY    SIDECASE   @@-@BAS   ECASE@CØ   MPARISON
+STØP
```

SCASE is created as a copy of BCASE for doing the side study.

ADHOC reports are redefined.

Results of the study are saved via +UNLOAD.

Since UNLOAD is done before DBM on DATABASE, report definition changes are only temporary.

Figure 12.50 Example side-case input.

REFINERY CRUDE RUNS   UNITS: KB/D   KB
===================

		JANUARY SIDECASE 2.5% SU	JANUARY BASECASE 3% SU	*TOTALS*
ARABIAN LIGHT	AL	26.010	-26.010	
ARABIAN HEAVY	AH	3.750	-31.329	-27.578
BREGA	BR	18.232	-1.305	16.927
TOTALS		47.993	-58.644	-10.651

REFINERY CRUDE RUNS   UNITS: K TONS   KT
===================

		JANUARY SIDECASE 2.5% SU	JANUARY BASECASE 3% SU	*TOTALS*
ARABIAN LIGHT	AL	110.000	-110.000	-120.438
ARABIAN HEAVY	AH	16.379	-136.816	68.666
BREGA	BR	73.959	-5.293	-51.772
TOTALS		200.338	-252.109	

INCENTIVE TO INCREASE PRODUCT DEMAND--$/TON
===

		JANUARY SIDECASE 2.5% SU	JANUARY BASECASE 3% SU	*TOTALS*
MOTOR GASOLINE	GS	-42.03	68.30	26.27
JET FUEL	JF	-33.01	3.49	-29.52
HEATING OIL	HO	-3.87	1.51	-2.36
FUEL OIL	FL	-13.47	-5.00	-18.47

```
*********************************************
* N.B.: ALL PRODUCT DEMANDS ARE FORMULATED AS FIXED. *
****REALIZATIONS HAVE BEEN INCLUDED IN FORMULATION****
****POSITIVE NUMBER==>INCENTIVE TO INCREASE DEMAND****
****NEGATIVE NUMBER==>INCENTIVE TO DECREASE DEMAND****
*********************************************
```

(a)

PROCESS UNIT CAPACITY SUMMARY   CASE: PIPSTIL D.

	UNIT FEED --KT/D F	USED CAPACITY --KB/D U	SPARE CAPACITY --KB/D S	MAXIMUM CAPACITY --KB/D M
JANUARY SIDECASE 2.5% SU	6.463	47.993	17.000	64.993
JANUARY BASECASE 3% SU	-8.133	-58.644	-6.349	-64.993
TOTALS	-1.670	-10.651	10.651	

PROCESS UNIT CAPACITY SUMMARY   CASE: CATCRAK C.

	UNIT FEED --KT/D F	USED CAPACITY --KB/D U	SPARE CAPACITY --KB/D S	MAXIMUM CAPACITY --KB/D M	VALUE OF CAPACITY --$/B I
JANUARY SIDECASE 2.5% SU	1.057	7.225	.	7.225	13.592
JANUARY BASECASE 3% SU	-1.057	-7.225	.	-7.225	-15.700
TOTALS					-2.107

PROCESS UNIT CAPACITY SUMMARY   CASE: REFORM F.

	UNIT FEED --KT/D F	USED CAPACITY --KB/D U	SPARE CAPACITY --KB/D S	MAXIMUM CAPACITY --KB/D M
JANUARY SIDECASE 2.5% SU	0.709	6.049	0.724	6.773
JANUARY BASECASE 3% SU	-0.709	-6.049	-0.724	-6.773
TOTALS				

NET REFINERY MATERIAL BALANCE --OOO TONS

		JANUARY SIDECASE 2.5% SU	JANUARY BASECASE 3% SU	*TOTALS*
CRUDE OIL FEEDS	P	-200.338	252.109	51.772
EXPORTS-IMPORTS (NET)	E	-58.643	9.793	-48.850
PRODUCT DEMANDS	D	255.000	-255.000	-2.921
MATERIAL LOSS	B	3.981	-6.902	0.000
TOTALS		0.000	0.000	0.000

(b)

Figure 12.51   (a) and (b) Example ADHOC reports for the side case.

331

It is worth noting several factors from this analysis:

- Brega crude runs increase as the sulfur specification decreases.
- Total crude runs decrease as the sulfur specification decreases.
- All products except gasoline become more expensive to manufacture.
- The value of catalytic cracker capacity decreases.

In the next section, we look at a technique that affords a more detailed view of how model activities vary as the sulfur specification changes.

12.10 A PARAMETRIC ANALYSIS

In Section 12.9, we looked at a side case based on changing the sulfur specification for fuel oil from 3% to 2.5%. In particular, we noticed a very large increase in Brega crude runs as the sulfur specification decreased and a significant decrease in total crude runs. It would be interesting to be able to analyze what happens to crude runs throughout this change in sulfur specification. Thus we might wish to have results not only at 3 and 2.5%, but also at 2.9, 2.8, 2.7, and 2.6%. Obviously, this could be done by running a whole series of side cases as we did for the 2.5% case in Section 12.9. However, parametric programming techniques allow us to do this in a way that uses both analyst and computer time more efficiently.

PLATOFORM offers the user the ability to do parametric analysis on right-hand-side coefficients, objective function coefficients, columns of the model, and rows of the model. To run the parametric analysis, the user must be familiar with the formulation of the model. [In fact, effective use of linear programming (LP), whatever software support system is used, requires the analyst to be somewhat familiar with the model's formulation.]

In our parametric example, we are interested in changing the fuel oil specification, which appears on a specification vector as shown below:

```
        QFL1WTUF
XSU.FLUF  -3.0
```

We want this coefficient to increase from -3.0 to -2.5 (a specification decrease implies a coefficient increase because of the minus sign). Also, we wish to see results at coefficient values of -3.0, -2.9, -2.8, -2.7, -2.6, and -2.5. All of this information must be passed to PLATOFORM, using the parametric-optimization syntax. We also provide case names for each of the solution points we wish to analyze. Figure 12.52 illustrates the input syntax to request a parametric optimization.

```
 +ØPT
 -SCASE          BCASEM1
 -CØMMENT        JANUARY    BASECASE    @3.0%@SU
 =PARACØL        1          5
      *RHS                  QFL1WTUF
      XSU.FLUF   .1
      *CASE
      BCASEM19   JANUARY    PARMTRIC    @2.9%@SU
      BCASEM18   JANUARY    PARMTRIC    @2.8%@SU
      BCASEM17   JANUARY    PARMTRIC    @2.7%@SU
      BCASEM16   JANUARY    PARMTRIC    @2.6%@SU
      BCASEM15   JANUARY    PARMTRIC    @2.5%@SU
 -PSØLUTN
```

Parametric procedures are called as an option under +OPT.

=PARACOL defines interval at which to report (1) and maximum number of reports (5).

*RHS defines column to be changed (QFL1WTUF) and changes on each row (.1 on XSU.FLUF).

*CASE defines case names and comments for SOL....C.

PSOLUTN requests solution print at each reporting interval.

Figure 12.52   Input to run parametric analysis for fuel oil sulfur.

Figure 12.53a, b, and c shows a set of ADHOC reports that result from this parametric run. Some changes have been made to the ADHOC reports, including the following:

- Definition of a fuel oil blending report.
- Definition of a recipe blending report.
- Change in the roles of cases and units in the CAPSUM report (as we did for the delta reports).

The ADHOC report "Incentive to Increase Product Demand—$/Ton" in Figure 12.53b is particularly interesting. Of course, it shows an increase in the cost of fuel oil at each stage. More interestingly, however, no changes take place in any other marginal values between 2.9 and 2.5%. This is an indication that there are no basis changes over this range. In fact, the entire parametric run had only one basis change (between 3.0 and 2.9%). The change at that basis was that light virgin naphtha was no longer burned as fuel, and blend 3 began to be used for jet fuel (see report "Recipe Blending Summary" in Figure 12.53a).

12.11 SUMMARY

In this chapter, we have looked at the development and usage of an LP model using the PLATOFORM system. We saw how the various components of the system fit together to give the user a powerful software support system for using LP as a planning tool.

Starting with a basic description of supply planning, we saw how a data base could be designed to support the data. We then saw how this data could be

FUEL OIL BLEND COMPOSITION BY SULFUR LEVEL
===

		JANUARY BASECASE 3.0% SU	JANUARY PARMTRIC 2.9% SU	JANUARY PARMTRIC 2.8% SU	JANUARY PARMTRIC 2.7% SU	JANUARY PARMTRIC 2.6% SU	JANUARY PARMTRIC 2.5% SU
VIRGIN HEATING OIL	VH	23.8870	22.9098	21.9745	21.0391	20.1038	19.1684
VACUUM DISTILAT	VD	21.7504	20.5474	19.3871	18.2268	17.0664	15.9061
BREGA VACUUM RESID	YB	0.3650	1.3119	2.2591	3.2063	4.1535	5.1006
ARAB.LT VACUUM RESID	YL	12.5172	12.5172	12.5172	12.5172	12.5172	12.5172
ARAB.HY VACUUM RESID	YH	31.6093	25.9957	20.4428	14.8898	9.3369	3.7840
FUEL OIL IMPORT 3.0% SU	FI	7.4915	14.3758	21.0772	27.7786	34.4800	41.1814
REFINRY FUEL EQUIV	RF	2.3795	2.3422	2.3422	2.3422	2.3422	2.3422
TOTALS		100.0000	100.0000	100.0000	100.0000	100.0000	100.0000

```
*************************************************
* FUEL OIL IMPORT ASSUMED TO ALWAYS BE 3% SULFUR *
*************************************************
```

RECIPE BLENDING SUMMARY
=========================

			JANUARY BASECASE 3.0% SU	JANUARY PARMTRIC 2.9% SU	JANUARY PARMTRIC 2.8% SU	JANUARY PARMTRIC 2.7% SU	JANUARY PARMTRIC 2.6% SU	JANUARY PARMTRIC 2.5% SU
JETFUEL	JF BL 2	JFB2	100.0000	98.4454	95.8702	93.2950	90.7198	88.1447
	JF BL 3	JFB3	.	1.5546	4.1298	6.7050	9.2802	11.8553
****
HEATOIL	VIRG HO	HOVH	71.1696	71.6671	72.4911	73.3152	74.1393	74.9633
	CATDIST	HOCD	28.8304	28.3329	27.5089	26.6848	25.8607	25.0367
****
REFFUEL	LVN	RFLN	100.0000	.	.	.	.	.
****

(a)

REFINERY CRUDE RUNS   UNITS: KB/D   KB
==================

	JANUARY BASECASE 3.0% SU	JANUARY PARMTRIC 2.9% SU	JANUARY PARMTRIC 2.8% SU	JANUARY PARMTRIC 2.7% SU	JANUARY PARMTRIC 2.6% SU	JANUARY PARMTRIC 2.5% SU
ARABIAN LIGHT AL	26.010	26.010	26.010	26.010	26.010	26.010
ARABIAN HEAVY AH	31.329	25.765	20.261	14.758	9.254	3.750
BREGA BR	1.305	4.689	8.075	11.461	14.846	18.232
TOTALS	58.644	56.465	54.347	52.229	50.111	47.993

REFINERY CRUDE RUNS   UNITS: K TONS   KT
==================

	JANUARY BASECASE 3.0% SU	JANUARY PARMTRIC 2.9% SU	JANUARY PARMTRIC 2.8% SU	JANUARY PARMTRIC 2.7% SU	JANUARY PARMTRIC 2.6% SU	JANUARY PARMTRIC 2.5% SU
ARABIAN LIGHT AL	110.000	110.000	110.000	110.000	110.000	110.000
ARABIAN HEAVY AH	136.816	112.519	88.484	64.449	40.414	16.379
BREGA BR	5.293	19.022	32.756	46.491	60.225	73.959
TOTALS	252.109	241.541	231.240	220.939	210.639	200.338

INCENTIVE TO INCREASE PRODUCT DEMAND---$/TON
==

	JANUARY BASECASE 3.0% SU	JANUARY PARMTRIC 2.9% SU	JANUARY PARMTRIC 2.8% SU	JANUARY PARMTRIC 2.7% SU	JANUARY PARMTRIC 2.6% SU	JANUARY PARMTRIC 2.5% SU
MOTOR GASOLINE GS	-68.30	-42.03	-42.03	-42.03	-42.03	-42.03
JET FUEL JF	-3.49	-33.01	-33.01	-33.01	-33.01	-33.01
HEATING OIL HO	-1.51	-3.87	-3.87	-3.87	-3.87	-3.87
FUEL OIL FL	5.00	1.31	-2.39	-6.08	-9.77	-13.47

```
*******************************************
* N.B.: ALL PRODUCT DEMANDS ARE FORMULATED AS FIXED. *
****REALIZATIONS HAVE BEEN INCLUDED IN FORMULATION****
****POSITIVE NUMBER==>INCENTIVE TO INCREASE DEMAND****
****NEGATIVE NUMBER==>INCENTIVE TO DECREASE DEMAND****
*******************************************
```

(b)

Figure 12.53 (a), (b) ADHOC reports for parametric study on fuel oil sulfur.

(continued on next page)

PROCESS UNIT CAPACITY SUMMARY

		UNIT FEED --KT/D (F)	USED CAPACITY --KB/D (U)	SPARE CAPACITY --KB/D (S)	MAXIMUM CAPACITY --KB/D (M)	VALUE OF CAPACITY --$/B (I)
PIPSTIL	BCASEM1 D.A	8.133	58.644	6.349	64.993	.
	BCASEM19 D.B	7.792	56.465	8.528	64.993	.
	BCASEM18 D.C	7.459	54.347	10.646	64.993	.
	BCASEM17 D.D	7.127	52.229	12.764	64.993	.
	BCASEM16 D.E	6.795	50.111	14.882	64.993	.
	BCASEM15 D.F	6.463	47.993	17.000	64.993	.
CATCRAK	BCASEM1 C.A	1.057	7.225	.	7.225	15.700
	BCASEM19 C.B	1.057	7.225	.	7.225	13.592
	BCASEM18 C.C	1.057	7.225	.	7.225	13.592
	BCASEM17 C.D	1.057	7.225	.	7.225	13.592
	BCASEM16 C.E	1.057	7.225	.	7.225	13.592
	BCASEM15 C.F	1.057	7.225	.	7.225	13.592
REFORM	BCASEM1 F.A	0.709	6.049	0.724	6.773	.
	BCASEM19 F.B	0.709	6.049	0.724	6.773	.
	BCASEM18 F.C	0.709	6.049	0.724	6.773	.
	BCASEM17 F.D	0.709	6.049	0.724	6.773	.
	BCASEM16 F.E	0.709	6.049	0.724	6.773	.
	BCASEM15 F.F	0.709	6.049	0.724	6.773	.

NET REFINERY MATERIAL BALANCE --000 TONS

		JANUARY BASECASE 3.0% SU	JANUARY PARMTRIC 2.9% SU	JANUARY PARMTRIC 2.8% SU	JANUARY PARMTRIC 2.7% SU	JANUARY PARMTRIC 2.6% SU	JANUARY PARMTRIC 2.5% SU
CRUDE OIL FEEDS	P	-252.109	-241.541	-231.240	-220.939	-210.639	-200.338
EXPORTS-IMPORTS (NET)	E	-9.793	-19.775	-29.492	-39.209	-48.926	-58.643
PRODUCT DEMANDS	D	255.000	255.000	255.000	255.000	255.000	255.000
MATERIAL LOSS	B	6.902	6.316	5.732	5.148	4.565	3.981
TOTALS		-0.000	-0.000	-0.000	0.000	0.000	0.000

(c)

Figure 12.53 continued   (c) ADHOC reports for parametric study on fuel oil sulfur.

336

looked at in a modular fashion and a matrix generator developed to fit the particular view of the data.

We saw how model optimization, which is mathematically the most complex piece of an LP application, is very simple for the user to call. This included advanced capabilities for carrying out parametric analysis. (The PLATOFORM system also includes very strong capabilities for mixed-integer and nonlinear optimization that were not covered in our discussion.)

We also looked at the ease of running side studies using PLATOFORM. There are also systems efficiency features in the generator's capability to do a partial regeneration rather than a regeneration of the entire matrix.

One of the most important aspects of an LP support system is its report writer capabilities. We looked at report writers, with particular attention given to the ADHOC report module of PLATOFORM. Examples showed the ease and speed with which reports can be developed and revised.

PLATOFORM is an essential part of Exxon's use of LP. It has enabled us to bring our data handling, model generation, and results analysis tools to a level comparable to existing optimization capabilities. In fact, PLATOFORM has given us the facilities to push forward our ability to optimize nonlinear and nonconvex formulations.

The PLATOFORM system has been in constant evolution over the past ten years. To continue to meet the needs of LP users, we expect the PLATOFORM system to continue its evolution in a controlled manner. In fact, a large project has been undertaken to define the needs that PLATOFORM must meet to continue to be a viable tool into the 1990s.

APPENDIX A. TABLES IN OPRINT

ALPHABETIC LIST OF TABLES IN OPRINT
==

TABLE	DESCRIPTION			INDEX NO.
BA....T	CRUDE OIL AVAILS ----KT	FAWLEY	MILFORD	0001
BE.UFUMT	IMPORT/EXPORT LIMITS--KT	MILFORD	FAWLEY	0004
BE.UMUFT	IMPORT/EXPORT LIMITS--KT	FAWLEY		0004
BS...UFT	PRODUCT DEMANDS ----KT	FAWLEY		0003
BU.CPUFT	PROCESS UNIT CAPACITIES	EXIST	FAWLEY	0005
BU.TAUFT	EXCEPTIONAL STRUCTURE	DOWN	FAWLEY	0005
EB.GBUFT	EXCEPTIONAL STRUCTURE	GBLEND	FAWLEY	0014
EB.PRUFT	EXCEPTIONAL STRUCTURE	PROCESS	FAWLEY	0010
FA.BL..T	CRUDE OIL PRICES ----$/	BARREL		0001
FC...UFT	CRUDE TRANSP PRICE-$/TON	FAWLEY		0007
FE.UMUFT	IMPORT/EXPORT PRICES-$/T	MILFORD	FAWLEY	0004
FS....T	PRODUCT SALES PRICES-$/T	EUROPE		0003
FU.GRUFT	CAPACITY&INVESTMENT COST	FAWLEY		0006
IA...Z	CRUDE OILS AVAILABLE			0001
IB.F.UFZ	COMPONENTS TO BLENDS	FUELS	FAWLEY	0015
IB.G.UFZ	COMPONENTS TO BLENDS	MOGAS	FAWLEY	0013
ID...UFZ	CRUDE PROCESSING OPTIONS	FAWLEY		0007
IG....Z	GENERATOR MODULE OPTIONS			0017
IM...UFZ	RECIPE BLENDING OPTIONS			0011
IP...UFZ	PROCESS UNIT OPTIONS	FAWLEY		0009
IQ.F.UFZ	QUALITY BLENDING OPTIONS	FUELS	FAWLEY	0015
IQ.G.UFZ	QUALITY BLENDING OPTIONS	MOGAS	FAWLEY	0013
IU...UFZ	CAPACITY&INVESTMENT OPTN	FAWLEY		0005
KD...UFZ	DISTILLATION CAP CONSUME	FAWLEY		0008
KQ.F...Z	BLENDING METHODOLOGY	FUELS	EUROPE	0015
KQ.G...Z	BLENDING METHODOLOGY	MOGAS	EUROPE	0013
P.DAYS.T	DAYS IN TIME PERIODS			0006
P2....Z	TIME PERIODS UNDER STUDY			0017
QB.F.UFT	BLENDING COMP QUALITIES	FUELS	FAWLEY	0016
QB.G.UFT	BLENDING COMP QUALITIES	MOGAS	FAWLEY	0014
QC....T	CRUDE OIL PROPERTIES			0002
SB.F.UFT	BLENDED PRODUCT SPECS	FUELS	FAWLEY	0016
SB.G.UFT	BLENDED PRODUCT SPECS	MOGAS	FAWLEY	0014
YD...UFT	CRUDE DISTILLATION YIELD	FAWLEY		0007
YM.JFUFT	PRODUCT BLENDING RECIPES	JETFUEL	FAWLEY	0011
YM.RFUFT	PRODUCT BLENDING RECIPES	REFFUEL	FAWLEY	0012
YP.C.UFT	PROCESS UNIT YIELDS	CATCRAK	FAWLEY	0009
YP.F.UFT	PROCESS UNIT YIELDS	REFORM	FAWLEY	0009

OPRINT   CRUDE OIL SUPPLY   LOCATION ..   ACTPROB BCASE   DATERUN 82122101   INDEX NO.   0001

IA.....Z   UPDATED 82122101   RESIDENT ON BCASE
========   ========

CRUDE OILS AVAILABLE

ARABIAN LIGHT   AL
ARABIAN HEAVY   AH
BREGA   BR

BA....T   UPDATED 82122101   RESIDENT ON BCASE
========   ========

CRUDE OIL AVAILS -----KT

              JANUARY   FEBRUARY   MARCH
              M1   M2   M3
ARABIAN LIGHT   AL   110.000   105.000   115.000
ARABIAN HEAVY   AH   165.000   155.000   170.000
BREGA   BR   80.000   70.000   80.000

FA.BL..T   UPDATED 82122101   RESIDENT ON BCASE
========   ========

CRUDE OIL PRICES -----$/ BARREL

                   ALL
                   PERIODS
ARABIAN LIGHT   AL   35.00
ARABIAN HEAVY   AH   34.00
BREGA   BR   42.00

341

DBM - PLATOFORM SYSTEM - VER 1715

82/12/21 20.42.04

OPRINT   CRUDE OIL SUPPLY   LOCATION ..   ACTPROB BCASE   DATERUN 82122101   INDEX NO.   0002

QC.....T   UPDATED 82122101   RESIDENT ON BCASE
========   =========

          CRUDE OIL PROPERTIES

                                    SPECFIC
                                    GRAVITY

                                      SG
ARABIAN LIGHT   AL   0.8580
ARABIAN HEAVY   AH   0.8860
BREGA   BR   0.8230

OPRINT   PRODUCT SUPPLY   LOCATION UF   ACTPROB BCASE   DATERUN 82122101   INDEX NO.   0003

BS...UFT   UPDATED 82122101   RESIDENT ON BCASE
=======   PRODUCT DEMANDS ------KT   FAWLEY   U.K.   ========

                                     JAN   FEB   MAR
                                     FIX BND   FIX BND   FIX BND

                                     M1FX   M2FX   M3FX
MOTOR GASOLINE   GS   40.000   35.000   45.000
JET FUEL   JF   20.000   20.000   20.000
HEATING OIL   HO   50.000   50.000   40.000
FUEL OIL   FL   145.000   130.000   145.000

FS.....T   UPDATED 82122101   RESIDENT ON BCASE
=======   PRODUCT SALES PRICES-$/T   EUROPE   ========

                                     ALL
                                     PERIODS

MOTOR GASOLINE   GS   -430.00
JET FUEL   JF   -300.00
HEATING OIL   HO   -315.00
FUEL OIL   FL   -250.00

DATE   82122101
LO1   ALL DEMANDS HAVE BEEN FORMAULATED AS FIXED REQUIREMENTS
LO2   REALIZATIONS WILL HAVE NO EFFECT ON OPTIMIZATION BUT
LO3   WILL CAUSE MARGINAL VALUES TO REFLECT REVENUES

```
              DBM - PLATOFORM SYSTEM - VER 1715                          82/12/21  20.42.04

OPRINT   EXTERNAL IMPORT/EXPORT   LOCATION UF        ACTPROB BCASE      DATERUN 82122101      INDEX NO.      0004

BE.UFUMT
========
         IMPORT/EXPORT LIMITS--KT    FAWLEY    U.K.      UPDATED 82122101                  RESIDENT ON BCASE
                                                         MILFORD HAVEN   U.K.              ========

                        JAN        FEB        MAR
                        FIX BND    FIX BND    FIX BND

REFORMATE        PF     M1FX       M2FX       M3FX
                        1.650      0.000      0.000

BE.UMUFT
========
         IMPORT/EXPORT LIMITS--KT    MILFORD HAVEN   U.K.      UPDATED 82122101             RESIDENT ON BCASE
                                                               FAWLEY   U.K.                ========

                        JAN        FEB        MAR
                        FIX BND    FIX BND    FIX BND

REFORMATE               M1FX       M2FX       M3FX
CATLYTC CRACKED NAPH-LS  PF  0.000  1.650      0.000
                         CL  0.580  0.000      0.000

FE.UMUFT
========
         IMPORT/EXPORT PRICES-$/T    MILFORD HAVEN   U.K.      UPDATED 82122101             RESIDENT ON BCASE
                                                               FAWLEY   U.K.                ========

                        ALL
                        PERIODS

FUEL OIL IMPORT 3.0% SU  FI         245.00
```

```
                    DBM -  PLATOFORM SYSTEM - VER 1715                      82/12/21  20.42.04

OPRINT    REFINERY CAPACITIES      LOCATION UF        ACTPROB BCASE      DATERUN 82122101    INDEX NO.      0005

IU...UFZ    CAPACITY&INVESTMENT OPTN  FAWLEY      U.K.           UPDATED 82122101      RESIDENT ON BCASE
========                                                                               ========

PIPSTIL TOTFEED    D.TF    C
REFORM  TOTFEED    F.TF    C
CATCRAK TOTFEED    C.TF    C

BU.CPUFT    PROCESS UNIT CAPACITIES    EXISTING CAPACITY     UPDATED 82122101      RESIDENT ON BCASE
========                                                        FAWLEY    U.K.     ========

                      ALL
                      PERIODS

PIPSTIL TOTFEED    D.TF    65.000
REFORM  TOTFEED    F.TF     7.500
CATCRAK TOTFEED    C.TF     8.000

DATE              82122101
ALL CPACITIES ARE GIVEN IN BARRELS PER STREAM DAY
LO1         WILL BE CONVERTED TO BARRELS PER PERIOD USING
LO2         TURN AROUND DATA AND NUMBER OF DAYS PER PERIOD
LO3

BU.TAUFT    PROCESS UNIT CAPACITIES    TURN   AROUND   DAYS    UPDATED 82122101      RESIDENT ON BCASE
========                                                        FAWLEY    U.K.      ========

                      JANUARY

                      M1
REFORM  ........   F.    3.000
CATCRAK ........   C.    3.000
```

345

DBM - PLATOFORM SYSTEM - VER 1715

OPRINT   REFINERY CAPACITIES   LOCATION UF   ACTPROB BCASE   DATERUN 82122101   INDEX NO.   0006

P.DAYS.T   UPDATED 82122101   RESIDENT ON BCASE
========   ========

DAYS IN TIME PERIODS

                        DAYS
JANUARY   M1   31
FEBRUARY   M2   28
MARCH   M3   31

FU.GRUFT   CAPACITY&INVESTMENT COST   UPDATED 82122101   RESIDENT ON BCASE
========   FAWLEY   U.K.   ========

                        OPERTNG
                        COST
                        $/B
                        OP
PIPSTIL TOTFEED   D.TF   0.15
REFORM   TOTFEED   F.TF   0.60
CATCRAK TOTFEED   C.TF   0.65

OPRINT   CRUDE OIL DISTILLATION   LOCATION UF   ACTPROB BCASE   DATERUN 82122101   INDEX NO.   0007

ID...UFZ   CRUDE PROCESSING OPTIONS   FAWLEY   U.K.   UPDATED 82122101   RESIDENT ON BCASE
========   ========

PIPSTIL

		D.
ARABIAN LIGHT	AL	Y
ARABIAN HEAVY	AH	Y
BREGA	BR	Y

YD...UFT   CRUDE DISTILLATION YIELD   FAWLEY   U.K.   UPDATED 82122101   RESIDENT ON BCASE
========   ========

		ARAB.LT PIPSTIL ALD.	ARAB.HY PIPSTIL AHD.	BREGA PIPSTIL BRD.
LIGHT VIRGIN NAPHTHA	LN	3.50	3.00	4.50
INTERMD VIRGIN NAPHTHA	IN	10.00	7.50	13.50
VIRGIN HEATING OIL	VH	39.00	30.00	43.00
VACUUM DISTILAT	VD	28.50	23.00	28.00
BREGA VACUUM RESID	YB			10.00
ARAB.LT VACUUM RESID	YL	16.50		
ARAB.HY VACUUM RESID	YH		33.50	
MATERAIL LOSS	ML	2.50	3.00	1.00

FC...UFT   CRUDE TRANSP PRICE-$/TON   FAWLEY   U.K.   UPDATED 82122101   RESIDENT ON BCASE
========   ========

		ALL PERIODS
ARABIAN LIGHT	AL	24.150
ARABIAN HEAVY	AH	24.150
BREGA	BR	10.050

347

DBM - PLATOFORM SYSTEM - VER 1715

82/12/21 20.42.04

OPRINT   CRUDE OIL DISTILLATION   LOCATION UF   ACTPROB BCASE   DATERUN 82122101   INDEX NO.   0008

KD...UFZ
========
DISTILLATION CAP CONSUME   FAWLEY   U.K.   UPDATED 82122101   RESIDENT ON BCASE
                                                                ========

PIPSTIL

TOTAL   FEED   TF   D.
                    D.

OPRINT   OTHER REFINERY PROCESSES   LOCATION UF   ACTPROB BCASE   DATERUN 82122101   INDEX NO.   0009

IP...UFZ   PROCESS UNIT OPTIONS   FAWLEY   U.K.   UPDATED 82122101   RESIDENT ON BCASE
========   ========

		CATLYTC CRACKER LO SEV	CATLYTC CRACKER HI SEV	REFORMER
		CL	CH	F.
				Y
INTERMD VIRGIN NAPHTHA	IN			Y
VIRGIN HEATING OIL	VH	Y	Y	
VACUUM DISTILAT	VD	Y	Y	

YP.C.UFT   PROCESS UNIT YIELDS   CATLYTC CRACKER   UPDATED 82122101   RESIDENT ON BCASE
========   FAWLEY   U.K.   ========

		VIRG HO CC---LS	VIRG HO CC---HS	VACDIST CC---LS	VACDIST CC---HS
		VHCL	VHCH	VDCL	VDCH
BUTANE	C4	2.0000	3.2500	5.0000	6.0000
VIRGIN HEATING OIL	VH	-100.0000	-100.0000		
CATLYTC CRACKED NAPH-LS	CL	27.5000		32.5000	
CATLYTC CRACKED NAPH-HS	CH		37.7500		45.0000
CATLYTC CRACKED DISTILAT	CD	68.0000	55.5000	58.5000	44.0000
VACUUM DISTILAT	VD			-100.0000	-100.0000
REFINRY FUEL EQUIV	RF	2.5000	3.5000	4.0000	5.0000

YP.F.UFT   PROCESS UNIT YIELDS   REFORMER   UPDATED 82122101   RESIDENT ON BCASE
========   FAWLEY   U.K.   ========

		IVN REFORM
		INF. REFORM
BUTANE	C4	2.0000
INTERMD VIRGIN NAPHTHA	IN	-100.0000
REFORMATE	PF	90.0000
REFINRY FUEL EQUIV	RF	8.0000

OPRINT   OTHER REFINERY PROCESSES LOCATION UF   ACTPROB BCASE   DATERUN 82122101   INDEX NO.   0010

EB.PRUFT   UPDATED 82122101   RESIDENT ON BCASE
========   EXCEPTIONAL STRUCTURE   SPECIAL PROCESSES   FAWLEY   U.K.   ========

PIN.F.UF   PVH.C*UF   PVD.C*UF   Y
-1.3569

CTF.F.UF   -1.1287   -1.0870   .
CTF.C.UF   .

DATE   82122101
LO1   CAPACITY CONSUMPTION IN CUBIC METERS CALCULATED USING
LO2   THE FOLLOWING ASSUMED FEED GRAVITIES
LO3   INTERMEDIATE VIRGIN NAPHTHA=.737
LO4   VIRGIN HEATING OIL   =.886
LO5   VACUUM DISTILLATE   =.920

350

OPKINT   RECIPE BLENDING   LOCATION UF   ACTPROB BCASE   DATERUN 82122101   INDEX NO.   0011

IM...UFZ
========   RECIPE BLENDING OPTIONS   FAWLEY   U.K.   UPDATED 82122101   RESIDENT ON BCASE
                                                                                ========

		JET FUEL	HEATING OIL	REFINRY FUEL EQUIV
		JF	HO	RF
BUTANE	C4			
LIGHT VIRGIN NAPHTHA	LN			Y
VIRGIN HEATING OIL	VH		1	Y
CATLYTC CRACKED DISTILAT	CD		1	
JETFUEL BLEND #1	B1	Y		
JETFUEL BLEND #2	B2	Y		
JETFUEL BLEND #3	B3	Y		

YM.JFUFT
========   PRODUCT BLENDING RECIPES   JET FUEL   UPDATED 82122101   RESIDENT ON BCASE
                                        FAWLEY   U.K.   ========

		JF BL 1 JETFUEL	JF BL 2 JETFUEL	JF BL 3 JETFUEL
		B1JF	B2JF	B3JF
LIGHT VIRGIN NAPHTHA	LN	-0.3000	-0.2000	-0.1000
INTERMD VIRGIN NAPHTHA	IN			-0.1000
VIRGIN HEATING OIL	VH	-0.7000	-0.8000	-0.8000
CATLYTC CRACKED DISTILAT	CD			
JET FUEL	JF	1.0000	1.0000	1.0000

DBM - PLATOFORM SYSTEM - VER 1715

82/12/21   20.42.04

OPRINT   RECIPE BLENDING   LOCATION UF   ACTPROB BCASE   DATERUN 82122101   INDEX NO.   0012

YM.RFUFT   UPDATED 82122101   RESIDENT ON BCASE
========   PRODUCT BLENDING RECIPES   REFINRY FUEL   EQUIV   FAWLEY   U.K.   ========

                        BUTANE   LVN
                        REFFUEL   REFFUEL

                        C4RF   LNRF
BUTANE   C4   -1.0000   -1.0000
LIGHT   VIRGIN NAPHTHA LN
REFINRY FUEL   EQUIV RF   1.1100   1.1070
MATERAIL   LOSS   ML   -0.1100   -0.1070

DATE   82122101
LO1   RF ENTRIES REPRESENT GREATER FUEL VALUE OF COMPONENTS
LO2   ML ENTRIES INCLUDED TO ALLOW MATERIAL BALANCE REPORTS

352

OPRINT   MOTOR GASOLINE BLENDING   LOCATION UF   ACTPROB BCASE   DATERUN 82122101   INDEX NO.   0013

IQ.G.UFZ   UPDATED 82122101   RESIDENT ON BCASE
========   QUALITY BLENDING OPTIONS   MOTOR   GASOLINE   FAWLEY   U.K.   ========

                                     MOTOR
                                     GASOLINE

                                     GS
RVP   MINSPEC   RVMN   X
% 212F MAXSPEC   D1MX   X
RON .4G MINSPEC   R4MN   X

IB.G.UFZ   UPDATED 82122101   RESIDENT ON BCASE
========   COMPONENTS TO BLENDS   MOTOR   GASOLINE   FAWLEY   U.K.   ========

                                     MOTOR
                                     GASOLINE

                                     GS
BUTANE   C4   X
LIGHT VIRGIN NAPHTHA   LN   X
REFORMATE   PF   X
CATLYTC CRACKED NAPH-LS   CL   X
CATLYTC CRACKED NAPH-HS   CH   X

KQ.G...Z   UPDATED 82122101   RESIDENT ON BCASE
========   BLENDING METHODOLOGY   MOTOR   GASOLINE   EUROPE   ========

                        BLENDING   INPUT   MODEL   SCALING
                        METHOD   DATA   DATA   FACTOR
                                    UNITS   UNITS
                                    FT   RT   SC
                        BA   100
PERCENT OFF AT 212 F D1   VO
REID   VAPOR PRESSURE RV   VO   ON1   RBN
RESERCH OCTNAE .4G/PB/L R4 VO

353

DBM - PLATOFORM SYSTEM - VER 1715

OPRINT   MOTOR GASOLINE BLENDING   LOCATION UF   ACTPROB BCASE   DATERUN 82122101   INDEX NO.   0014

SB.G.UFT   BLENDED PRODUCT SPECS   MOTOR GASOLINE   UPDATED 82122101   RESIDENT ON BCASE
========   FAWLEY   U.K.   ========

	RVP MINSPEC	% 212F MAXSPEC	RON .4G MINSPEC
	RVMN	D1MX	R4MN
MOGAS ALL TP GS..	10.0000	60.0000	99.3700

QB.G.UFT   BLENDING COMP QUALITIES   MOTOR GASOLINE   UPDATED 82122101   RESIDENT ON BCASE
========   FAWLEY   U.K.   ========

		SPECFIC GRAVITY	PERCENT OFF AT 212 F	REID VAPOR PRESSURE	RESERCH OCTNAE .4G/PB/L
		SG	D1	RV	R4
BUTANE	C4	0.5700	100.0000	75.0000	101.6100
LIGHT VIRGIN NAPHTHA	LN	0.6500	95.0000	12.0000	86.3000
REFORMATE	PF	0.8650	35.0000	6.0000	102.4300
CATLYTC CRACKED NAPH-LS	CL	0.7300	60.0000	7.0000	94.9400
CATLYTC CRACKED NAPH-HS	CH	0.7500	64.0000	9.0000	99.0500

EB.GBUFT   EXCEPTIONAL STRUCTURE   MOGAS BLENDING   UPDATED 82122101   RESIDENT ON BCASE
========   FAWLEY   U.K.   ========

	QGS1VOUF	QGS2VOUF	QGS3VOUF
FAT1..J.	3.1200		
FAT2..J.		3.1200	
FAT3..J.			3.1200

DATE   82122101
LO1   REPRESENTS LEAD ADDITION COST FOR MOTOR GASOLINE
LO2 $7.80 PER GRAM OF LEAD
LO3   .4 GRAMS OF LEAD PER LITER OF GASOLINE

354

OPRINT   FUEL OIL BLENDING   LOCATION UF   ACTPROB BCASE   DATERUN 82122101   INDEX NO.   OO15

```
IQ.F.UFZ     QUALITY BLENDING OPTIONS   FUEL   OILS      UPDATED 82122101      RESIDENT ON BCASE
========                                                 FAWLEY      U.K.      ========

                                FUEL OIL

                                   FL
SULFUR   MAXSPEC   SUMX              X
VBN      MAXSPEC   W1MX              X
```

```
IB.F.UFZ     COMPONENTS TO BLENDS       FUEL   OILS      UPDATED 82122101      RESIDENT ON BCASE
========                                                 FAWLEY      U.K.      ========

                                FUEL OIL

                                         FL
VIRGIN   HEATING OIL         VH           X
CATLYTC  CRACKED DISTILAT    CD           X
VACUUM   DISTILAT            VD           X
BREGA    VACUUM  RESID       YB           X
ARAB.LT  VACUUM  RESID       YL           X
ARAB.HY  VACUUM  RESID       YH           X
FUEL OIL IMPORT 3.0% SU      FI           X
REFINRY  FUEL    EQUIV       RF           X
```

```
KQ.F...Z     BLENDING METHODOLOGY       FUEL   OILS      UPDATED 82122101      RESIDENT ON BCASE
========                                                 EUROPE               ========
```

			BLENDING METHOD	INPUT DATA UNITS	MODEL DATA UNITS	INPUT QUALITY MNEMONIC
			BA	FT	RT	QC
WEIGHT	PERCENT SULFUR	SU	WT			
VISCSTY	BLENDNG NUMBER	W1	WT	CS	VBN	V1

```
                    DBM - PLATOFORM SYSTEM - VER 1715                    82/12/21  20.42.04

OPRINT    FUEL OIL BLENDING         LOCATION UF      ACTPROB BCASE       DATERUN 82122101    INDEX NO.    0016

SB.F.UFT           BLENDED PRODUCT SPECS    FUEL    OILS        UPDATED 82122101          RESIDENT ON BCASE
========                                                                FAWLEY      U.K.            ========

                                    SULFUR      VISC CS
                                    MAXSPEC     MAXSPEC

FUEL     ALL TP      FL..           SUMX        V1MX
                                    3.0000      500.0000

QB.F.UFT           BLENDING COMP QUALITIES    FUEL    OILS     UPDATED 82122101          RESIDENT ON BCASE
========                                                                FAWLEY    .   U.K.            ========

                             WEIGHT      VISCSTY     VISCSTY
                             PERCENT     CENTSTKS    BLENDNG
                             SULFUR                  NUMBER
                             SU          V1          W1
VIRGIN   HEATING OIL     VH  1.0000      2.8740      .
CATLYTC  CRACKED DISTILAT CD 1.5000      5.6300      .
VACUUM   DISTILAT        VD  1.7000      7.4200      .
BREGA    VACUUM RESID    YB  0.6000      .           44.6000
ARAB.LT  VACUUM RESID    YL  4.0000      .           48.0000
ARAB.HY  VACUUM RESID    YH  5.0000      .           51.4000
FUEL OIL IMPORT 3.0% SU  FI  3.0000      500.0000    .
REFINRY  FUEL    EQUIV   RF  3.5000      .           44.0000
```

356

```
DBM  -  PLATOFORM  SYSTEM  -  VER 1715                              82/12/21   20.42.04

OPRINT   MATRIX GENERATOR CONTROL  LOCATION  ..      ACTPROB  BCASE     DATERUN 82122101     INDEX NO.    0017

IG......Z                                                    UPDATED 82122101      RESIDENT ON BCASE
========                                                                          ========
          GENERATOR MODULE OPTIONS


                                    EUROPE    MILFORD    FAWLEY
                                              HAVEN
                                              U.K.       U.K.
                                    ..        UM         UF

CRUDE AVABILITIES            CA       X
EXTERNAL IMPORT/EXPORT       BO                 X
CAPACITY&INVESTMENT          CI                            X
PIPESTILL                    PE                            X
SPECIAL PROCESSES            PR                            X
MISC.BLENDS(RECIPES)         RE                            X
MOGAS BLENDING               GB                            X
FUEL BLENDING                FB                            X
DEMANDS                      DM                            X


P2......Z                    TIME PERIODS UNDER STUDY               UPDATED 82122101      RESIDENT ON BCASE
========                                                                          ========

                                  1-CH
                                  PERIOD
                                  CODE
                                  T1

JANUARY             M1            1
FEBRUARY            M2            2
MARCH               M3            3
```

357

APPENDIX B. ADHOC REPORTS

ADHOC REPORTS   FIRST QUARTER BASECASE REPORTS

REFINERY CRUDE RUNS   UNITS: KB/D   KB
====================

		CASE BCASEM1 JANUARY	CASE BCASEM2 FEBRUARY	CASE BCASEM3 MARCH
ARABIAN LIGHT	AL	26.010	27.488	27.193
ARABIAN HEAVY	AH	31.329	5.884	27.029
BREGA	BR	1.305	7.289	5.821
TOTALS		58.644	40.661	60.043

REFINERY CRUDE RUNS   UNITS: K TONS   KT
====================

		CASE BCASEM1 JANUARY	CASE BCASEM2 FEBRUARY	CASE BCASEM3 MARCH
ARABIAN LIGHT	AL	110.000	105.000	115.000
ARABIAN HEAVY	AH	136.816	23.208	118.039
BREGA	BR	5.293	26.707	23.614
TOTALS		252.109	154.915	256.652

INCENTIVE TO INCREASE PRODUCT DEMAND---$/TON
==

		CASE BCASEM1 JANUARY	CASE BCASEM2 FEBRUARY	CASE BCASEM3 MARCH
MOTOR GASOLINE	GS	-68.30	64.40	-663.32
JET FUEL	JF	-3.49	-49.77	18.56
HEATING OIL	HO	-1.51	-34.77	125.12
FUEL OIL	FL	5.00	5.00	60.12

```
************************************************
* N.B.: ALL PRODUCT DEMANDS ARE FORMULATED AS FIXED. *
****REALIZATIONS HAVE BEEN INCLUDED IN FORMULATION****
****POSITIVE NUMBER==>INCENTIVE TO INCREASE DEMAND****
****NEGATIVE NUMBER==>INCENTIVE TO DECREASE DEMAND****
************************************************
```

ADHOC REPORTS   FIRST QUARTER BASECASE REPORTS

PROCESS UNIT CAPACITY SUMMARY   CASE: JANUARY
==

		UNIT FEED ---KT/D F	USED CAPACITY ---KB/D U	SPARE CAPACITY ---KB/D S	MAXIMUM CAPACITY ---KB/D M	VALUE OF CAPACITY ---$/B I
PIPSTIL	D.	8.133	58.644	6.349	64.993	15.700
CATLYTC CRACKER	C.	1.057	7.225		7.225	.
REFORMER	F.	0.709	6.049	0.724	6.773	.

PROCESS UNIT CAPACITY SUMMARY   CASE: FEBRUARY
==

		UNIT FEED ---KT/D F	USED CAPACITY ---KB/D U	SPARE CAPACITY ---KB/D S	MAXIMUM CAPACITY ---KB/D M	VALUE OF CAPACITY ---$/B I
PIPSTIL	D.	5.533	40.661	24.332	64.993	4.186
CATLYTC CRACKER	C.	1.170	7.999		7.999	.
REFORMER	F.	0.514	4.385	3.115	7.499	.

PROCESS UNIT CAPACITY SUMMARY   CASE: MARCH
==

		UNIT FEED ---KT/D F	USED CAPACITY ---KB/D U	SPARE CAPACITY ---KB/D S	MAXIMUM CAPACITY ---KB/D M	VALUE OF CAPACITY ---$/B I
PIPSTIL	D.	8.279	60.043	4.950	64.993	66.607
CATLYTC CRACKER	C.	1.170	7.999		7.999	.
REFORMER	F.	0.753	6.425	1.074	7.499	.

ADHOC REPORTS   FIRST QUARTER BASECASE REPORTS

NET REFINERY MATERIAL BALANCE---000 TONS
===

		CASE BCASEM1 JANUARY	CASE BCASEM2 FEBRUARY	CASE BCASEM3 MARCH	*TOTALS*
CRUDE OIL FEEDS	P	-252.109	-154.915	-256.652	-663.677
EXPORTS-IMPORTS (NET)	E	-9.793	-83.673	.	-93.466
PRODUCT DEMANDS	D	255.000	235.000	250.000	740.000 .
MATERIAL LOSS	B	6.902	3.588	6.652	17.143
TOTALS		-0.000	0.000	-0.000	0.000

MATERIAL BALANCE AROUND PROCESS UNITS--000 TONS   CASE: JANUARY
=== =============

		PIPSTIL	CATLYTC CRACKER	REFORMER	*TOTALS*
		D.	C.	F.	
CRUDE OIL FEED	XX	-252.109			-252.109
BUTANE	C4		1.966	0.440	2.405
LIGHT VIRGIN NAPHTHA	LN	8.193			8.193
INTERMD VIRGIN NAPHTHA	IN	21.976		-21.976	0.000
VIRGIN HEATING OIL	VH	86.221			86.221
REFORMATE	PF			19.778	19.778
CATLYTC CRACKED NAPH-HS	CH		14.743		14.743
CATLYTC CRACKED DISTILAT	CD		14.415		14.415
VACUUM DISTILAT	VD	64.300	-32.762		31.538
BREGA VACUUM RESID	YB	0.529			0.529
ARAB.LT VACUUM RESID	YL	18.150			18.150
ARAB.HY VACUUM RESID	YH	45.833			45.833
REFINRY FUEL EQUIV	RF	6.907	1.638	1.758	3.396
MATERIAL LOSS	.	-0.000	-0.000	-0.000	6.907
TOTALS		-0.000	-0.000	-0.000	-0.000

MATERIAL BALANCE AROUND PROCESS UNITS---000 TONS   CASE: FEBRUARY
===

	FEED	PIPSTIL	CATLYTC CRACKER	REFORMER	*TOTALS*
		D.	C.	F.	
CRUDE OIL	XX	-154.915			-154.915
BUTANE	C4		1.962	0.288	2.249
LIGHT VIRGIN NAPHTHA	LN	5.573			5.573
INTERMD VIRGIN NAPHTHA	IN	15.846		-14.386	1.460
VIRGIN HEATING OIL	VH	59.396			59.396
REFORMATE	PF			12.948	12.948
CATLYTC CRACKED NAPH-LS	CL		0.134		0.134
CATLYTC CRACKED NAPH-HS	CH		14.557		14.557
CATLYTC CRACKED DISTILAT	CD		14.475		14.475
VACUUM DISTILAT	VD	42.741	-32.762		9.979
BREGA VACUUM RESID	YB	2.671			2.671
ARAB.LT VACUUM RESID	YL	17.325			17.325
ARAB.HY VACUUM RESID	YH	7.775			7.775
REFINRY FUEL EQUIV	RF	3.588	1.634	1.151	2.785
MATERIAL LOSS	..				3.588
TOTALS		-0.000	-0.000	-0.000	-0.000

MATERIAL BALANCE AROUND PROCESS UNITS---OOO TONS CASE: MARCH
===

			PIPSTIL	CATLYTC CRACKER	REFORMER	*TOTALS*
			D.	C.	F.	
CRUDE OIL FEED	XX		-256.652			-256.652
BUTANE	C4			2.176	0.467	2.643
LIGHT VIRGIN NAPHTHA	LN		8.629			8.629
INTERMD VIRGIN NAPHTHA	IN		23.541		-23.342	0.199
VIRGIN HEATING OIL	VH		90.415			90.415
REFORMATE	PF				21.007	21.007
CATLYTC CRACKED NAPH-HS	CH			16.322		16.322
CATLYTC CRACKED DISTILAT	CD			15.960		15.960
VACUUM DISTILAT	VD		66.536	-36.272		30.264
BREGA VACUUM RESID	YB		2.361			2.361
ARAB.LT VACUUM RESID	YL		18.975			18.975
ARAB.HY VACUUM RESID	YH		39.543			39.543
REFINRY FUEL EQUIV	RF			1.814	1.867	3.681
MATERIAL LOSS	..		6.652			6.652
TOTALS			-0.000	-0.000	-0.000	-0.000

PROCESS UNIT ECONOMIC BALANCES---000 $ CASE: JANUARY
=================================

		PIPSTIL	CATLYTC CRACKER	REFORMER	*TOTALS*
		D.	C.	F.	
CRUDE OIL FEED	XX	-65138.45			-65138.45
CRUDE TRANS COST	YY	-6013.81			-6013.81
***********		**********	**********	**********	**********
BUTANE	C4		1092.02	244.17	1336.18
LIGHT VIRGIN NAPHTHA	LN	2059.85			2059.85
INTERMD VIRGIN NAPHTHA	IN	11408.85		-11408.85	0.00
VIRGIN HEATING OIL	VH	27289.49			27289.49
REFORMATE	PF			10877.91	10877.91
CATLYTC CRACKED NAPH-HS	CH		7185.11		7185.11
CATLYTC CRACKED DISTILAT	CD		4562.50		4562.50
VACUUM DISTILAT	VD	18742.06	-9549.38		9192.68
BREGA VACUUM RESID	YB	175.10			175.10
ARAB.LT VACUUM RESID	YL	3797.83			3797.83
ARAB.HY VACUUM RESID	YH	7951.81			7951.81
REFINRY FUEL EQUIV	RF		372.05	399.30	771.35
***********		**********	**********	**********	**********
CAPACTY CONSMPN COST	ZZ	-272.73	-3662.30	-112.53	-4047.55
***********		**********	**********	**********	**********
TOTALS		-0.00	-0.00	-0.00	-0.00

PROCESS UNIT ECONOMIC BALANCES---OOO $ CASE: FEBRUARY
==

			PIPSTIL	CATLYTC CRACKER	REFORMER	*TOTALS*
			D.	C.	F.	
CRUDE	OIL	FEED	XX	-43593.27	.	.	-43593.27
CRUDE	TRANS	COST	YY	-3364.63	.	.	-3364.63
****
BUTANE			C4		684.94	100.47	785.40
LIGHT. VIRGIN NAPHTHA			LN	1949.28			1949.28
INTERMD VIRGIN NAPHTHA			IN	5542.46		-5031.85	510.61
VIRGIN HEATING OIL			VH	20775.05			20775.05
REFORMATE			PF			4754.92	4754.92
CATLYTC CRACKED NAPH-LS			CL		48.34		48.34
CATLYTC CRACKED NAPH-HS			CH		5247.96		5247.96
CATLYTC CRACKED DISTILAT			CD		5062.93		5062.93
VACUUM DISTILAT			VD	13458.18	-10316.02		3142.16
BREGA VACUUM RESID			YB	819.01			819.01
ARAB.LT VACUUM RESID			YL	3365.99			3365.99
ARAB.HY VACUUM RESID			YH	1218.72			1218.72
REFINRY FUEL EQUIV			RF		355.13	250.14	605.26
****
CAPACTY CONSMPN COST			ZZ	-170.79	-1083.28	-73.67	-1327.74
****
TOTALS				0.00	-0.00	0.00	0.00

PROCESS UNIT ECONOMIC BALANCES---000 $ CASE: MARCH

===

		PIPSTIL	CATLYTC CRACKER	REFORMER	*TOTALS*
		D.	C.	F.	
CRUDE OIL FEED	XX	-66830.85			-66830.85
CRUDE TRANS COST	YY	-5865.20			-5865.20
****
BUTANE	C4		2616.02	561.15	3177.17
LIGHT VIRGIN NAPHTHA	LN	5588.73			5588.73
INTERMD VIRGIN NAPHTHA	IN	26024.00		-25803.90	220.09
VIRGIN HEATING OIL	VH	17168.46			17168.46
REFORMATE	PF			25007.70	25007.70
CATLYTC CRACKED NAPH-HS	CH		17576.25		17576.25
CATLYTC CRACKED DISTILAT	CD		3030.49		3030.49
VACUUM DISTILAT	VD	12634.07	-6887.48		5746.60
BREGA VACUUM RESID	YB	448.39			448.39
ARAB.LT VACUUM RESID	YL	3603.05			3603.05
ARAB.HY VACUUM RESID	YH	7508.57			7508.57
REFINRY FUEL EQUIV	RF		344.37	354.58	698.95
****
CAPACTY CONSMPN COST	ZZ	-279.23	-16679.66	-119.53	-17078.42
****
TOTALS		0.00	0.00	-0.00	0.00

APPENDIX **C.** SUPPLEMENTARY DISPLAY OF ACTPROB DATABASE

This appendix is intended for the more serious reader who is interested in pursuing PLATOFORM's data structure in more detail. The section contains a printout of the contents of ACTPROB DATABASE. The material included is as follows:

Table Name(s)	Page(s)	Purpose in PLATOFORM
G E N T A B . C	370	Generic cataloging of table sets
D I C T . . . C	371	Cataloging of dictionaries
All dictionaries	372–378	Translation of entity codes
Z R . A H . . Z	379–380	ADHOC report definitions
C O L C	381	Column name translation control for SOL-PRINT
R O W C	382	Row name translation control for SOLPRINT
Z X . * * . . Z	383	SOLPRINT auxiliary translation rules
Z Z . . . * * Z	383–386	Control tables for OPRINT

1212   TABLE   GENTAB.C HAS   34 ROWS   10 COLUMNS

GENTAB.C	S1	S2	H1	H2	T1	F1	M1	D1	D2	D3
BA....T	CRUDES.D		PERNAM2D		S	D3	XXXAABBX	CRUDE OI	L AVAILS	-----KT
BE.LOLOT	PRODUCTD		PERNAM2D	MAXMIN.D	G	D3	XXXXXAAX	IMPORT/E	XPORT LI	MITS--KT
BS..LOT	PRODUCTD		PERNAM2D	MAXMIN.D	G	D3	XXXAABBX	PRODUCT	DEMANDS	-----KT
BU.CILOT	UNITS..D	CAPTYPED	PERNAM2D		G	D3	XXXAABBX	PROCESS	UNIT CAP	ACITIES
EB.GMLOT	*		*		G	D4	XXXAAXXX	EXCEPTIO	NAL STRU	CTURE
FA.DA..T	CRUDES.D		PERNAM2D		G	D2	XXXXXAAX	CRUDE OI	L PRICES	----$/
FC...LOT	CRUDES.D		PERNAM2D		G	D3	XXXAABBX	CRUDE TR	ANSP PRI	CE-$/TON
FE.LOLOT	PRODUCTD		PERNAM2D		G	D2	XXXXXAAX	IMPORT/E	XPORT PR	ICES-$/T
FS..LOT	PRODUCTD		PERNAM2D		G	D2	XXXAABBX	PRODUCT	SALES PR	ICES-$/T
FU.CILOT	UNITS..D	CAPTYPED	CSTTYPED		G	D2	XXXAABBX	CAPACITY	&INVESTM	ENT COST
IA.....Z	CRUDES.D		*		S	A8		CRUDE OI	LS AVAIL	ABLE
IB.PCLOZ	PRODUCTD		PRODUCTD		G	A8	XXXAABBX	COMPONEN	TS TO BL	ENDS
ID...LOZ	CRUDES.D		UNITS..D		S	A8	XXXXXAAX	CRUDE PR	OCESSING	OPTIONS
IG.....Z	GENRMODD		LOCATN.D		G	A8		GENERATO	R MODULE	OPTIONS
IM...LOZ	PRODUCTD		PRODUCTD		S	A8	XXXXXAAX	RECIPE B	LENDING	OPTIONS
IP...LOZ	PRODUCTD		UNITS..D		G	A8	XXXXXAAX	PROCESS	UNIT OPT	IONS
IQ.PCLOZ	QUALITYD	MAXMIN.D	PRODUCTD		G	A8	XXXAABBX	QUALITY	BLENDING	OPTIONS
IU...LOZ	UNITS..D	CAPTYPED	PERNAM2D		G	A8	XXXXXAAX	CAPACITY	&INVESTM	ENT OPTN
KD...LOZ	CAPTYPED		UNITS..D		G	A8	XXXXXAAX	DISTILLA	TION CAP	CONSUME
KQ.PCLOZ	QUALITYD		BLEND..D		G	A8	XXXAABBX	BLENDING	METHODO	LOGY
LQ.....Z	PRODCLSD		PRODCLSD		S	A8		QUALITY	TABLES T	O USE
P.DAYS.T	PERNAM2D		*		S	DO		DAYS IN	TIME PER	IODS
P2.....Z	PERNAM2D		PERNAM2D		S	A8		TIME PER	IODS UND	ER STUDY
QB.PCLOT	PRODUCTD		PRODUCTD		G	D4	XXXAABBX	BLENDING	COMP QU	ALITIES
QC....T	CRUDES.D		PRODUCTD		S	D4		CRUDE OI	L PROPER	TIES
SB.PCLOT	PRODUCTD	PERNAM2D	PRODUCTD	MAXMIN.D	G	D4	XXXAABBX	BLENDED	PRODUCT	SPECS
SG.PCLOT	QUALITYD		QUALITYD		G	D4	XXXAABBX	BLENDING	TOLERAN	CES
YD...LOT	PRODUCTD		MISC...D	UNITS..D	G	D2	XXXXXAAX	CRUDE DI	STILLATI	ON YIELD
YL...LOT	PRODCLSD		MISC...D		G	D2	XXXXXAAX	PRODUCT	BLENDING	LOSSES
YM.PRLOT	PRODUCTD		PRODUCTD	PRODUCTD	G	D2	XXXAABBX	PRODUCT	BLENDING	RECIPES
YP.UNLOT	PRODUCTD		PRODUCTD	UNITS..D	G	D4	XXXAABBX	PROCESS	UNIT YIE	LDS
ZR.RC..Z	*		*		G	A8	XXXAAXXX	REPORT W	RITER DE	FINITION
ZX.SP..Z	*		*		G	A8	XXXXAAXX	SOLPRINT	AUXILIA	RY TABLE
ZZ...OPZ	*		*		G	A8	XXXXXAAX	OPRINT C	ONTROL T	ABLE

1212   TABLE   DICT...C HAS   21 ROWS   4 COLUMNS

DICT...C	D1	D2	D3	C1
BALANCED	BALANCE	CATEGORI	ES	B
BLEND..D	BLENDING	METHODO	LOGY	BL
CAPINV.D	CPAICITY	INVESTM	ENT TYPE	CI
CAPREP.D	CPAICITY	REPORT	HEADS	C
CAPTYPED	CAPACITY	TYPES		CT
CRUDES.D	CRUDE OI	LS		CR
CSTTYPED	COST TYP	ES		CS
DAUNIT.D	DATA UNI	TS		DA
GENRMODD	GENERATO	R SUBMOD	ULES	GM
LOCATN.D	LOCATION	S		LO
MAXMIN.D	MAXIMUMS	AND MIN	IMUMS	MM
MISC...D	MISCELLA	NEOUS CO	DES	MS
OPRINT.D	OPRINT C	LASSIFIC	ATIONS	OP
PERNAM1D	1-CH TIM	E PERIOD	CODES	P
PERNAM2D	2-CH TIM	E PERIOD	CODES	P2
PRODCLSD	PRODUCT	CLASSES		PC
PRODUCTD	PRODUCTS			PR
QUALITYD	QUALITIE	S		QU
REPCLS.D	REPORT W	RITER CL	ASSES	RC
SOLPRNTD	SOLPRINT	CONTROL	CLASSES	SP
UNITS..D	PROCESS	UNITS &	OPTIONS	UN

DICTIONARY BALANCED   BALANCE CATEGORIES

CODE	MNEMONIC M1	DESCRIPTOR D1	D2	D3	TAGS T1	T2
P	CRUDE	CRUDE	OIL	FEEDS		
E	EXP/IMP	EXPORTS-IMPORTS	(NET)			
D	DEMANDS	PRODUCT DEMANDS				
B	LOSS	MATERIAL	LOSS			

DICTIONARY BLEND..D   BLENDING METHODOLOGY

CODE	MNEMONIC M1	DESCRIPTOR D1	D2	D3	TAGS T1	T2
BA	METHOD	BLENDINGMETHOD				
FT	IN UNIT	INPUT	DATA	UNITS		
RT	MO UNIT	MODEL	DATA	UNITS		
QC	IN QUAL	INPUT	QUALITY	MNEMONIC		
SC	SCALE	SCALING FACTOR				

DICTIONARY CAPINV.D   CPAICITY INVESTMENT TYPE

CODE	MNEMONIC M1	DESCRIPTOR D1	D2	D3	TAGS T1	T2
CP	EXIST	EXISTING CAPACITY				
TA	DOWN	TURN	AROUND	DAYS		
GR						

DICTIONARY DAUNIT.D   DATA UNITS

CODE	MNEMONIC	DESCRIPTOR			TAGS	
	M1	D1	D2	D3	T1	T2
BL	BARREL	BARREL				
..	TON	TON				
KB	KB/D	OOO	BARRELS	PER DAY		
KT	K TONS	OOO	TONS	PER PRD		

DICTIONARY GENRMODD   GENERATOR SUBMODULES

CODE	MNEMONIC	DESCRIPTOR			TAGS	
	M1	D1	D2	D3	T1	T2
CA	CRAVAIL	CRUDE AVABILITIES				
BO	EXTERN	EXTERNAL IMPORT/EXPORT				
CI	CAPINV	CAPACITY&INVESTMENT				
PE	PSTLL	PIPESTILL				
PR	PROCESS	SPECIAL PROCESSES				
RE	RECIPE	MISC.BLENDS (RECIPES)				
GB	GBLEND	MOGAS BLENDING				
FB	FBLEND	FUEL BLENDING				
DM	DEMAND	DEMANDS				

DICTIONARY LOCATN.D   LOCATIONS

CODE	MNEMONIC	DESCRIPTOR			TAGS	
	M1	D1	D2	D3	T1	T2
J.	EXXON	EXXON				
..	EUROPE	EUROPE				
UM	MILFORD	MILFORD HAVEN	U.K.			
UF	FAWLEY	FAWLEY	U.K.			

DICTIONARY MAXMIN.D   MAXIMUMS AND MINIMUMS

CODE	MNEMONIC	DESCRIPTOR			TAGS	
	M1	D1	D2	D3	T1	T2
MN	MINSPEC	MINIMUM	PRODUCT	SPEC		
MX	MAXSPEC	MAXIMUM	PRODUCT	SPEC		
FX	FIX BND	FIXED	BOUND			

DICTIONARY MISC...D   MISCELLANEOUS CODES

CODE	MNEMONIC	DESCRIPTOR			TAGS	
	M1	D1	D2	D3	T1	T2
T1	1CH-CODE	1-CH	PERIOD	CODE		

DICTIONARY OPRINT.D   OPRINT CLASSIFICATIONS

CODE	MNEMONIC	DESCRIPTOR			TAGS	
	M1	D1	D2	D3	T1	T2
CS	CRUDE	CRUDE OIL SUPPLY				
PS	PRODUCT	PRODUCT SUPPLY				
EX	EXTERNL	EXTERNAL IMPORT/EXPORT				
RC	CAPACTY	REFINERY CAPACITIES				
CD	DISTILL	CRUDE OIL DISTILLATION				
PR	PROCESS	OTHER REFINERY PROCESSES				
RE	RECIPE	RECIPE BLENDING				
GB	MOGAS	MOTOR GASOLINE BLENDING				
FB	FUEL	FUEL OIL BLENDING				
GC	CONTROL	MATRIX GENERATOR CONTROL				

DICTIONARY PERNAM1D   1-CH TIME PERIOD CODES

CODE	MNEMONIC	DESCRIPTOR			TAGS	
	M1	D1	D2	D3	T1	T2
.	ALL TP	ALL	PERIODS			
1	JAN	JANUARY				
2	FEB	FEBRUARY				
3	MAR	MARCH				

DICTIONARY PERNAM2D   2-CH TIME PERIOD CODES

CODE	MNEMONIC	DESCRIPTOR			TAGS	
	M1	D1	D2	D3	T1	T2
..	ALL TP	ALL	PERIODS			
M1	JAN	JANUARY				
M2	FEB	FEBRUARY				
M3	MAR	MARCH				

DICTIONARY PRODCLSD   PRODUCT CLASSES

CODE	MNEMONIC	DESCRIPTOR			TAGS	
	M1	D1	D2	D3	T1	T2
G.	MOGAS	MOTOR	GASOLINE			
F.	FUELS	FUEL	OILS			

375

DICTIONARY PRODUCTD   PRODUCTS

CODE	MNEMONIC	DESCRIPTOR			TAGS	
	M1	D1	D2	D3	T1	T2
XX	CRUDE	CRUDE	OIL	FEED		
YY	CRTRAN	CRUDE	TRANS	COST	#	
C4	BUTANE	BUTANE				
LN	LVN	LIGHT	VIRGIN	NAPHTHA		
IN	IVN	INTERMD	VIRGIN	NAPHTHA		
VH	VIRG HO	VIRGIN	HEATING	OIL		
PF	REFMT	REFORMATE				
CL	CATNAPL	CATLYTC	CRACKED	NAPH-LS		
CH	CATNAPH	CATLYTC	CRACKED	NAPH-HS		
CD	CATDIST	CATLYTC	CRACKED	DISTILAT		
VD	VACDIST	VACUUM	DISTILAT			
YB	BR RESID	BREGA	VACUUM	RESID		
YL	AL RESID	ARAB.LT	VACUUM	RESID		
YH	AH RESID	ARAB.HY	VACUUM	RESID		
GS	MOGAS	MOTOR	GASOLINE			
SG	MOGAS	MOTOR	GASOLINE			
B1	JF BL 1	JETFUEL	BLEND #1			
B2	JF BL 2	JETFUEL	BLEND #2			
B3	JF BL 3	JETFUEL	BLEND #3			
JF	JETFUEL	JET FUEL				
HO	HEATOIL	HEATING	OIL			
FI	FO IMP	FUEL OIL	IMPORT	3.0% SU		
FL	FUEL	FUEL OIL				
RF	REFFUEL	REFINRY	FUEL	EQUIV	#	
FR	REFFUEL	REFINRY	FUEL			
ML	LOSS	MATERAIL		LOSS		
..	LOSS	MATERIAL		LOSS		
ZZ	CAPCOST	CAPACTY	CONSMPN	COST	#	

DICTIONARY QUALITYD   QUALITIES

CODE	MNEMONIC	DESCRIPTOR			TAGS	
	M1	D1	D2	D3	T1	T2
SU	SULFUR	WEIGHT	PERCENT	SULFUR		
V1	VISC CS	VISCSTY	CENTSTKS			
W1	VBN	VISCSTY	BLENDNG	NUMBER		
GS	GRAVITY	SPECFIC	GRAVITY			
SG	GRAVITY	SPECFIC	GRAVITY			
D1	% 212F	PERCENT	OFF AT	212 F		
RV	RVP	REID	VAPOR	PRESSURE		
R4	RON .4G	RESERCH	OCTNAE	.4G/PB/L		
VO	VOLUME	VOLUME				
WT	WEIGHT	WEIGHT				
..						

DICTIONARY REPCLS.D   REPORT WRITER CLASSES

CODE	MNEMONIC	DESCRIPTOR			TAGS	
	M1	D1	D2	D3	T1	T2
AH	ADHOC	ADHOC	REPORTS			

DICTIONARY SOLPRNTD   SOLPRINT CONTROL CLASSES

CODE	MNEMONIC	DESCRIPTOR			TAGS	
	M1	D1	D2	D3	T1	T2
R.	ROWS	ROWS				
C.	COLUMNS	COLUMNS				
XX	AUXILARY	AUXILARY				

DICTIONARY UNITS..D   PROCESS UNITS & OPTIONS

CODE	MNEMONIC	DESCRIPTOR			TAGS	
	M1	D1	D2	D3	T1	T2
D.	PIPSTIL	PIPSTIL				
C.	CATCRAK	CATLYTC CRACKER				
CL	CC---LS	CATLYTC CRACKER	LO SEV			
CH	CC---HS	CATLYTC CRACKER	HI SEV			
F.	REFORM	REFORMER				

TPRINT

ACTPROB DATABASE   DATERUN 82122101

ZR.AH..Z
========

REPORT WRITER DEFINITION ADHOC REPORTS

UPDATED 82122101

RESIDENT ON BCASE
========

	CRUDES	CAPSUM	PRODVALS	MATBAL	UNITMB	UNITDB
MASK1	P**.D***	P**.**UF	D**...**	P**.D***		
MAPS1	1234KT78	F2345.78				XX345
STYP1			D			
SCLE1			-1	-1		
MASK2	P**.D***	C**.****		D**...**		
MAPS2	123KB78	U2345678			XX345	YY345
SCLE2	.159	.159				
MASK3		C**.****		E**.UF**		
MAPS3		M2345678				
MAPT3		S2345678				
SCLE3		.159				
STYP3		XU				
MASK4		C**.****		E**.**UF		
MAPS4		S2345678		12347856		ZZ345
SCLE4		-.159		-1		
MASK5		C**.****		B...ML..		
MAPS5		12345678				
SCLE5		-6.29				
STYP5		D		RA		
VSCMSK1	O**OOOOO	OOO*OOOO	***O****	***O****	****O***	****O***
VSCTAB1	QC....T	*SCALE				
VSCCOL1	GS	DP				
VSCDSL1	*****KB**	I********				
VSCMSK2	OOO*OOOO	OOO*OOOO	OPROOOOO	BOOOOOOO	PROOOOOO	PROOOOOO
VSCTAB2	*SCALE					
VSCCOL2	DP					
VSCSLC2	****KB**					
CASEMASK	****O****	***O****	***O****	***O****	****O***	****O***
STUBMASK	OCROOOOO	OOOOUNOO	OPROOOOO	BOOOOOOO	PROOOOOO	PROOOOOO
HEADMASK	OOO®OO3O	OOO®OOOO	OOO®OOOO	OOO®OOOO	OOUNOOOO	OOUNOOOO
SPLTMASK	OOOODAOO	OOO®OOOO	OOOOOOOO	OOOOOOOO	OOOO®OOO	OOOO®OOO
POINT	3	3	2	3	3	2
TOTAL	STUB			X	X	X
NEWPAGE		X		X	X	X
TITLE	LO3	LO4	LO5	LO6	LO7	LO8
TITLE1	LO2	LO1	LO1	LO1	LO1	LO1
TRAIL1			LO9			
TRAIL2			L10			
TRAIL3			L11			
TRAIL4			L12			
TRAIL5			L13			
TRAIL6			LO9			
*PLAN	1	3	2	4		
MASM1		MATRIX			MATRIX	
DOTC1		X			X	

```
                    DBM  -  PLATOFORM SYSTEM  -  VER 1715

        CRUDES    CAPSUM   PRODVALS   MATBAL    UNITMB      UNITDB

DOTR1                                                        X
COLM1                                            P**.****    P**.D***
ROWM1                                            B**.****    A**.....
KEYM1                                            TUE.        TUE.
MASM2                                            MATRIX      MATRIX
DOTC2                                            X           X
DOTR2                                                        X
COLM2                                            P**.D***    P**.D***
ROWM2                                            A**.....    FAT*..J.
KEYM2                                            TUE.        TUE.
MASM3                                                        MATRIX
DOTC3                                                        X
DOTR3                                                        X
COLM3                                                        P**.****
ROWM3                                                        B**..**
KEYM3                                                        TUE.
MASM4                                                        MATRIX
DOTC4                                                        X
DOTR4                                                        X
COLM4                                                        P**.****
ROWM4                                                        C**.****
KEYM4                                                        TUE.
SCALE                                                        -1
ULINE                                                        11
*BALANCE                                         1           2


DATE   82122101
       ======================================================
LO1                      REFINERY CRUDE RUNS
LO2              ====================
LO3          PROCESS UNIT CAPACITY SUMMARY                        UNITS:
LO4    INCENTIVE TO INCREASE PRODUCT DEMAND--$/TON                 CASE:
LO5        NET REFINERY MATERIAL BALANCE--OOO TONS
LO6    MATERIAL BALANCE AROUND PROCESS UNITS--OOO TONS     CASE:
LO7          PROCESS UNIT ECONOMIC BALANCES---OOO $        CASE:
LO8    ******************************************************
LO9    * N.B.: ALL PRODUCT DEMANDS ARE FORMULATED AS FIXED. *
L10    ****REALIZATIONS HAVE BEEN INCLUDED IN FORMULATION****
L11    ****POSITIVE NUMBER==>INCENTIVE TO INCREASE DEMAND****
L12    ****NEGATIVE NUMBER==>INCENTIVE TO DECREASE DEMAND****
L13
```

UTILITY - PLATOFORM SYSTEM - VER 2408.F1

1212   TABLE   COL....C HAS   9 ROWS   8 COLUMNS

COL....C	F1	F2	F3	F4	F5	F6	F7	F8
AVAILBTY	A	CRUDES.D	&	*	*	&	LOCATN.D	&
BLEND	B	PRODUCTD	&	*	PRODUCTD	&	LOCATN.D	&
CAPACITY	C	CAPTYPED	&	*	UNITS..D	&	LOCATN.D	&
DEMAND	D	PRODUCTD	&	*	*	&	LOCATN.D	&
EXTERNAL	E	PRODUCTD	&	*	LOCATN.D	&	LOCATN.D	&
RECIPE	M	PRODUCTD	&	*	PRODUCTD	&	LOCATN.D	&
DISTILL	P	CRUDES.D	&	*	#12.S1	&	LOCATN.D	&
PROCESS	P	PRODUCTD	&	*	UNITS..D	&	LOCATN.D	&
SPECIFY	Q	PRODUCTD	&	PERNAM1D	QUALITYD	&	LOCATN.D	&

UTILITY - PLATOFORM SYSTEM - VER 2408.F1

82/12/21 20.42.04

1212   TABLE   ROW....C HAS   7 ROWS   8 COLUMNS

ROW....C	F1	F2	F3	F4	F5	F6	F7	F8
AVAILBTY	A	CRUDES.D	&	*	*	&	LOCATN.D	&
ACCNTING	B	#21.S1	&	*	PRODUCTD	&	LOCATN.D	&
BALANCE	B	PRODUCTD	&	*	QUALITYD	&	LOCATN.D	&
CAPACITY	C	CAPTYPED	&	*	UNITS..D	&	LOCATN.D	&
MIN@SPEC	N	QUALITYD	&	*	PRODUCTD	&	LOCATN.D	&
MAX@SPEC	X	QUALITYD	&	*	PRODUCTD	&	LOCATN.D	&
FINANCL	F	CSTTYPED	&	PERNAM1D	*	&	LOCATN.D	&

ACTPROB DATABASE   DATERUN 82122101

TPRINT

ZX.C...Z
========   SOLPRINT AUXILIARY TABLE COLUMNS   UPDATED 82122101   RESIDENT ON DATABASE
                                                                                       ========

          TABLE   DICT
S1   ZX.XX..Z UNITS..D

ZX.R...Z
========   SOLPRINT AUXILIARᵛ TABLE ROWS   UPDATED 82122101   RESIDENT ON DATABASE
                                                                                       ========

          TABLE   DICT
S1   ZX.XX..Z *

ZX.XX..Z
========   SOLPRINT AUXILIARY TABLE AUXILARY   UPDATED 82122101   RESIDENT ON DATABASE
                                                                                       ========

          ROWS   COLS
S1   ..   D

ZZ...CDZ
========   OPRINT CONTROL TABLE   CRUDE OIL DISTILLATION   UPDATED 82122101   RESIDENT ON DATABASE
                                                                                       ========

          STYLE
          DDD
ID...UFZ
YD...UFT
FC...UFT
KD...UFZ

```
DBM - PLATOFORM SYSTEM - VER 1715                                    82/12/21  20.42.04

                              ACTPROB DATABASE        DATERUN 82122101
TPRINT

ZZ...CSZ         OPRINT CONTROL TABLE    CRUDE OIL SUPPLY      UPDATED 82122101    RESIDENT ON DATABASE
=======                                                                           ========

          STYLE
          DDD
IA......Z
BA......T
FA.**...T
QC.....T

ZZ...EXZ         OPRINT CONTROL TABLE    EXTERNAL IMPORT/EXPORT UPDATED 82122101   RESIDENT ON DATABASE
=======                                                                           ========

          STYLE
          DDD
BE.*****T
FE.*****T

ZZ...FBZ         OPRINT CONTROL TABLE    FUEL OIL BLENDING     UPDATED 82122101    RESIDENT ON DATABASE
=======                                                                           ========

          STYLE
          DDD
IQ.F.UFZ
IB.F.UFZ
KQ.F...Z
SB.F.UFT
QB.F.UFT
```

DBM - PLATOFORM SYSTEM - VER 1715

ACTPROB DATABASE   DATERUN 82122101

TPRINT

ZZ...GBZ   OPRINT CONTROL TABLE   MOTOR GASOLINE BLENDING   UPDATED 82122101   RESIDENT ON DATABASE
========   ========

STYLE
DDD

IQ.G.UFZ
IB.G.UFZ
KQ.G...Z
SB.G.UFT
QB.G.UFT
EB.GBUFT

ZZ...GCZ   OPRINT CONTROL TABLE   MATRIX GENERATOR CONTROL   UPDATED 82122101   RESIDENT ON DATABASE
========   ========

STYLE
DDD

IG.....Z
P2.....Z

ZZ...PRZ   OPRINT CONTROL TABLE   OTHER REFINERY PROCESSES   UPDATED 82122101   RESIDENT ON DATABASE
========   ========

STYLE
DDD

IP...UFZ
YP.**UFT
EB.PRUFT

ZZ...PSZ   OPRINT CONTROL TABLE   PRODUCT SUPPLY   UPDATED 82122101   RESIDENT ON DATABASE
========   ========

STYLE
DDD

BS...UFT
FS.....T

385

DBM - PLATOFORM SYSTEM - VER 1715   82/12/21 20.42.04

                    ACTPROB DATABASE   DATERUN 82122101

TPRINT

ZZ...RCZ   OPRINT CONTROL TABLE   REFINERY CAPACITIES   UPDATED 82122101   RESIDENT ON DATABASE
========   ========

STYLE
DDD
IU...UFZ
BU.**UFT
P.DAYS.T
FU.**UFT

ZZ...REZ   OPRINT CONTROL TABLE   RECIPE BLENDING   UPDATED 82122101   RESIDENT ON DATABASE
========   ========

STYLE
DDD
IM...UFZ
YM.**UFT

INDEX